Engineering Materials

This series provides topical information on innovative, structural and functional materials and composites with applications in optical, electrical, mechanical, civil, aeronautical, medical, bio- and nano-engineering. The individual volumes are complete, comprehensive monographs covering the structure, properties, manufacturing process and applications of these materials. This multidisciplinary series is devoted to professionals, students and all those interested in the latest developments in the Materials Science field, that look for a carefully selected collection of high quality review articles on their respective field of expertise.

Indexed at Compendex (2021) and Scopus (2022)

Rakesh Kumar Sonker · Kedar Singh ·
Rajendra Sonkawade
Editors

Low-Dimensional Multifunctional Hybrid Nanostructured Materials

Energy and Environmental Applications

Springer

Editors
Rakesh Kumar Sonker
Department of Physics, Motilal Nehru College
University of Delhi
New Delhi, Delhi, India

Kedar Singh
School of Physical Sciences
Jawaharlal Nehru University
New Delhi, Delhi, India

Rajendra Sonkawade
Department of Physics
Shivaji University
Kolhapur, Maharashtra, India

ISSN 1612-1317 ISSN 1868-1212 (electronic)
Engineering Materials
ISBN 978-981-96-7951-5 ISBN 978-981-96-7952-2 (eBook)
https://doi.org/10.1007/978-981-96-7952-2

This Springer imprint is published by the registered company Springer Nature Singapore Pte Ltd.
The registered company address is: 152 Beach Road, #21-01/04 Gateway East, Singapore 189721, Singapore

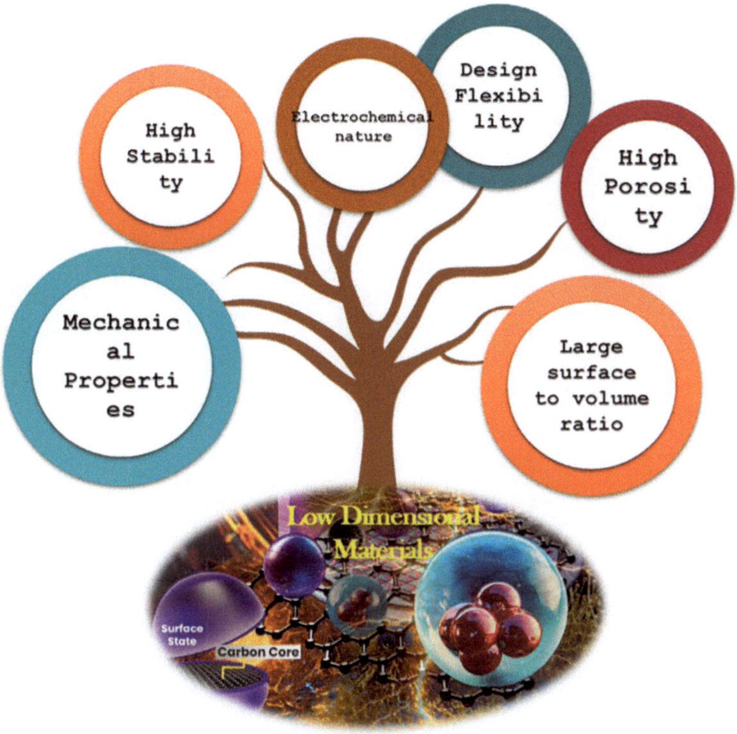
High
Stabili
ty
Electrochemical
nature
Design
Flexibi
lity
High
Porosi
ty
Mechanic
al
Properti
es
Large
surface
to volume
ratio
Low Dimensional
Materials
Surface
State
Carbon Core

Preface

Addressing the urgent need for sustainable energy and environmental solutions, low-dimensional, multifunctional hybrid nanomaterials are gaining prominence. Materials such as graphene oxide (GO), GO-metal oxide composites, metal oxide composites, and carbon nanotube-based hybrids offer unique capabilities for efficient energy harvesting, storage, and conversion, alongside environmental remediation through adsorption, catalysis, and electro- or photo catalytically generated reactive oxygen species. As the energy sector evolves towards renewable, efficient, cost-effective, convenient, and safe sources for a sustainable future, hydrogen and batteries are increasingly viewed as essential energy carriers for a low-carbon economy, poised to reshape the energy landscape alongside the growth of renewables. Despite this shift, the abundance and established infrastructure of fossil fuels suggest their continued importance in the energy mix. The adaptability and design flexibility of these advanced nanomaterials make them invaluable for confronting global challenges in both sustainable energy and environmental sustainability.

Low-Dimensional and Multifunctional Hybrid Nanostructured Materials delves into the fundamental materials crucial for energy and environmental design, fabrication methods for energy devices, and the environmental implications of solar and hydrogen storage. The book offers a detailed analysis of smart nanomaterials currently used in energy and environmental applications, including carbon-based materials, MOFs, GO, MXENEs, low-dimensional materials, metal oxide composites, TMDs, and metallopolymer nanomaterials. It highlights recent progress in energy storage materials like low-dimensional materials, chalcogenides, CNTs, MOFs, and hybrid nanomaterials, showcasing their potential for material scientists in energy storage research. This book provides an in-depth exploration of the opportunities and challenges within energy and environmental applications, utilizing relevant illustrations to enhance understanding. The chapters offer comprehensive discussions on current and future energy technologies, challenges, and fundamental concepts, serving as a valuable resource for researchers, students, academicians, and professionals. It covers low dimensional and multifunctional hybrid materials, including popular characterization techniques across its 13 chapters, which thoroughly explore energy and environmental applications. Chapter "Low-Dimensional and Multifunctional Hybrid Materials for Energy Storage: An Introduction" introduces low-dimensional and multifunctional hybrid materials, emphasizing how combining nanomaterials of different dimensions can overcome limitations, and reviews their applications in batteries and supercapacitors. Chapter "Characterization Technique for Energy Storage Nanomaterials" focuses on nanomaterial characterization, highlighting the importance of their high surface-area-to-volume ratio and exploring the structural, chemical, physical, and electrical properties of low-dimensional nanomaterials for advanced energy storage. Chapter "Low Dimensional Nanomaterials for Hydrogen Storage" examines the impact of economic and population growth on energy demand, the dominance of fossil fuels, and the environmental costs of greenhouse gas emissions. It also explores recent advancements in high-entropy alloy-based metal hydrides for hydrogen production and storage. Chapter "Li-Ion Batteries for Energy Storage" details progress in Lithium-ion Batteries (LIBs) for energy storage, driven by electric vehicle demand, discussing recent developments and future directions for performance enhancement. Chapter "Recent Investigation of Na- and K-Ion Batteries" covers the latest developments in Sodium (Na) and Potassium (K) ion batteries using transition metal oxides and their hybrid/composite materials. Chapter "Flexible Nanogenerators Based on Low Dimensional Nanomaterials" explores flexible nanogenerators and their smart material applications (metal oxides, MXENEs, carbon-based materials), highlighting their high output power and self-powered sensing capabilities for renewable energy and advanced sensors. Chapter "Metal–Organic Framework (MOF)-Derived Flexible Energy Storage" discusses the intense interest in flexible MOFs as multifunctional nanostructures for electrochemical energy storage, concluding with current limitations and strategies for advancement. Chapter "Graphene-Based Multifunctional Hybrid Nanomaterial for Energy Storage" reviews the potential of graphene in energy storage devices, focusing on graphene-based hybrids in batteries and supercapacitors, and analyzing the challenges and practical implications. Chapter "Two-Dimensional Nanomaterials for Energy Storage" explores the growing energy demand and its impact on 2D nanomaterial synthesis research, highlighting the unique properties of 2D materials and electrodes based on 2D heterostructures for enhanced performance. Chapter "Graphene Oxide-Based Hybrid Composite Films for Fuel Cell Applications" examines the synthesis, properties, and applications of fuel cells, focusing on recent advancements in graphene-based materials that enhance efficiency, particularly in SOFCs and PEMFCs. Chapter "Environmental Impact of Nanostructured Materials for Energy Storage Devices" explores the intersection of environmental challenges, energy demands, and sustainable practices, showcasing the role of smart materials in energy storage and eco-friendly solutions. Chapter "Evolution of Energy Storage System and Its Life-Saving Applications" discusses the latest advancements in smart materials research and the transformative evolution of energy storage systems across various sectors, emphasizing their crucial role in supporting renewable energy and ensuring reliable power. Finally, Chapter "MXene-Based Functional Materials for Supercapacitors Energy Device" explores the potential of MXenes as cutting-edge materials for next-generation supercapacitors, leveraging their exceptional properties to enhance energy storage capabilities. Despite their promise, MXenes encounter significant challenges, which researchers are actively tackling through innovative material modifications and advancements in synthesis techniques and device integration.

This book aims to inspire readers to explore the exciting field of low-dimensional and multifunctional hybrid nanostructured materials. Feedback and suggestions are welcomed for future editions.

New Delhi, India — Rakesh Kumar Sonker
New Delhi, India — Kedar Singh
Kolhapur, India — Rajendra Sonkawade

Contents

About the Editors

Dr. Rakesh Kumar Sonker has received his Ph.D. degree in 2016 from Department of Applied Physics, Babasaheb Bhimrao Ambedkar University, Lucknow, India. Subsequently he joined the Department of Physics, Acharya Narendra Dev College, University of Delhi, as Assistant Professor. Presently, he is working as an Assistant Professor in the Department of Physics, Motilal Nehru College, University of Delhi, India. Also, Dr. Sonker was selected in 2017 for National Post-doctoral Fellowship of Department of Science and Technology, selected for BRICS Young Scientist Forum 2022, India. His current interests of research include the synthesis of functional 2D/3D nanomaterial, smart nanofibers/nanoporous materials, Quantum Dots, metallopolymers, etc., characterizations and their energy storage device and sensor applications.

Prof. Kedar Singh received his Ph.D. degree in Physics 2003, from the University of Rajasthan, Jaipur, India. Subsequently he joined the Department of Physics, Institute of Science, Banaras Hindu University, Varanasi, as Assistant Professor. Presently he is Professor of Physics and Dean, School of Physical Sciences, Jawaharlal Nehru University, Delhi, India. His current research interests are in condensed matter physics, magnetic nano material, thermoelectric materials and semiconducting thin films, gas/bio sensors, DMS-quantum dots, core-shell materials, plasmonic nanocrystals, chalcogenide glasses and their nano composites for energy applications etc.

Prof. Rajendra Sonkawade received his Ph.D. degrees in Physics 2006, from the Hemwati Nandan Bahuguna University, Srinagar (Garhwal), Uttarakhand India. Subsequently, he worked at Inter University Accelerator Centre (formerly Nuclear Science Centre), New Delhi research centre of UGC, New Delhi under MHRD, Government of India, for around 18 years as a senior scientist. He has also worked as Professor at Department of Applied Physics, Babasaheb Bhimrao Ambedkar University (Central University), Lucknow, India. Presently he is working as a Professor at Department of Physics, Shivaji University, Kolhapur, Maharashtra, India. His current research interests are in metal oxide thin films and nanostructures for gas sensing radiation physics, supercapacitor and polymer science etc.

Low-Dimensional and Multifunctional Hybrid Materials for Energy Storage: An Introduction

Divya Rani, Prashant Tripathi, Priyanka, and Pushpendra Kumar

Abstract The global energy demand is rapidly increasing, driving extensive research efforts to develop highly efficient energy storage devices. Among these, supercapacitors and batteries are the two most widely used energy storage systems (ESS). The performance of these ESSs is directly influenced by the properties of their key components, including electrodes, separators, and electrolytes. Recent studies highlight the crucial role of low-dimensional materials in enhancing the performance of batteries, supercapacitors, and hybrid ESS, owing to their larger surface area, superior ionic mobility, and enhanced electronic conductivity. However, while nanoparticles as individual units offer significant advantages, they also exhibit certain limitations. These shortcomings can be effectively addressed by integrating nanomaterials of different dimensionalities, each possessing distinct physical and chemical properties. This chapter provides a comprehensive overview of low-dimensional materials, including 0D, 1D, and 2D structures, along with their composites and hybrids in combination with conventional electrode and current collector materials for batteries and supercapacitors. It begins with a discussion on the working mechanisms of supercapacitors and batteries, followed by an in-depth analysis of research advancements in pristine 0D, 1D, and 2D materials, as well as their composites and hybrid structures for energy storage applications. This chapter serves as a valuable reference for researchers engaged in the development of next-generation energy storage devices.

1 Introduction

The extensive use of fossil fuels leads to excessive CO_2 emissions, making the development of renewable energy sources inevitable. However, these renewable energy sources are intermittent, necessitating efficient energy storage systems (ESS)

D. Rani · Priyanka · P. Kumar (✉)
School of Physical Science, Jawaharlal Nehru University, New Delhi, India
e-mail: pushpendra@mail.jnu.ac.in

P. Tripathi
Department of CSE, Faculty of Engineering and Technology, SGT University, Gurugram, Haryana, India

R. K. Sonker et al. (eds.), *Low-Dimensional Multifunctional Hybrid Nanostructured Materials*, Engineering Materials, https://doi.org/10.1007/978-981-96-7952-2_1

to balance supply and demand. ESS plays a crucial role in reducing peak energy consumption, mitigating power outages, and powering electric vehicles. Energy storage systems can be classified based on their energy density and power density. Among them, batteries and supercapacitors represent two key categories. Supercapacitors excel in applications requiring rapid energy transfer, such as buses, trains, elevators, and memory backup systems. In contrast, batteries store and deliver energy more gradually, making them suitable for devices like laptops, smartphones, and electric vehicles. While batteries cannot provide instantaneous energy surges, supercapacitors are unable to sustain a steady energy supply over long durations. Consequently, both technologies are essential depending on specific application requirements. In certain scenarios, a combination of both storage mechanisms is required, particularly in electric vehicles (EVs) [1]. EVs demand sudden bursts of energy for starting, acceleration, and deceleration while requiring a stable energy flow to maintain constant velocity. Hybrid energy storage systems (HESS) integrate the characteristics of both supercapacitors and batteries to meet these diverse energy demands [2]. Given the ongoing energy crisis, researchers are exploring lower-dimensional materials with high energy density, superior power output, and enhanced cycling stability.

The performance of electrochemical storage systems is significantly influenced by surface sites, particle size, and interfacial properties. Lower-dimensional materials offer a larger surface area, improved ionic mobility, and superior electronic conductivity compared to conventional electrode materials, making them highly promising for next-generation batteries, supercapacitors, and hybrid ESS [3]. Despite their potential, only a few lower-dimensional materials have been incorporated into conventional ESS so far. Notable examples include multi-walled carbon nanotube additives and carbon-coated silicon nanoparticles. Through nanoscale engineering, electrochemical reactions can be tailored to specific needs. For instance, the pore size in carbon–sulfur composites determines the formation of lithium polysulfide chains [4, 5]. Similarly, nano-structured hybrid architectures, such as carbon-silicon and carbon–sulfur composites, help address challenges associated with conversion and alloying electrodes, paving the way for advanced energy storage solutions [3].

This chapter explores recent advancements in low-dimensional materials for energy storage systems. The progress made in developing these materials and their composites with conventional materials is discussed in detail. Before delving into specific advancements, a comprehensive review of the underlying mechanisms of supercapacitors and batteries is essential.

2 Mechanism

2.1 Mechanism for Supercapacitor

A capacitor stores electrical energy by accumulating charges on closely spaced electrodes separated by an insulating material. A supercapacitor is an advanced form of capacitor with significantly higher capacitance. Unlike conventional capacitors, which utilize a solid dielectric between their electrodes, supercapacitors employ a liquid electrolyte that serves as an electronically insulating medium while facilitating ion migration. Supercapacitors store electrical charge through two primary mechanisms: electric double-layer capacitance (EDLC) and pseudocapacitance [6].

2.1.1 Mechanism for EDLC

The mechanism of an electric double-layer capacitor (EDLC) as presented in Fig. 1, closely resembles that of a conventional dielectric capacitor, as initially described by the Helmholtz model. According to this model, two layers of opposite charges form at the electrode/electrolyte interface at an atomic-scale distance, without any charge transfer occurring. The capacitance in such a system is exceptionally high due to the minimal separation between these layers ($d \ll 10^{-9}$ m), which is on the atomic scale:

$$C = \frac{k \in_0 A}{d} \tag{1}$$

The Gouy-Chapman model builds upon the Helmholtz model, which describes ions as point charges whose concentration near an electrode follows the Boltzmann

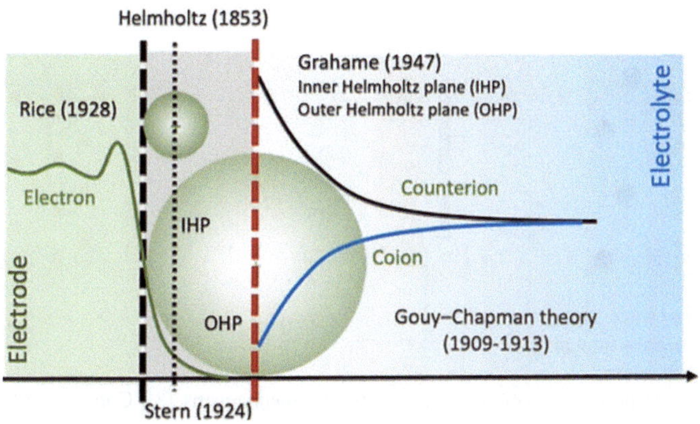

Fig. 1 Various explanations for the EDLC mechanism [6]. Reproduced with permission @ copyright (2022) American Chemical Society

distribution law. However, the Helmholtz model fails to accurately represent the highly charged layer of ions near the electrode surface. To address this limitation, Stern proposed a revised model that integrates both the Helmholtz and Gouy-Chapman theories. In Stern's model, the ion distribution is divided into two distinct regions: the Stern layer and the diffuse layer [7]. Activated carbon, graphene, and carbon nanotubes are among the most widely used materials for electric double-layer supercapacitors due to their high surface area and excellent conductivity.

2.1.2 Mechanism for Pseudo-capacitor

Charge storage in pseudo-capacitive supercapacitors occurs through faradaic reactions at the electrode surface (Fig. 2). These supercapacitors are further classified into three subcategories:

1. Underpotential Deposition Supercapacitors
2. Surface Redox Pseudo-Capacitive Supercapacitors
3. Intercalation Pseudo-Capacitive Supercapacitors.

- ***Underpotential Deposition Supercapacitors***

Underpotential deposition (UPD) refers to the electrodeposition of a metal onto another at a potential less negative than its equilibrium potential. This technique is employed in underpotential deposition supercapacitors for charge storage [7].

- ***Surface Redox Pseudo-Capacitive Supercapacitors***

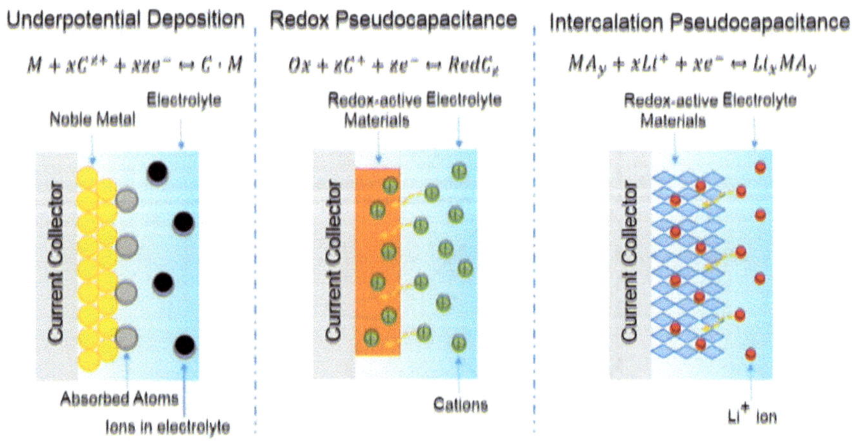

Fig. 2 Different pseudo-capacitance charge storage mechanisms [8]. Copyright © (2022) John Wiley and Sons

In these supercapacitors, charge storage occurs via electrochemical adsorption of ions onto the electrode surface through redox reactions. This process enhances capacitance by facilitating rapid and reversible faradaic reactions [8].

- ***Intercalation Pseudo-Capacitive Supercapacitors***

These supercapacitors store charge by intercalating ions between atomic layers of the electrode material through redox reactions. For example, MXene-based supercapacitors accommodate ions between MXene sheets. While this mechanism resembles that of batteries, the key difference is that intercalation is limited to the electrode surface, often referred to as superficial intercalation. Since faradaic charge transfer is involved, intercalation pseudo-capacitive supercapacitors exhibit lower power density compared to EDLCs [8].

2.2 Mechanism for Batteries

Electrochemical cells are of two types:

i. **Galvanic cell**: It engenders an electric current from a redox reaction.
ii. **Electrolytic cell**: Electrolytic cells utilize an electric current to effectuate chemical reactions.

A secondary electrochemical cell functions as both a galvanic and an electrolytic cell. During discharge, it operates as a galvanic cell, converting chemical energy into electrical energy. Conversely, during charging, it behaves as an electrolytic cell, requiring external energy to drive the reverse reaction. The fundamental redox reactions underpinning these processes enable the cell to store and release energy efficiently. Similarly, rusting of iron is also a redox reaction. In this process, iron undergoes oxidation by losing electrons to oxygen in the presence of moisture or water, forming hydrated ferric oxide. Simultaneously, oxygen is reduced, completing the redox cycle that leads to corrosion [8].

$$4Fe + 3O_2 + 2xH_2O \rightarrow 2Fe_2O_3 \cdot xH_2O$$

Electron transfer can be harnessed effectively if electrons are directed through an external circuit, this is precisely the function of electrochemical cells. These cells consist of two electrodes, the cathode and anode, which possess different electrochemical potentials and are separated by a separator while both remain in contact with an electrolyte. To facilitate electron collection, the electrodes are connected to current collectors. The choice of the current collector depends on the electrode material and its electrochemical potential, ensuring that the collector remains stable and does not participate in side reactions. Typically, different materials are used for different electrodes based on their operating potential. For instance, copper (Cu) is commonly used as a current collector for graphite anodes, while aluminum (Al) is employed for lithium cobalt oxide (LCO) cathodes as presented in Fig. 3a [9].

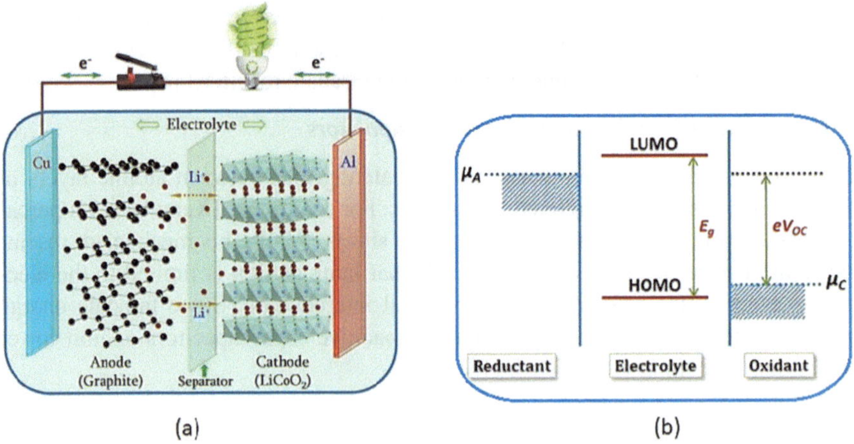

Fig. 3 **a** Typical Li-ion cell [9], **b** electrochemical window of electrolyte [10]. Reproduced with permission @ copyright (2016) Elsevier, reproduced with permission @ copyright (2013) American Chemical Society

Oxidation and reduction reactions in electrochemical cells are two complementary processes that occur simultaneously at different electrodes. In a Li-ion battery, for instance, lithiated graphite undergoes oxidation, while cobalt oxide undergoes reduction during discharge, driven by the negative Gibbs free energy difference between the reactants and products, indicating a spontaneous reaction. During charging, the reverse process occurs when an external potential is applied.

$$Li_xC_6 \rightarrow xLi^+ + xe^- + 6C \quad \text{Anodic reaction}$$

$$Li_{1-x}CoO_2 + xLi^+ + xe^- \rightarrow LiCoO_2 \quad \text{Cathodic reaction}$$

$$Li_xC_6 + Li_{1-x}CoO_2 \rightarrow LiCoO_2 + C_6 \quad \text{Overall cell reaction}$$

The electrolyte serves as the medium through which Li-ions shuttle between the electrodes. Therefore, it must exhibit high ionic conductivity while remaining an electronic insulator and should not participate in redox reactions. To ensure this, the lowest unoccupied molecular orbital (LUMO) of the solvent should be higher than the chemical potential of the anode, and the highest occupied molecular orbital (HOMO) should be lower than the chemical potential of the cathode, thereby preventing unwanted redox reactions, as illustrated in Fig. 3b [10]. However, the stability of the electrochemical window is not solely dictated by the HOMO and LUMO levels. It also depends on factors such as the Gibbs free energy difference between reactants and products [11], dissociation capacity, steric hindrance, viscosity, and several other parameters. While these factors play a crucial role, they fall beyond the scope of this chapter. To ensure battery safety, a separator is placed between the electrodes.

Functioning like a membrane, it acts as an electronic insulator while permitting the passage of electrolytes and Li-ions. In solid-state batteries, both the electrolyte and separator are replaced by a single component known as the solid electrolyte [12].

3 Zero-Dimensional and Their Multifunctional Hybrid Materials in ESS

Zero-dimensional (0D) particles possess all three dimensions within the range of 2–10 nm. Their ultra-small size results in a high surface-to-volume ratio, providing a substantial number of active sites that make them ideal candidates for energy storage and high-power density applications. Furthermore, the functional groups attached to their surface not only enhance their physicochemical properties but also enable the formation of advanced composites tailored for energy storage [13].

3.1 Zero-Dimensional and Their Multifunctional Hybrid Materials in Supercapacitors

Carbon-based materials are widely used in electric double-layer supercapacitors. Many groups have explored the integration of carbon quantum dots (CQDs) and graphene quantum dots (GQDs) with specific metals, metal oxides, hydroxides, and conductive polymers, presenting a promising approach for supercapacitor applications. Various composites, such as NiO/CQDs, CuO/GQDs, RuO_2/CQDs, NiS/CQDs, and $Ni(OH)_2$/CQDs, have been investigated for their potential in supercapacitors [13]. Metal oxide nanoparticles typically function as electrochemical pseudocapacitors (PCSs), storing energy through redox reactions of electro-active species. The combination of PCSs with good electrical conductors and EDL-type materials, such as carbon-based quantum dots, enhances specific capacitance. For instance, NiO/CQDs exhibit specific capacitance values of 1572 F g^{-1} and 956 F g^{-1} at current densities of 2 A g^{-1} and 10 A g^{-1}, respectively. After 1000 galvanostatic charge–discharge (GCD) cycles, a capacity retention of 93% is maintained at a current density of 2 A g^{-1}. The strong interaction between CQDs and NiO, coupled with their distinctive one-dimensional nanostructure and extensive specific surface area, contributes to the remarkable electrochemical properties of the CQDs/NiO electrode [14]. Reduced CQDs decorated with RuO_2 (RCQDs/RuO_2), with a high Ru loading of 41.9%, address the need for high power density. The RCQDs framework in RCQDs/RuO_2 facilitates charge transfer, and ionic mobility, and prevents RuO_2 nanoparticle aggregation. Consequently, a high specific capacitance of 460 F g^{-1} is achieved at 50 A g^{-1}, with excellent stability demonstrated by 93% capacity retention after 5000 GCD cycles at 5 A g^{-1} [15]. Yibing Xie and colleagues synthesized TiO_2 nanotubes on a Ti sheet via anodization and subsequently deposited a

CQDs-PPy layer using pulse voltammetry. The resulting CQDs-PPy/TiO_2 nanotubes serve as effective supercapacitor electrode materials. Due to their highly electrochemically active surface, these nanotubes retain 89% of their capacity at 20 A g^{-1} after 2000 GCD cycles, enabling the utilization of most of the PPy active sites. At 5 A g^{-1}, CQDs-PPy/TiO_2 nanotubes exhibit a specific capacitance of 533 F g^{-1} [16]. Removing the TiO_2 template results in a CQDs-PPy hybrid, which Xie et al. identified as a potential electrode material for flexible solid-state supercapacitors. However, this flexible supercapacitor exhibits a relatively low specific capacitance of 13 F g^{-1} even at a very low current density of 0.005 A g^{-1} [16]. Recently, MXene quantum dots have gained attention as electrode materials for supercapacitors and as composite materials in battery separators, particularly in Li–S batteries [17, 18]. Liu et al. prepared two-dimensional $Ti_3C_2T_x$ nanosheets (TCNs) containing zero-dimensional $Ti_3C_2T_x$ quantum dots (TCQDs) through microwave hydrothermal treatment [17]. The resulting TCQDs-TCNs' structure exhibits an areal capacity of 803 mF cm^{-2} at 1 mA cm^{-2}, attributed to TCQDs' abundant active sites and functional groups at the edges that enhance electrolyte ion capture. Meanwhile, TCNs provide highly electrochemically active matrices for efficient and rapid charge transfer. Zhao et al. developed a GQDs-integrated electro-spun carbon nanofiber (CNF) structure (AGRCNF) through carbonization and chemical activation (Fig. 4). This freestanding supercapacitor electrode achieves a specific capacitance of 335 F g^{-1} at 1 A g^{-1} [19]. Beyond these materials, various researchers have investigated transition metal-based quantum dots for energy storage system (ESS) applications [20].

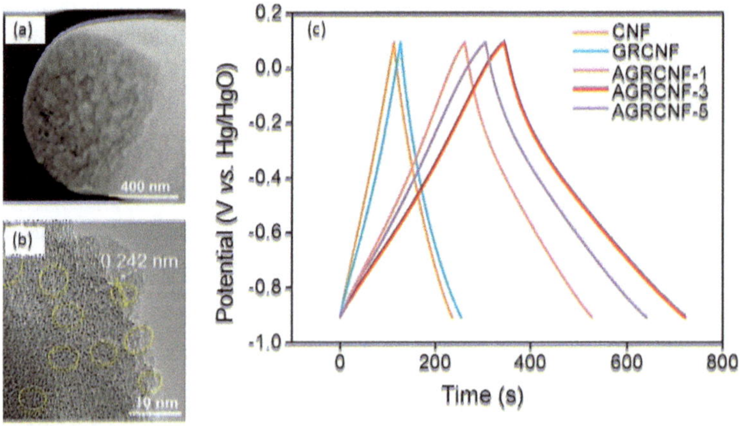

Fig. 4 **a**, **b** SEM, HRTEM images of AGRCNF-3, **c** GCD curves at 1 A/g of CNF, GRCNF and AGRCNF-X [19]. Reproduced with permission @ copyright (2020) American Chemical Society

3.2 *Zero-Dimensional and Their Multifunctional Hybrid Materials in Batteries*

These nanoparticles exhibit three-dimensional quantum confinement, as most of them have a radius ranging from 1 to 10 nm. Quantum dots are commonly incorporated into composite materials due to the presence of functional groups on their surface, which enhance their ability to interact with other components within the composite [21]. Fullerene (C_{60}) is a zero-dimensional material with a cage-like structure composed of hexagonal and pentagonal carbon rings. Its dimensions vary based on the number of these rings, which are interconnected by single and double carbon bonds [22]. Pure C_{60} nanoparticles (C_{60} NPs) can be synthesized through heat-induced evaporation of hand-ground C_{60} (C_{60} HG) powder at 800 °C. Figure 5 depicts the TEM/HRTEM images of BPQDs, PCNF/S/BPQDs, PCNF/S/BPQDs cathode, and the specific capacity of PCNF/S/BPQDs cathode over 200 GCD cycles [23].

Yin et al. investigated the application of fullerene in a Li||C_{60} NPs half-cell using $LiClO_4$ in dioxolane (DOL) and dimethyl ether (DME) as the electrolyte. This configuration exhibited a capacity of 359 mAh g^{-1} at a current density of 5 A g^{-1} [24]. The C_{60} coating on silicon anodes, particularly on silicon nanowire anodes, enhances structural integrity and mitigates undesirable reactions with the electrolyte [24]. Fullerene-based composites have demonstrated significant improvements in both capacity and cycle life when employed in supercapacitors. Specifically, a composite material comprising polyaniline, phenyl-C_{60}, methyl ester, and butyric acid achieved a specific capacitance of 2200 F g^{-1} under a current density of 2 A g^{-1} [25]. Carbon-based dots are also categorized as zero-dimensional particles. These dots consist of sp^2- and sp^3-hybridized carbon atoms, with their respective ratios varying across different carbon-based quantum dots. Among them, carbon nanodots (CNDs), carbon quantum dots (CQDs), and graphene quantum dots (GQDs) are notable. Different synthesis techniques lead to distinct atomic configurations, classifying CNDs as amorphous nanodots, while CQDs and GQDs exhibit crystalline structures. Graphene quantum dots (GQDs) play a crucial role in improving electrolyte characteristics

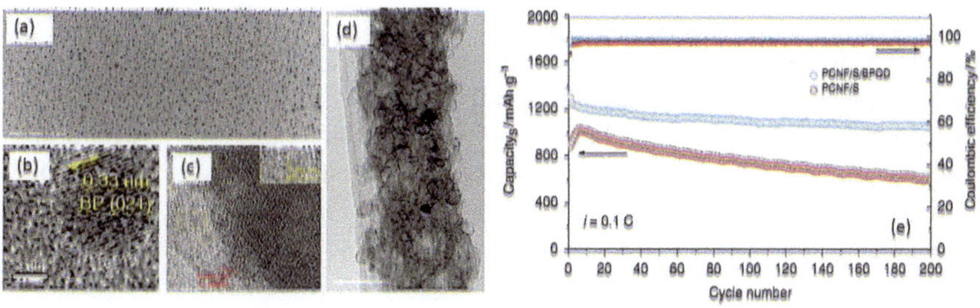

Fig. 5 **a** TEM image of BPQDs, **b** HRTEM image of BPQDs, **c** HRTEM image of PCNF/S/BPQDs, **d** TEM image of PCNF/S/BPQDs cathode, **e** specific capacity of PCNF/S/BPQDs cathode over 200 GCD cycles [23] where (**a**–**d**) 100 nm, 2 nm, 2 nm, 100 nm bar length. Reproduced with permission @ copyright (2018) Springer Nature

in lithium-sulfur (Li–S) batteries by providing heterogeneous nucleation sites that promote uniform lithium deposition and prevent dendrite formation. A significant concentration of GQDs at the electrode–electrolyte interface modulates the local electric field, ensuring dendrite-free lithium plating, as confirmed by in-situ Raman spectroscopy [24]. The GQD-modified electrolyte in a Li||C/S battery exhibits stable cycling performance, maintaining a capacity of 830 mAh g^{-1} at a rate of 0.5C over 200 cycles [26]. Carbon-based quantum dots can also serve as anode materials, applicable to both lithium-ion and sodium-ion batteries. In a half-cell configuration (CQD||Li) utilizing a 1 M $LiPF_6$ solution in a 1:1 molar ratio of ethylene carbonate (EC) and diethyl carbonate (DEC), galvanostatic charge–discharge (GCD) cycling was performed at multiple C-rates (0.5C, 1C, 10C, 20C). Initial cycles exhibited irreversible behavior, attributed to electrolyte decomposition during solid electrolyte interphase (SEI) formation and irreversible lithium-ion storage in CQDs. However, after several cycles, the interlayer spacing expanded, enhancing the contact area and wettability between the electrolyte and CQDs. This led to progressive improvements in coulombic efficiency and specific capacity stability. At a 5C rate, the CQD anode delivered 674 mAh g^{-1} [27]. Lithium and sulfur are considered among the most promising materials for addressing energy storage challenges. However, the commercialization of Li–S batteries faces significant hurdles, particularly the lithium polysulfide shuttling effect. To mitigate this issue, Ding et al. developed a polypropylene (PPy) separator coated with multiwalled carbon nanotubes and TiO_2 quantum dots (MWCNT@TiO_2QDs), achieving long-term cycling stability over 600 cycles at 0.5C [28]. Similarly, Xu and colleagues designed a cathode composite (PCNF/S/BPQD), incorporating black phosphorus quantum dots (BPQDs) with sulfur and porous carbon nanofibers (PCNFs). This composite effectively suppresses polysulfide shuttling through the catalytic activity of BPQD edge sites and the physical confinement provided by PCNF [23].

4 One Dimensional and Their Multifunctional Hybrid Materials in ESS

One-dimensional (1D) materials are characterized by a single dimension that extends beyond the nanoscale range. Due to their high aspect ratio, 1D nanostructures not only enhance electrical connectivity but also mitigate volume expansion issues through their inherently porous structure [29]. Various synthesis techniques are employed for their fabrication, including anodization, hydrothermal/solvothermal methods, electrochemical deposition, electrospinning, and vapour-phase approaches.

4.1 One Dimensional and Their Multifunctional Hybrid Materials in Supercapacitors

Conventional pseudocapacitive materials (CPMs), including conducting polymers, transition metal nitrides, selenides, oxides, and dichalcogenides, commonly suffer from volume expansion during charge–discharge cycles [30]. As previously mentioned, the porosity of one-dimensional (1D) architectures in various materials helps mitigate this issue [29]. Additionally, 1D structures offer advantages such as shortened electronic pathways and reduced ion diffusion lengths, enhancing electrochemical performance [29]. Most EDL supercapacitors are typically composed of carbon-based materials, such as carbon nanotubes (CNTs), carbon nanofibers (CNFs), and multi-walled carbon nanotubes (MWCNTs). While CNTs possess a high surface area, their capacity increases when activated by KOH due to an expansion of their surface area [31]. Tao et al. developed symmetric supercapacitors using 6 M KOH as the electrolyte and electrodes composed of 95% chemical vapour deposition (CVD)-grown CNFs and 5% polytetrafluoroethylene (PTFE) as a binder. Their device exhibited a specific capacitance of 297 F g^{-1} [32]. Similarly, Sun and colleagues synthesized bamboo-like carbon nanofibers (BCNFs) using electrospinning, followed by thermal treatment and etching, which resulted in BCNFs with a core structure resembling bamboo [33] (see Fig. 6 for more details.).

Its unique architecture allows it to be folded up to 180° without cracking. The electrochemical capacitance of this free-standing BCNF electrode was evaluated using a three-electrode setup. The results revealed a specific capacitance of 236 F g^{-1} at 5 A g^{-1}, and after 5000 GCD cycles, the device retained nearly 100% of its initial capacity at a current density of 10 A g^{-1}. The capacitance versus current density graph indicates that beyond 50 A g^{-1}, the specific capacitance remains

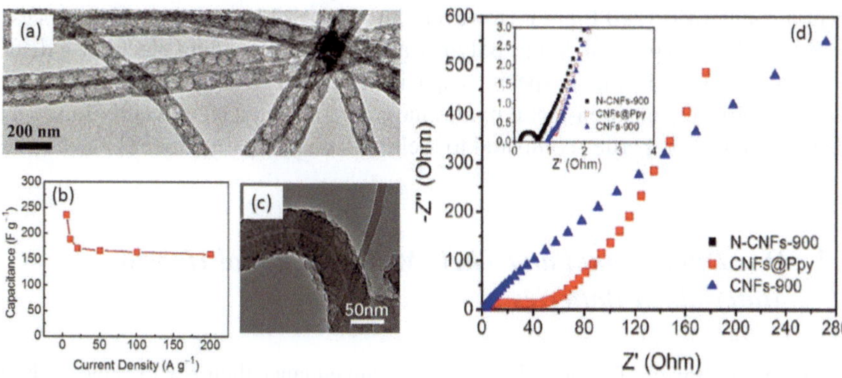

Fig. 6 **a, b** TEM image of BCNFs and specific capacity at different current densities of BCNFs respectively [33]. **c, d** TEM image of N-CNFs and EIS measurement of N-CNFs [34]. Reproduced with permission @ copyright (2015) American Chemical Society, reproduced with permission @ copyright (2012) American Chemical Society

nearly constant up to 200 A g^{-1}. This phenomenon is attributed to the presence of both large and small mesopores, which effectively trap ions even at high current densities. Using free-standing BCNFs, Sun and colleagues developed an all-solid-state flexible supercapacitor that retained 96% of its original capacity after 10,000 GCD cycles at 67 mA cm^{-3} [33]. The pseudocapacitive behaviour of carbonaceous materials can be enhanced through element doping. Chen et al. demonstrated this by carbonizing CNFs coated with polypyrrole in a nitrogen atmosphere to produce nitrogen-doped porous carbon nanofibers (N-CNFs) [34]. The N-CNFs exhibit a roughly rectangular cyclic voltammetry curve with minor humps, indicative of the combined effects of pseudocapacitance and electric double-layer capacitance. A three-electrode setup was used to measure the electrochemical impedance spectroscopy of the N-CNFs electrode. The results showed lower impedance compared to CNFs@PPy and CNFs-900. The specific capacitance was 202 F g^{-1} in the first GCD cycle and remained at 196 F g^{-1} after 3000 GCD cycles at 1 A g^{-1}. This excellent stability is attributed to the uniformly distributed mesopores and micropores in the N-CNFs, which facilitate rapid ion transport. Chen et al. also demonstrated that annealing temperature significantly influences pore size [34]. Beyond carbonaceous and doped-carbonaceous materials, scientists have explored conducting polymers and carbon-based 1D material composites, which exhibit both electric double-layer capacitance and pseudocapacitive properties. One such example is a polyaniline (PANI)-porous carbon nanofiber (PCNF) composite, synthesized via the polymerization of aniline on PCNFs [35]. The PCNFs contribute to the formation of the electric double layer, while PANI participates in redox reactions. Consequently, although the capacity retention of PANI-PCNFs is slightly lower than that of EDLC-based supercapacitors, its specific capacitance is higher. To optimize the electrochemical performance of PANI-PCNFs, polymerization duration was varied, and the resulting composites were analyzed. The composite synthesized with a 12 h polymerization time exhibited superior specific capacitance (238 F g^{-1}) compared to the 6 h polymerization counterpart (220 F g^{-1}). However, during the 12 h polymerization process, excessive PANI deposition led to polymer aggregation, impeding electrolyte diffusion and causing a decline in specific capacitance and current density [35]. Furthermore, porous 1D transition-metal oxides have demonstrated significantly improved electrochemical properties compared to their bulk counterparts [36–38].

4.2 One Dimensional and Their Multifunctional Hybrid Materials in Batteries

The incorporation of 1D materials in batteries can enhance their performance at high C-rates and alleviate issues related to volume expansion in alloying-type electrode materials. Through template-directed synthesis, $LiMn_2O_4$ nanotubes are fabricated using MWCNTs as a template. Due to the specific orientation of graphene sheets in the MWCNT template, the resulting $LiMn_2O_4$ nanotubes retain a well-defined crystal orientation, as confirmed by X-ray diffraction (XRD) analysis.

The specific capacity of aligned $LiMn_2O_4$ nanotubes in an aqueous electrolyte (0.5 M Li_2SO_4) is 102 mAh g^{-1} at 1 Ag^{-1}, corresponding to a 9.1C rate. These nanotubes enable ultrafast charging and discharging of the cell in just 6 s (600C) while retaining 54% of their initial capacity. This remarkable capacity retention at high current densities is attributed to the 10 nm thin $LiMn_2O_4$ layer on MWCNTs, which facilitates efficient lithium intercalation and deintercalation. Additionally, the superior conductivity of aqueous electrolytes, approximately twice that of organic electrolytes, enhances performance at high C rates [40]. Furthermore, Zhang and colleagues synthesized 1D nanoporous $LiNi_{0.5}Mn_{1.5}O_4$ tubes, which exhibit a specific capacity of 109 mAh g^{-1} at 20C, significantly outperforming bulk $LiNi_{0.5}Mn_{1.5}O_4$, which delivers only 80 mAh g^{-1} at the same rate. At high current densities, Li-ion accumulation on the bulk electrode surface obstructs diffusion pathways, whereas, in nanosized 1D electrodes, ion accumulation is minimized, allowing for more reversible rapid charging [41]. Similarly, 1D porous nanotubes of $LiFePO_4$ and $Li_3V_2(PO_4)_3$ demonstrate outstanding electrochemical performance as electrode materials [42, 43].

For alloying electrodes, Si implanted or encapsulated in carbon-based porous nanofibers effectively mitigates pulverization and volume expansion. Park and colleagues synthesized Si nanotubes coated with carbon to stabilize the Si-electrolyte interface via solid electrolyte interphase (SEI) formation [44]. Wu et al. developed double-walled Si–SiO_x nanotubes, where Si forms the inner layer and SiO_x constitutes the outer layer. This architecture confines lithiation-induced expansion toward the hollow interior of the tube. The interaction between lithium ions and SiO_x leads to SEI formation on the outer surface ($SiO_x + Li = Si + xLi_2O$), enhancing structural stability. As a result, the material retains 90% coulombic efficiency even after 6000 cycles at a 12C rate [45]. Mesoporous Si nanowires (MSNWs), synthesized via metal-assisted chemical etching, exhibit microporosity, which is critical for long-term stability [46]. Yu and coworkers fabricated SnO_2 nanoparticles (2–3 nm in size) embedded in CNTs. During the conversion reaction, transition metal oxides transform into nano-sized metallic particles dispersed within lithium oxide matrices. Since SnO_2 nanoparticles are confined within CNTs (S-CNTs), phase changes during charge/discharge cycling are limited, enhancing cycle stability. This composite retains 26% of its initial capacity after 620 charge/discharge cycles at 4 A g^{-1} [47]. A simple hydrothermal method enables the dispersion of porous Fe_3O_4/graphene nanowires onto amorphous vanadium oxide matrices, forming Fe_3O_4/VO_x/GNWs composites [39] as shown in Fig. 7. This unique structure delivers a specific capacity of 808 mAh g^{-1} at 2 A g^{-1}. Graphene nanowires enhance ion diffusion and electron transport while also accommodating the volume expansion of iron oxide. Additionally, the porous nature of iron oxide increases the electrode–electrolyte interaction area. Several 1D materials have also been investigated as electrode materials for sodium-ion batteries, including TiO_2 nanotubes, $Na_2Ti_3O_7$ nanotubes, and $MgTi_2O_5$/C nanorods [48–51].

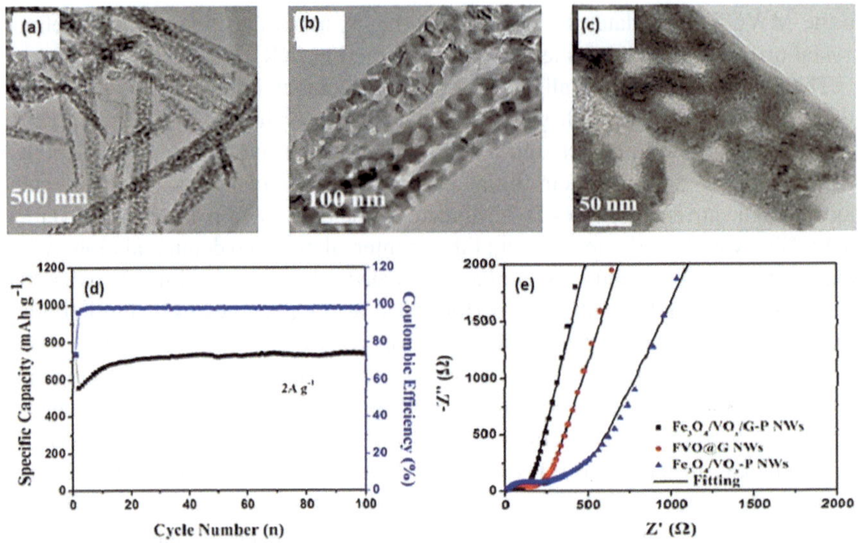

Fig. 7 **a**, **b** TEM image of Fe_3O_4/VO_x/GNWs, **c** TEM image of Fe_3O_4/VO_x/GNWs after first discharge, **d** cycling performance of Fe_3O_4/VO_x/GNWs, **e** EIS measurement of Fe_3O_4/VO_x/GNWs [39]. Reproduced with permission @ copyright (2014) American Chemical Society

5 Two-Dimensional and Their Multifunctional Hybrid Materials in ESS

Two-dimensional (2D) materials are characterized by a single nanoscale dimension, granting them unique properties. These materials exhibit excellent electrical conductivity, an exceptionally high surface area, and remarkable electrochemical characteristics. Their layered structure provides stability across chemical, electrochemical, and thermal conditions [52]. However, the widespread application of 2D materials is hindered by their inherent tendency to restack, which limits their performance in certain applications [53].

5.1 2D and Their Multifunctional Hybrid Materials in Supercapacitors

Two-dimensional (2D) materials, such as MXenes, are highly promising electrode materials for ESS. MXenes are typically synthesized by selectively etching the A-layer from MAX phases using chemical etching methods. For instance, the removal of aluminum (Al) from Ti_3AlC_2 through chemical etching, followed by dimethyl sulfoxide (DMSO) intercalation, results in well-separated Ti_3C_2 sheets, enhancing their electrochemical performance [54].

KOH-treated Ti_3C_2 exhibits a volumetric capacitance of 215 F cm^{-3} at 5 A g^{-1} [54]. However, functional group interactions cause MXene layers to re-stack, limiting their effectiveness as electrode materials in supercapacitors. To address this challenge, Ayman and colleagues developed a composite of MXene and cobalt ferrite nanoparticles (CoF NP/MXene) (Fig. 8). These nanoparticles act as spacers between MXene layers, preventing restacking. The CoF NP/MXene composite achieves a specific capacitance of 1268 F g^{-1} at 1 A g^{-1} [55]. The presence of cobalt ferrite nanoparticles introduces redox peaks, which are typically absent in pure MXene-based supercapacitors. Beyond MXenes, the electrochemical properties of MoS_2 and polyaniline (PANI) composites have been studied by Wang and colleagues. PANI is a well-known pseudocapacitive electrode material for supercapacitors, but the instability of its redox-active sites leads to structural degradation over repeated charge/discharge cycles. MoS_2, on the other hand, has a large surface area suitable for electric double-layer capacitors (EDLC) and contains Mo centers capable of oxidation states ranging from +2 to +6, contributing to its pseudocapacitive behavior. In the MoS_2/PANI composite, PANI chains are intercalated between MoS_2 sheets, enhancing mechanical stability during redox reactions while improving specific capacitance by combining pseudocapacitive and EDLC properties. The composite achieves a specific capacitance of 390 F g^{-1} at 0.8 A g^{-1}. Additionally, the MoS_2 content plays a critical role in conductivity, charge transfer, and stability [56]. Due to their inherent pseudocapacitive properties, transition metal hydroxides such as $Ni(OH)_2$ and $Co(OH)_2$ are widely recognized for their high capacitance, following transition metal oxides and sulfides. Vidhya and colleagues demonstrated that Co–$Ni(OH)_2$ exhibits even higher specific capacity and ionic diffusion rates than pure metal hydroxides, as confirmed by GCD and electrochemical impedance spectroscopy (EIS) tests [57]. At 1.5 A g^{-1}, Co–$Ni(OH)_2$ delivers a specific capacitance of 1366 F g^{-1} with a retention rate of 96.26% after 2000 cycles. Integrating transition metal sulfides with carbon-based materials is an effective strategy to enhance specific capacitance and charge transfer properties. Tu et al. synthesized a WS_2/reduced graphene oxide (rGO) composite using a molten salt method. In this composite, W^{6+}/W^{4+} redox reactions and oxygen-containing functional groups on the rGO surface store energy via faradaic reactions, while the rGO-electrolyte interface facilitates EDLC storage. Additionally, rGO enhances charge transfer by increasing the hybrid's conductivity. The WS_2/rGO composite prepared via the molten salt method achieves a specific capacitance of 1355 F g^{-1} at 2 A g^{-1} [58]. Tomy et al. [59] have provided a comprehensive review of other 2D materials employed in supercapacitor applications, including black phosphorus, polymer composites, and carbon-based materials.

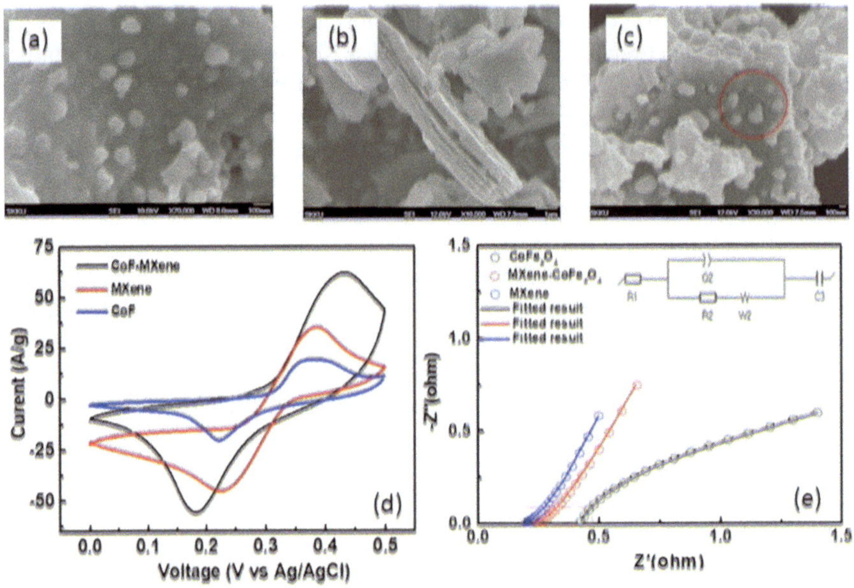

Fig. 8 **a** SEM image of CoF NP, **b** SEM image of exfoliated MXene layers, **c** SEM image of CoF NPs in MXene layers, **d** CV curve of CoF-MXene composite, MXene and CoF, **e** EIS measurement of CoF-MXene composite, MXene and CoF [55]. Reproduced with permission @ copyright (2020) American Chemical Society

5.2 Two-Dimensional and Their Multifunctional Hybrid Materials in Batteries

In batteries, redox reactions occur within the bulk of the electrode, leading to phase transformations and volumetric expansion of electrode materials, which in turn contribute to the degradation of battery performance. These challenges can be mitigated by utilizing 2D materials as electrode components. For instance, 2D materials can help address issues such as pulverization in alloy-type electrodes, polysulfide shuttling in lithium-sulfur batteries, and the poor reversibility of lithium peroxide formation in Li-air batteries [60].

5.2.1 Electrode Composite with 2D Materials and Their Coating on the Electrode

Efficient ion transport in 2D materials enhances their ionic conductivity, and their electrical properties can be tailored to meet specific requirements. In widely used electrode materials such as LCO, NMC, LFP, Si, Sn, and LTO, poor electrical conductivity contributes significantly to polarization. For instance, graphene exhibits an exceptional charge carrier mobility of 200,000 cm^{-2} Vs and an electrical conductivity of 10^6 S cm^{-1}. Besides graphene, acetylene black and carbon black are extensively used as electrode additives in batteries [61]. Incorporating just 5% graphene into cathode composites can significantly enhance diffusion coefficients. This improvement is attributed to the formation of long-range electron transport pathways and high electron conductivity [59]. Furthermore, electrode corrosion can be mitigated by coating the electrode with suitable 2D materials. Rajmjoo et al. applied an outer layer of rGO and an inner layer of SiO_2 to NMC (532). It has been observed that this double-layer coating effectively prevents HF-induced dissolution of transition metals [62]. Another cathode material, $Li_3V_2(PO_4)_3$ (LVP), has a very low intrinsic conductivity of just 10^{-8} S cm^{-1}. Zhao et al. developed a composite of graphene and LVP, which not only enhances electrical conductivity but also mitigates vanadium dissolution during Li-ion electrodeposition [63].

5.2.2 Application of 2 D Materials to Prevent Polysulfide Shuttling

The challenge of low energy density in lithium–sulfur (Li–S) batteries can be addressed through advanced material design; however, these batteries face significant obstacles, particularly polysulfide shuttling. To mitigate this issue, Fan and colleagues coated a polypropylene (PP) separator with a reduced graphene oxide-sodium lignosulfonate (rGO-SL) composite. The rGO-SL/PP separator offers a crucial advantage by introducing negatively charged sulfonic groups, which effectively repel polysulfides while permitting the free passage of Li^+ ions. When incorporated into a full cell, this separator achieved a 74% coulombic efficiency over 1000 charge–discharge cycles at a 2C rate [64]. Another strategy involves optimizing cathode host materials in Li–S batteries. Porous carbon structures are commonly employed to confine sulfur and mitigate polysulfide dissolution. Li and co-workers developed a few-layer phosphorene-carbon nanofiber (FLP-CNF) framework as a cathode host, leveraging the strong interaction between lithium polysulfides and phosphorene. This framework not only enhances sulfur retention but also facilitates rapid redox reactions between sulfur and lithium sulfide (Li_2S), improving electrochemical performance [65]. Additionally, extensive research has explored the application of two-dimensional materials in addressing polysulfide shuttling, as detailed by Rojaee [60].

5.2.3 2D Materials Usage in Current Collectors

When selecting a material for the current collector, it is crucial to consider its susceptibility to corrosion and compatibility with various electrolytes. Shi et al. addressed these limitations by replacing conventional copper and aluminum current collectors with a graphene membrane. Using vacuum filtration, they integrated the active material into the graphene membrane, which significantly increased the contact area between the electrode and the collector. This modification resulted in a considerable reduction in electrode polarization compared to traditional aluminum and copper current collectors [66]. Similarly, Lei et al. employed a pulse laser deposition technique to coat graphene paper with V_2O_5, utilizing it as a cathode, while lithiated graphene served as the anode [67]. Additionally, MXene sheets have gained attention as alternative current collectors. Wang et al. demonstrated the use of d-$Ti_3C_2T_x$ as a current collector for both the cathode and anode, leveraging its broad electrochemical window, which prevents involvement in redox reactions [68]. The specific capacitance and rate performance of batteries utilizing $Ti_3C_2T_x$-based current collectors were found to be comparable to those employing conventional Cu and Al collectors. However, a key advantage of $Ti_3C_2T_x$ MXene-based collectors is their contribution to reducing device thickness and weight [68]. Rojaee has provided a comprehensive discussion on the enhanced applications of 2D materials in ESS, highlighting their potential in next-generation battery technologies [60].

6 Hybrid Energy Storage Systems

Hybrid energy storage systems (HESS) integrate two or more energy storage technologies to optimize performance. These combinations can include: (1) Fuel-cell and battery, (2) Battery and supercapacitor and (3) Hydrogen and supercapacitor [69]. Battery-supercapacitor HESS has been extensively employed in electric vehicles. Under heavy load conditions, the system shifts to supercapacitor mode, while under light load conditions, it operates in battery mode [70]. Instead of using separate battery and supercapacitor units, an innovative approach is to integrate both functionalities into a single unit by combining battery-type and supercapacitor-type materials within a single cell. For instance, the cathode material VO_2 has been combined with an activated carbon anode to achieve such integration. Sahoo and colleagues developed a rGO@VO_2 composite as the cathode and utilized activated carbon on carbon cloth (AC@CC) as the anode. The electrochemical performance of the rGO@VO_2||AC@CC cell is shown in Fig. 9 [71]. Luo et al. synthesized Nb-doped NMC and fabricated an electrode using Nb-NMC, graphene, Super P, and PVDF in a 72:8:10:10 ratio. This capacitor-battery hybrid system exhibited stable cycling at a 5C rate, retaining 73.9% of its initial capacity after 300 cycles. The incorporation of graphene and other carbon-based additives, featuring 1D to 3D configurations and functional groups such as C=O and C–O, contributed to excellent electrical conductivity, reaching approximately 10 S cm^{-1} [72]. Hu et al. developed an organic electrolyte-based zinc-ion battery that demonstrated high energy density and long cycle life, combining the characteristics of both a battery (high energy density) and a supercapacitor (extended lifespan). Their design incorporated bipolar conjugated microporous polymers onto rGO (BCGM@rGO) as the cathode material. The Zn||BCGM@rGO cell retained 90% of its initial capacity after 25,000 cycles at 0.2 A/g [73]. Similarly, Shi and his team constructed MXene-based coaxial Zn-ion hybrid fiber supercapacitors, achieving a capacitance of 101 mF cm^{-2} at 200 mV s^{-1}. After 5000 cycles at 20 mA cm^{-2}, the device maintained 83.58% of its initial capacity [74]. For a comprehensive review of HESS, refer to Zheng's work [75].

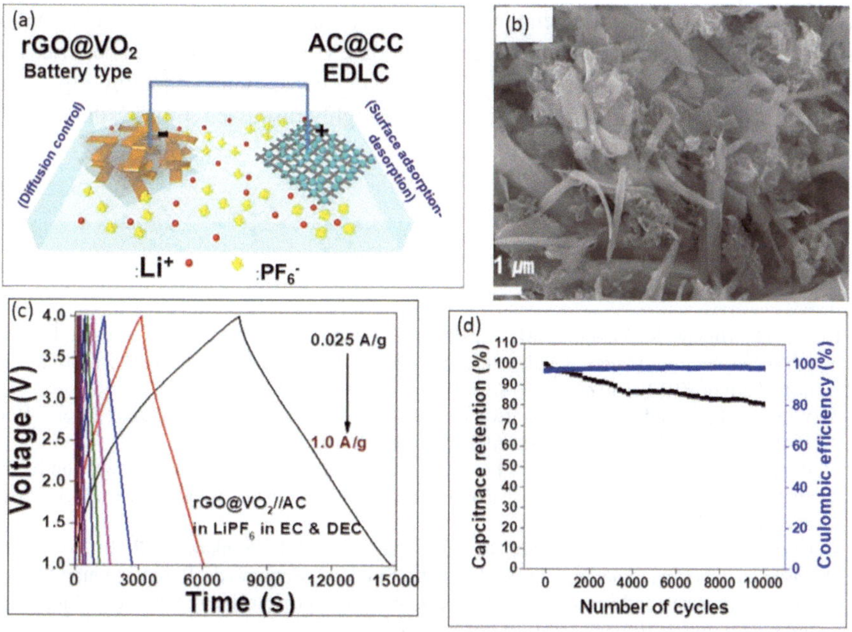

Fig. 9 **a** Schematic representation of battery-supercapacitor hybrid. **b** FESEM image of rGO@VO_2 cathode. **c, d** GCD at different current density and cycling performance at 2 A g^{-1} of AC@CC||rGO@VO_2 cell [71]. Reproduced with permission @ copyright (2019) American Chemical Society

7 Summary

The development of ESS with high energy density, high power density, and extended cycle life has become a global priority due to growing environmental concerns and the increasing demand for renewable energy. This has further amplified the need for efficient ESS. In recent decades, nanomaterials have significantly advanced energy storage technologies by offering innovative solutions to existing challenges. Their unique physical and chemical properties, which vary across different dimensions, play a crucial role in enhancing ESS performance. The integration of these nanomaterial dimensions further optimizes energy storage capabilities. This book chapter begins with a fundamental introduction to the need for ESS, with a specific focus on supercapacitors and batteries, along with a discussion of some real-time applications. Before delving into low-dimensional and multifunctional hybrid materials as electrode materials for supercapacitors and batteries, we provide a detailed analysis of the working mechanisms of these storage devices. First, the mechanisms of supercapacitors are explored, emphasizing EDLC and pseudocapacitance. This is followed by an in-depth discussion of battery mechanisms. The contributions of various research groups are reviewed, highlighting how 0D materials offer exceptional opportunities for developing composite materials incorporating metal oxides, hydroxides, and conductive polymers, which enhance both energy and power density. Additionally, studies on 1D materials designed to accommodate volume expansion in batteries are examined. The role of 2D materials in increasing capacity and enabling rapid and stable charge–discharge processes is also thoroughly discussed. Finally, the chapter concludes with a brief overview of HESS.

References

1. Tie, D., Huang, S., Wang, J., Ma, J., Zhang, J., Zhao, Y.: Hybrid energy storage devices: advanced electrode materials and matching principles. Energy Storage Mater. **21**, 22–40 (2019). https://doi.org/10.1016/j.ensm.2018.12.018
2. Winter, M., Brodd, R.J.: What are batteries, fuel cells, and supercapacitors? Chem. Rev. **104**(10), 4245–4270 (2004). https://doi.org/10.1021/cr020730k
3. Pomerantseva, E., Bonaccorso, F., Feng, X., Cui, Y., Gogotsi, Y.: Energy storage: the future enabled by nanomaterials. Science **366**(6468), eaan8285 (2019). https://doi.org/10.1126/science.aan8285
4. Ghongade, S.D., Waikar, M.R., Sonker, R.K., Chakarvarti, S.K., Sonkawade, R.G.: Gas sensors based on hybrid nanomaterial. In: Sonker, R.K., Singh, K., Sonkawade, R. (eds.), Smart Nanostructure Materials and Sensor Technology, pp. 261–283. Springer Nature Singapore, Singapore. https://doi.org/10.1007/978-981-19-2685-3_13
5. Yang, J., Wang, G., Teixeira, A.P., Silva, G.G., Hansen, Z., Jamal, M.J., Mathew, K., Xiong, J., Zhou, T., Mackowiak, M., Fleming, P.D., Wu, Q.: A biomass-based cathode for long-life lithium-sulfur batteries. Electrochem. Commun. **140**, 107325 (2022). https://doi.org/10.1016/j.elecom.2022.107325
6. Wu, J.: Understanding the electric double-layer structure, capacitance, and charging dynamics. Chem. Rev. **122**(12), 10821–10859 (2022). https://doi.org/10.1021/acs.chemrev.2c00097
7. Patra, A., Namsheer, K., Jose, J.R., Sahoo, S., Chakraborty, B., Rout, C.S.: Understanding the charge storage mechanism of supercapacitors: in situ/operando spectroscopic approaches and theoretical investigations. J. Mater. Chem. A **9**(46), 25852–25891 (2021). https://doi.org/10.1039/D1TA07401F
8. Sun, J., Luo, B., Li, H.: A review on the conventional capacitors, supercapacitors, and emerging hybrid ion capacitors: past, present, and future. Adv. Energy Sustain. Res. **3**(6), 2100191 (2022). https://doi.org/10.1002/aesr.202100191
9. Liu, C., Neale, Z.G., Cao, G.: Understanding electrochemical potentials of cathode materials in rechargeable batteries. Mater. Today **19**(2), 109–123 (2016). https://doi.org/10.1016/j.mattod.2015.10.009
10. Goodenough, J.B., Park, K.-S.: The Li-ion rechargeable battery: a perspective. J. Am. Chem. Soc. **135**(4), 1167–1176 (2013). https://doi.org/10.1021/ja3091438
11. Peljo, P., Girault, H.H.: Electrochemical potential window of battery electrolytes: the HOMO–LUMO misconception. Energy Environ. Sci. **11**(9), 2306–2309 (2018). https://doi.org/10.1039/C8EE01286E
12. Wang, L., Li, J., Lu, G., Li, W., Tao, Q., Shi, C., Jin, H., Chen, G., Wang, S.: Fundamentals of electrolytes for solid-state batteries: challenges and perspectives. Front. Mater. **7**, 111 (2020). https://doi.org/10.3389/fmats.2020.00111
13. Breczko, J., Wysocka-Żołopa, M., Grądzka, E., Winkler, K.: Zero-dimensional carbon nanomaterials for electrochemical energy storage. ChemElectroChem **11**(10), e202300752 (2024). https://doi.org/10.1002/celc.202300752
14. Xu, J., Xue, Y., Cao, J., Wang, G., Li, Y., Wang, W., Chen, Z.: Carbon quantum dots/nickel oxide (CQDs/NiO) nanorods with high capacitance for supercapacitors. RSC Adv. **6**(7), 5541–5546 (2016). https://doi.org/10.1039/C5RA24192H

15. Zhu, Y., Ji, X., Pan, C., Sun, Q., Song, W., Fang, L., Chen, Q., Banks, C.E.: A carbon quantum dot decorated RuO_2 network: outstanding supercapacitances under ultrafast charge and discharge. Energy Environ. Sci. **6**(12), 3665 (2013). https://doi.org/10.1039/c3ee41776j
16. Xie, Y., Du, H.: Electrochemical capacitance of a carbon quantum dots–polypyrrole/titania nanotube hybrid. RSC Adv. **5**(109), 89689–89697 (2015). https://doi.org/10.1039/C5RA16538E
17. Liu, P., Liu, H., Zhang, T., Chen, L., Guo, W., Gu, T., Yu, F., Liu, Y., Wang, G.: $Ti_3C_2T_x$ quantum dot-modified $Ti_3C_2T_x$ nanosheets freestanding films for flexible solid-state pseudocapacitors. Chem. Eng. J. **477**, 146913 (2023). https://doi.org/10.1016/j.cej.2023.146913
18. Yang, K., Li, C., Qi, H., Dai, Y., Cui, Y., He, Y.: Developing a MXene quantum dot-based separator for Li–S batteries. J. Mater. Chem. A **11**(19), 10425–10434 (2023). https://doi.org/10.1039/D3TA01082A
19. Zhao, J., Zhu, J., Li, Y., Wang, L., Dong, Y., Jiang, Z., Fan, C., Cao, Y., Sheng, R., Liu, A., Zhang, S., Song, H., Jia, D., Fan, Z.: Graphene quantum dot reinforced electrospun carbon nanofiber fabrics with high surface area for ultrahigh rate supercapacitors. ACS Appl. Mater. Interfaces **12**(10), 11669–11678 (2020). https://doi.org/10.1021/acsami.9b22408
20. Li, Y., Shi, G., Xu, C., Zhang, Y., Liu, B., Huo, P.: A review of the improved electrochemical property achieved by the incorporation of transition metal-based quantum dots applied in supercapacitors. J. Power Sources **619**, 235200 (2024). https://doi.org/10.1016/j.jpowsour.2024.235200
21. Xie, Q., Perez-Cordero, E., Echegoyen, L.: Electrochemical detection of $C_{60}{}^{6-}$ and $C_{70}{}^{6-}$: enhanced stability of fullerides in solution. J. Am. Chem. Soc. **114**(10), 3978–3980 (1992). https://doi.org/10.1021/ja00036a056
22. Dresselhaus, M.S., Dresselhaus, G., Eklund, P.C.: Fullerenes. J. Mater. Res. **8**(8), 2054–2097 (1993). https://doi.org/10.1557/JMR.1993.2054
23. Xu, Z.-L., Lin, S., Onofrio, N., Zhou, L., Shi, F., Lu, W., Kang, K., Zhang, Q., Lau, S.P.: Exceptional catalytic effects of black phosphorus quantum dots in shuttling-free lithium sulfur batteries. Nat. Commun. **9**(1), 4164 (2018). https://doi.org/10.1038/s41467-018-06629-9
24. Yin, L., Cho, J., Kim, S.J., Jeon, I., Jeon, I., Park, M., Park, M., Jeong, S., Lee, D.H., Seo, D., Cho, C.: Abnormally high-lithium storage in pure crystalline C_{60} nanoparticles. Adv. Mater. **33**(43), 2104763 (2021). https://doi.org/10.1002/adma.202104763
25. Ramadan, A., Anas, M., Ebrahim, S., Soliman, M., Abou-Aly, A.I.: Polyaniline/fullerene derivative nanocomposite for highly efficient supercapacitor electrode. Int. J. Hydrogen Energy **45**(32), 16254–16265 (2020). https://doi.org/10.1016/j.ijhydene.2020.04.093
26. Hu, Y., Chen, W., Lei, T., Jiao, Y., Wang, H., Wang, X., Rao, G., Wang, X., Chen, B., Xiong, J.: Graphene quantum dots as the nucleation sites and interfacial regulator to suppress lithium dendrites for high-loading lithium-sulfur battery. Nano Energy **68**, 104373 (2020). https://doi.org/10.1016/j.nanoen.2019.104373
27. Javed, M., Saqib, A.N.S., Ata-ur-Rehman, Ali, B., Faizan, M., Anang, D.A., Iqbal, Z., Abbas, S.M.: Carbon quantum dots from glucose oxidation as a highly competent anode material for lithium and sodium-ion batteries. Electrochim. Acta **297**, 250–257 (2019). https://doi.org/10.1016/j.electacta.2018.11.167
28. Ding, H., Zhang, Q., Liu, Z., Wang, J., Ma, R., Fan, L., Wang, T., Zhao, J., Ge, J., Lu, X., Yu, X., Lu, B.: TiO_2 quantum dots decorated multi-walled carbon nanotubes as the multifunctional separator for highly stable lithium sulfur batteries. Electrochim. Acta **284**, 314–320 (2018). https://doi.org/10.1016/j.electacta.2018.07.167
29. Wu, S., Wu, H., Zou, M., Shi, X., Yuan, Y., Bai, W., Cao, A.: Short-range ordered graphitized-carbon nanotubes with large cavity as high-performance lithium-ion battery anodes. Carbon **158**, 642–650 (2020). https://doi.org/10.1016/j.carbon.2019.11.036
30. Gurav, S.R., Waikar, M.R., Rasal, A.S., Sonker, R.K., Sonkawade, R.G.: Current development and challenges in textile-based flexible supercapacitors. In: Sonker, R.K. (ed.) Smart and Flexible Energy Devices, pp. 539–550. CRC Press, Boca Raton, FL (2022)
31. Xu, B., Wu, F., Su, Y., Cao, G., Chen, S., Zhou, Z., Yang, Y.: Competitive effect of KOH activation on the electrochemical performances of carbon nanotubes for EDLC: balance between

porosity and conductivity. Electrochim. Acta **53**(26), 7730–7735 (2008). https://doi.org/10.1016/j.electacta.2008.05.033

32. Tao, X.Y., Zhang, X.B., Zhang, L., Cheng, J.P., Liu, F., Luo, J.H., Luo, Z.Q., Geise, H.J.: Synthesis of multi-branched porous carbon nanofibers and their application in electrochemical double-layer capacitors. Carbon **44**(8), 1425–1428 (2006). https://doi.org/10.1016/j.carbon.2005.11.024
33. Sun, Y., Sills, R.B., Hu, X., Seh, Z.W., Xiao, X., Xu, H., Luo, W., Jin, H., Xin, Y., Li, T., Zhang, Z., Zhou, J., Cai, W., Huang, Y., Cui, Y.: A bamboo-inspired nanostructure design for flexible, foldable, and twistable energy storage devices. Nano Lett. **15**(6), 3899–3906 (2015). https://doi.org/10.1021/acs.nanolett.5b00738
34. Chen, L.-F., Zhang, X.-D., Liang, H.-W., Kong, M., Guan, Q.-F., Chen, P., Wu, Z.-Y., Yu, S.-H.: Synthesis of nitrogen-doped porous carbon nanofibers as an efficient electrode material for supercapacitors. ACS Nano **6**(8), 7092–7102 (2012). https://doi.org/10.1021/nn302147s
35. Kumar, U., Sikarwar, S., Sonker, R.K., Yadav, B.C.: Carbon nanotube: synthesis and application in solar cell. J. Inorg. Organomet. Polym. Mater. **26**, 1231–1242 (2016)
36. Rahul, Singh, S., Singh, P.K., Sharma, J.P., Kakroo, S., Sonker, R., Khan, Z.H.: Encompassing environment synthesis, characterization and photovoltaic utilization of cadmium sulphide quantum dots. Mater. Today: Proc. **34**, 767–770 (2021)
37. Jiang, H., Ma, J., Li, C.: Hierarchical porous $NiCo_2O_4$ nanowires for high-rate supercapacitors. Chem. Commun. **48**(37), 4465 (2012). https://doi.org/10.1039/c2cc31418e
38. Shen, L., Che, Q., Li, H., Zhang, X.: Mesoporous $NiCo_2O_4$ nanowire arrays grown on carbon textiles as binder-free flexible electrodes for energy storage. Adv. Funct. Mater. **24**(18), 2630–2637 (2014). https://doi.org/10.1002/adfm.201303138
39. An, Q., Lv, F., Liu, Q., Han, C., Zhao, K., Sheng, J., Wei, Q., Yan, M., Mai, L.: Amorphous vanadium oxide matrixes supporting hierarchical porous Fe_3O_4/graphene nanowires as a high-rate lithium storage anode. Nano Lett. **14**(11), 6250–6256 (2014). https://doi.org/10.1021/nl5025694
40. Tang, W., Hou, Y., Wang, F., Liu, L., Wu, Y., Zhu, K.: $LiMn_2O_4$ nanotube as cathode material of second-level charge capability for aqueous rechargeable batteries. Nano Lett. **13**(5), 2036–2040 (2013). https://doi.org/10.1021/nl400199r
41. Zhang, X., Cheng, F., Yang, J., Chen, J.: $LiNi_{0.5}Mn_{1.5}O_4$ porous nanorods as high-rate and long-life cathodes for Li-ion batteries. Nano Lett. **13**(6), 2822–2825 (2013). https://doi.org/10.1021/nl401072x
42. Niu, C., Meng, J., Wang, X., Han, C., Yan, M., Zhao, K., Xu, X., Ren, W., Zhao, Y., Xu, L., Zhang, Q., Zhao, D., Mai, L.: General synthesis of complex nanotubes by gradient electrospinning and controlled pyrolysis. Nat. Commun. **6**, 7402 (2015). https://doi.org/10.1038/ncomms8402
43. Zhu, C., Yu, Y., Gu, L., Weichert, K., Maier, J.: Electrospinning of highly electroactive carbon-coated single-crystalline $LiFePO_4$ nanowires. Angew. Chem. Int. Ed. **50**(28), 6278–6282 (2011). https://doi.org/10.1002/anie.201005428
44. Park, M.-H., Kim, M.G., Joo, J., Kim, K., Kim, J., Ahn, S., Cui, Y., Cho, J.: Silicon nanotube battery anodes. Nano Lett. **9**(11), 3844–3847 (2009). https://doi.org/10.1021/nl902058c
45. Wu, H., Chan, G., Choi, J.W., Ryu, I., Yao, Y., McDowell, M.T., Lee, S.W., Jackson, A., Yang, Y., Hu, L., Cui, Y.: Stable cycling of double-walled silicon nanotube battery anodes through solid-electrolyte interphase control. Nat. Nanotechnol. **7**(5), 310–315 (2012). https://doi.org/10.1038/nnano.2012.35
46. Li, X., Yan, C., Wang, J., Graff, A., Schweizer, S.L., Sprafke, A., Schmidt, O.G., Wehrspohn, R.B.: Stable silicon anodes for lithium-ion batteries using mesoporous metallurgical silicon. Adv. Energy Mater. **5**(4), 1401556 (2015). https://doi.org/10.1002/aenm.201401556
47. Yu, J., Wei, H.-Y., Yang, Z.-Y., Cai, J.-X.: Ultrafine SnO_2 nanoparticles encapsulated in high-conductivity graphited carbon nanotubes as anodes for high electrochemistry performance lithium-ion batteries. J. Electron. Mater. **48**(11), 7250–7257 (2019). https://doi.org/10.1007/s11664-019-07542-7

48. Xiong, H., Slater, M.D., Balasubramanian, M., Johnson, C.S., Rajh, T.: Amorphous TiO_2 nanotube anode for rechargeable sodium ion batteries. J. Phys. Chem. Lett. **2**(20), 2560–2565 (2011). https://doi.org/10.1021/jz2012066
49. Wang, W., Yu, C., Lin, Z., Hou, J., Zhu, H., Jiao, S.: Microspheric $Na_2Ti_3O_7$ consisting of tiny nanotubes: an anode material for sodium-ion batteries with ultrafast charge-discharge rates. Nanoscale **5**(2), 594–599 (2013). https://doi.org/10.1039/C2NR32661B
50. Xie, F., Deng, Y., Xie, Y., Xu, H., Chen, G.: Ultra-small nanoparticles of $MgTi_2O_5$ embedded in carbon rods with superior rate performance for sodium ion batteries. Chem. Commun. **51**(17), 3545–3548 (2015). https://doi.org/10.1039/C4CC10203G
51. Ghongade, S.D., Patil, M.D., Waikar, M.R., Sonkawade, A.R., Bagwan, A.M., Shinde, S.J., Moholkar, A.V., Sonkawade, R.G.: Unveiling elegant in-situ properties: structure and vibrations of the polymer solution synthesised $BaFe_{12}O_{19}$. Surf. Interfaces **53**, 105005 (2024). https://doi.org/10.1016/j.surfin.2024.105005
52. Panchal, K., Kumar, D.: 2D nanomaterials for advanced supercapacitor application. In: Gupta, R. (ed.) Handbook of Energy Materials, pp. 1–31. Springer Nature Singapore, Singapore (2022). https://doi.org/10.1007/978-981-16-4480-1_15-1
53. Wang, J., Malgras, V., Sugahara, Y., Yamauchi, Y.: Electrochemical energy storage performance of 2D nanoarchitectured hybrid materials. Nat. Commun. **12**(1), 3563 (2021). https://doi.org/10.1038/s41467-021-23819-0
54. Dall'Agnese, Y., Lukatskaya, M.R., Cook, K.M., Taberna, P.-L., Gogotsi, Y., Simon, P.: High capacitance of surface-modified 2D titanium carbide in acidic electrolyte. Electrochem. Commun. **48**, 118–122 (2014). https://doi.org/10.1016/j.elecom.2014.09.002
55. Ayman, I., Rasheed, A., Ajmal, S., Rehman, A., Ali, A., Shakir, I., Warsi, M.F.: $CoFe_2O_4$ nanoparticle-decorated 2D MXene: a novel hybrid material for supercapacitor applications. Energy Fuels **34**(6), 7622–7630 (2020). https://doi.org/10.1021/acs.energyfuels.0c00959
56. Wang, J., Wu, Z., Hu, K., Chen, X., Yin, H.: High conductivity graphene-like MoS_2/polyaniline nanocomposites and its application in supercapacitor. J. Alloys Compd. **619**, 38–43 (2015). https://doi.org/10.1016/j.jallcom.2014.09.008
57. Vidhya, M.S., Ravi, G., Yuvakkumar, R., Velauthapillai, D., Thambidurai, M., Dang, C., Saravanakumar, B.: Nickel-cobalt hydroxide: a positive electrode for supercapacitor applications. RSC Adv. **10**(33), 19410–19418 (2020). https://doi.org/10.1039/D0RA01890B
58. Tu, C.-C., Lin, L.-Y., Xiao, B.-C., Chen, Y.-S.: Highly efficient supercapacitor electrode with two-dimensional tungsten disulfide and reduced graphene oxide hybrid nanosheets. J. Power Sources **320**, 78–85 (2016). https://doi.org/10.1016/j.jpowsour.2016.04.083
59. Tomy, M., Ambika Rajappan, A., Vm, V., Thankappan Suryabai, X.: Emergence of novel 2D materials for high-performance supercapacitor electrode applications: a brief review. Energy Fuels **35**(24), 19881–19900 (2021). https://doi.org/10.1021/acs.energyfuels.1c02743
60. Rojaee, R., Shahbazian-Yassar, R.: Two-dimensional materials to address the lithium battery challenges. ACS Nano **14**(3), 2628–2658 (2020). https://doi.org/10.1021/acsnano.9b08396
61. Gwon, H., Kim, H.-S., Lee, K.U., Seo, D.-H., Park, Y.C., Lee, Y.-S., Ahn, B.T., Kang, K.: Flexible energy storage devices based on graphene paper. Energy Environ. Sci. **4**(4), 1277 (2011). https://doi.org/10.1039/c0ee00640h
62. Razmjoo Khollari, M.A., Paknahad, P., Ghorbanzadeh, M.: Improvement of the electrochemical performance of a nickel rich $LiNi_{0.5}Co_{0.2}Mn_{0.2}$ cathode material by reduced graphene oxide/SiO_2 nanoparticle double-layer coating. New J. Chem. **43**(6), 2766–2775 (2019). https://doi.org/10.1039/C8NJ05835K
63. Zhao, D., Feng, Y., Wang, Y., Xia, Y.: Electrochemical performance comparison of $LiFePO_4$ supported by various carbon materials. Electrochim. Acta **88**, 632–638 (2013). https://doi.org/10.1016/j.electacta.2012.10.101
64. Fan, P., Ye, C., Xu, L.: One-dimensional nanostructured electrode materials based on electrospinning technology for supercapacitors. Diam. Relat. Mater. **134**, 109803 (2023). https://doi.org/10.1016/j.diamond.2023.109803
65. Li, L., Chen, L., Mukherjee, S., Gao, J., Sun, H., Liu, Z., Ma, X., Gupta, T., Singh, C.V., Ren, W., Cheng, H., Koratkar, N.: Phosphorene as a polysulfide immobilizer and catalyst in

high-performance lithium-sulfur batteries. Adv. Mater. **29**(2), 1602734 (2017). https://doi.org/10.1002/adma.201602734

66. Shi, Y., Wen, L., Zhou, G., Chen, J., Pei, S., Huang, K., Cheng, H.-M., Li, F.: Graphene-based integrated electrodes for flexible lithium-ion batteries. 2D Mater. **2**(2), 024004 (2015). https://doi.org/10.1088/2053-1583/2/2/024004
67. Lei, T., Chen, W., Lv, W., Huang, J., Zhu, J., Chu, J., Yan, C., Wu, C., Yan, Y., He, W., Xiong, J., Li, Y., Yan, C., Goodenough, J.B., Duan, X.: Inhibiting polysulfide shuttling with a graphene composite separator for highly robust lithium-sulfur batteries. Joule **2**(10), 2091–2104 (2018). https://doi.org/10.1016/j.joule.2018.07.022
68. Wang, C.-H., Kurra, N., Alhabeb, M., Chang, J.-K., Alshareef, H.N., Gogotsi, Y.: Titanium carbide (MXene) as a current collector for lithium-ion batteries. ACS Omega **3**(10), 12489–12494 (2018). https://doi.org/10.1021/acsomega.8b02032
69. Shirinda, K., Kanzumba, K.: A review of hybrid energy storage systems in renewable energy applications. Int. J. Smart Grid Clean Energy **11**(2), 99–108 (2022). https://doi.org/10.12720/sgce.11.2.99-108
70. Yang, B., Wang, J., Zhang, X., Wang, J., Shu, H., Li, S., He, T., Lan, C., Yu, T.: Applications of battery/supercapacitor hybrid energy storage systems for electric vehicles using perturbation observer based robust control. J. Power Sources **448**, 227444 (2020). https://doi.org/10.1016/j.jpowsour.2019.227444
71. Sahoo, R., Lee, T.H., Pham, D.T., Luu, T.H.T., Lee, Y.H.: Fast-charging high-energy battery-supercapacitor hybrid: anodic reduced graphene oxide–vanadium(IV) oxide sheet-on-sheet heterostructure. ACS Nano **13**(9), 10776–10786 (2019). https://doi.org/10.1021/acsnano.9b05605
72. Luo, Z., Li, J., Zhang, S., Chen, X., Kumar, P., Wang, J., Zhu, X., Meng, W., Qiu, J., Ming, H.: The layer oxide cathode exhibits optimal capacitance cooperative effects in capacitor batteries. ACS Appl. Energy Mater. **8**(2), 1292–1307 (2025). https://doi.org/10.1021/acsaem.4c02750
73. Hu, C., Chen, Y., Song, Z., Miao, L., Duan, H., Lv, Y., Xie, L., Liu, M., Gan, L.: Bipolar conjugated microporous polymer anchoring graphene hybrids for high-performance zinc-organic batteries. J. Mater. Chem. A **12**(21), 12818–12825 (2024). https://doi.org/10.1039/D4TA01952K
74. Shi, B., Li, L., Chen, A., Jen, T.-C., Liu, X., Shen, G.: Continuous fabrication of $Ti_3C_2T_x$ MXene-based braided coaxial zinc-ion hybrid supercapacitors with improved performance. Nano-Micro Lett. **14**(1), 34 (2022). https://doi.org/10.1007/s40820-021-00757-6
75. Zheng, C., Wang, Y., Liu, Z., Sun, T., Kim, N., Jeong, J., Cha, S.W.: A hybrid energy storage system for an electric vehicle and its effectiveness validation. Int. J. Precis. Eng. Manuf.-Green Technol. **8**(6), 1739–1754 (2021). https://doi.org/10.1007/s40684-020-00304-5

Characterization Technique for Energy Storage Nanomaterials

Gayatri R. Chodankar, Suman A. Sawant, Maqsood R. Waikar, Rakesh K. Sonker, and Rajendra G. Sonkawade

Abstract Effective energy storage devices necessitate the development of superior electrode materials as well as an in-depth knowledge of characterization techniques. Currently, smart materials (metallic, polymeric, and nanocomposite) are considered fascinating and leading candidates in the field of energy storage applications. To test the applicability of such materials in the field of energy storage field, an in-depth knowledge of electrochemical characterization techniques is mandatory, since they aid to comprehend a substance's fundamental electrochemistry. So, this chapter explores a range of electrochemical characterization techniques (like voltammetric techniques, galvanometric techniques, impedance measurement techniques, and additional prominent techniques) applied in electrochemical investigations. These methods collectively enable an understanding phenomenon taking place at the electrode–electrolyte boundary in an electrochemical cell. Key findings reveal that the electrochemical performance of smart materials is intrinsically linked to their structural and compositional characteristics, which can be accurately evaluated using advanced characterization techniques. Furthermore, the integration of smart nanomaterials into solid-state and flexible devices, combined with emerging trends such as real-time, in-situ characterization, multifunctional platforms, and sustainable material development, is poised to accelerate the design of next-generation energy storage systems that are more efficient, durable, and eco-friendly. This chapter highlighted methods that is applicable for analyse the electrochemical behaviour of smart materials in the field of energy storage devices.

Keywords Smart materials · Energy storage devices · Characterization techniques

G. R. Chodankar · S. A. Sawant · M. R. Waikar · R. G. Sonkawade (✉)
Radiation and Materials Research Laboratory, Department of Physics, Shivaji University, Kolhapur, Maharashtra, India
e-mail: rgs_phy@unishivaji.ac.in

R. K. Sonker
Department of Physics, Motilal Nehru College, University of Delhi, Delhi, India

R. K. Sonker et al. (eds.), *Low-Dimensional Multifunctional Hybrid Nanostructured Materials*, Engineering Materials, https://doi.org/10.1007/978-981-96-7952-2_2

1 Background and Significance

The evolution of sustainable and efficient energy storage technologies (ESTs) is reshaping the global energy landscape. As the demand for renewable energy escalates, the necessity for ground-breaking EST innovations intensifies. A crucial aspect of this development involves designing high-performance electrode materials to improve energy density (E_d), charge–discharge rates, stability, and longevity. Among the most promising candidates for these applications are smart materials-metallic, polymeric, and nanocomposite materials-which exhibit unique properties tailored for ESTs [1, 2]. Smart materials are particularly valuable due to their high thermal and electrical conductivity, environmental stability, and superior mechanical properties. These attributes make them ideal for use in batteries, supercapacitors (SCs), and capacitors. For instance, metallic nanomaterials facilitate efficient electron transfer, polymeric materials offer flexibility and lightweight properties, and nanocomposites integrate multiple advantageous features in a single material [3, 4].

Although smart materials hold great promise for energy storage, their practical application requires rigorous electrochemical characterization. Various characterization techniques shed light on important factors such as charge storage dynamics, material resilience, and long-term degradation, all of which contribute to the advancement of ESTs. Characterization techniques provide insights into key properties such as charge storage mechanisms, material stability, and aging processes, which are essential for advancing ESTs [5].

In this chapter we deliberate the electrochemical, structural and morphological characterization methods used in EST in order to delve into the different smart material features. These techniques help assess critical physical parameters, including resistance, charge transfer dynamics, capacitance, and ion diffusion. By employing multiple characterization methods, researchers can gain a comprehensive understanding of material performance, enabling the optimization of smart materials for efficient and scalable energy storage solutions.

The key electrochemical techniques discussed in this chapter include: Voltammetric methods, galvanostatic charge–discharge (GCD), electrochemical impedance spectroscopy (EIS) and additional strategies such as potentiostatic methods, linear sweep voltammetry (LSV), and chronopotentiometry. These different methods provide a unique perspective of the electrochemical behaviour of materials and allow the researchers to characterize key parameters, i.e., capacitance, resistance, charge transfer and diffusion processes [6]. By understanding these properties, the materials are tailorable to fulfil the particular necessities of different ESDs. Additionally, these methods help pinpoint performance bottlenecks, like degradation mechanisms or the charge transfer inefficiency, consequently improving EST's durability and productiveness.

1.1 Objectives of Characterization Techniques

The primary objectives of characterization techniques are to analyse the structural, chemical, electrochemical, and thermal properties of materials to ensure high performance, safety, and sustainability. A detailed understanding of the internal structure, particle distribution, and chemical composition is essential, as these factors influence ion transport and storage capacity. Electrochemical performance is evaluated through parameters such as capacity retention and E_d, while thermal stability assessments ensure safe operation under diverse conditions.

Advanced characterization techniques also support the development of smart nanomaterials with self-healing and adaptive properties, extending device longevity. Additionally, they help maintain quality control as materials transition from laboratory research to industrial-scale production. Ensuring reliability under extreme conditions is particularly crucial for applications such as electric vehicles. Furthermore, interdisciplinary approaches in material science and sustainability assessments contribute to the global effort to develop eco-friendly, high-performance ESTs [7].

1.2 Role of Smart Nanomaterials in Energy Storage Devices

Smart nanomaterials play a vital role in improving the longevity, performance, and safety of energy storage devices. Leveraging their unique nanoscale properties-such as high surface area, tunable conductivity, and responsiveness to external stimuli-these materials significantly enhance energy density, charge/discharge rates, and cycle life in batteries and SCs. For example, carbon nanotubes and graphene maximize ion–electron interactions, while porous nanostructures facilitate seamless ion transfer, crucial for high-energy applications. Moreover, smart nanomaterials mitigate device degradation and mechanical stress during cycling, extending cycle life. Self-healing nanomaterials and flexible structures enhance durability, while thermal management properties ensure safety by preventing short circuits and thermal runaway. In next-generation battery technologies such as solid-state and sodium-ion batteries, smart nanomaterials contribute to improved stability, conductivity, and adaptability to varying operating conditions. Their integration is driving innovation in ESTs, enabling more efficient and sustainable energy storage solutions. As research progresses, smart nanomaterials will continue to be at the forefront of advancements in flexible, wearable, and high-performance energy storage technologies [8, 9].

2 Electrochemical Techniques

Electrochemical techniques constitute a broad spectrum of techniques used to study chemical reactions where there is electron exchange between molecules. On the other hand, these techniques are also important in the field of analytical chemistry, material science, and biochemistry. Fundamentally, they involve the application of an electrical potential to a chemical system, often by means of electrodes that are immersed in an electrolyte, a process known as electrolysis. Voltammetry experiments tests are performed on a solution in an electrode-equipped cell. Three electrodes are used in the electrochemical cell where the experiment is conducted which is represented in Fig. 1.

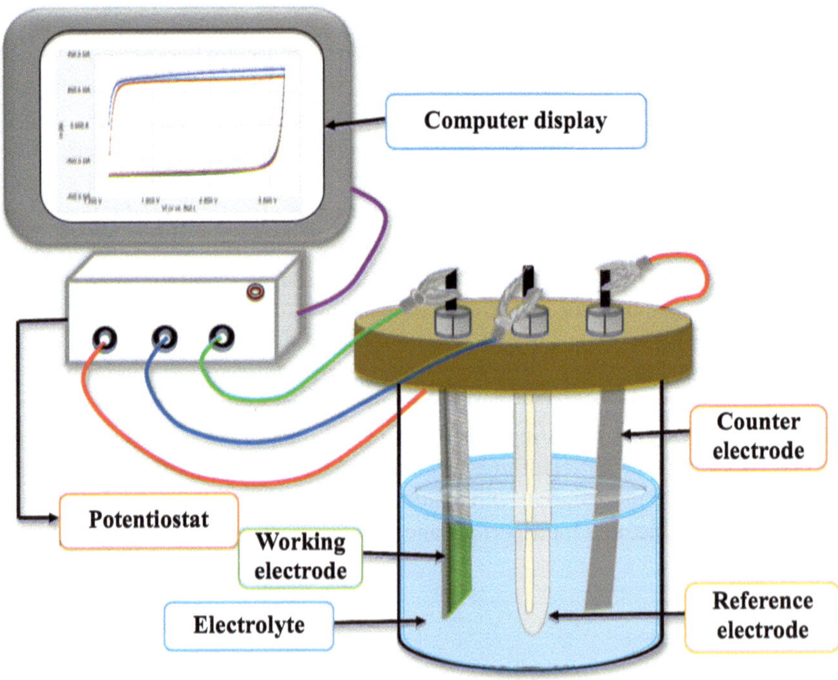

Fig. 1 Schematic of the electrochemical setup used to evaluate the electrochemical properties of electrode materials in energy storage devices (ESDs), highlighting the components involved in a three-electrode system

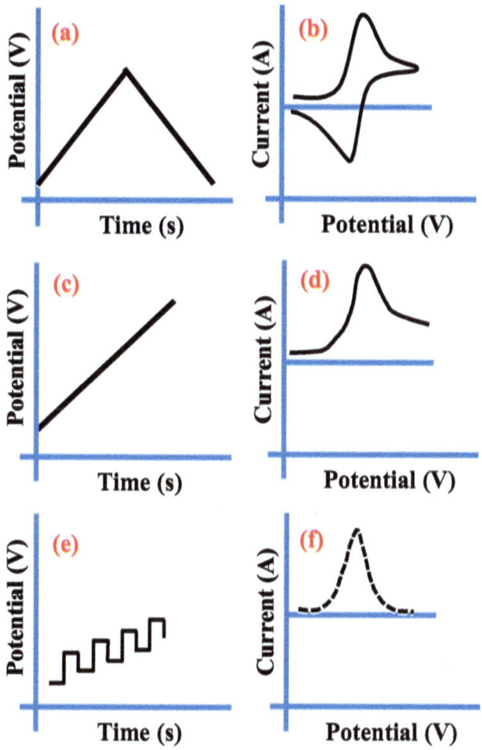

Fig. 2 The relationship between different possible potential programs and the corresponding electrochemical technique: **a, b** cyclic voltammetry; **c, d** linear sweep voltammetry and **e, f** differential pulse voltammetry

2.1 *Voltammetry Techniques*

The electrochemical properties of smart nanomaterials used in ESDs are characterized with voltammetry techniques. These techniques analyse energy storage mechanisms, charge transfer processes, and reaction kinetics, making them essential for evaluating batteries, SCs and capacitors. In this section, most common voltammetry techniques are discussed in detail, coupled to cover their relevance to energy storage nanomaterials and represented in Fig. 2a–h [10, 11].

2.2 *Overview of Voltammetry Techniques*

Voltammetry refers to a group of electrochemical techniques that measure the current as a function of the applied voltage (or potential) across an electrochemical cell. These techniques involve the use of a working electrode immersed in an electrolyte solution, with a counter electrode and reference electrode to complete the cell. As the potential is assorted, electrochemical reactions occur at the electrode surface, resulting in a current that can be measured. The shape and magnitude of the current–voltage curves (voltammograms) provide information about the material's redox processes, ion diffusion, and capacitance [11].

(A) **Cyclic Voltammetry (CV)**

In CV curve, the x-axis represents applied potential (*V*) while y-axis is the obtained current (*I*) in response to the applied potential (*V*). CV basically works on the principle of sweeping the potential between the lower and upper potential limit Fig. 2a, b. When the lower/upper limit is reached reversing the potential sweep is occurred to get a cyclic motion. The sweep of potential between upper/lower limits is a function of time which is referred as 'scan rate'. In CV tool, two conventions such as US convention and IUPAC convention which are commonly used [12], both the conventions are pictorial represented in Fig. 3a, b.

- **Key Features**
  - **Charge Storage Mechanism**: SCs exhibit nearly rectangular CV curves, while batteries display distinct redox peaks characteristic of material are obtained as shown in Fig. 3c, d.

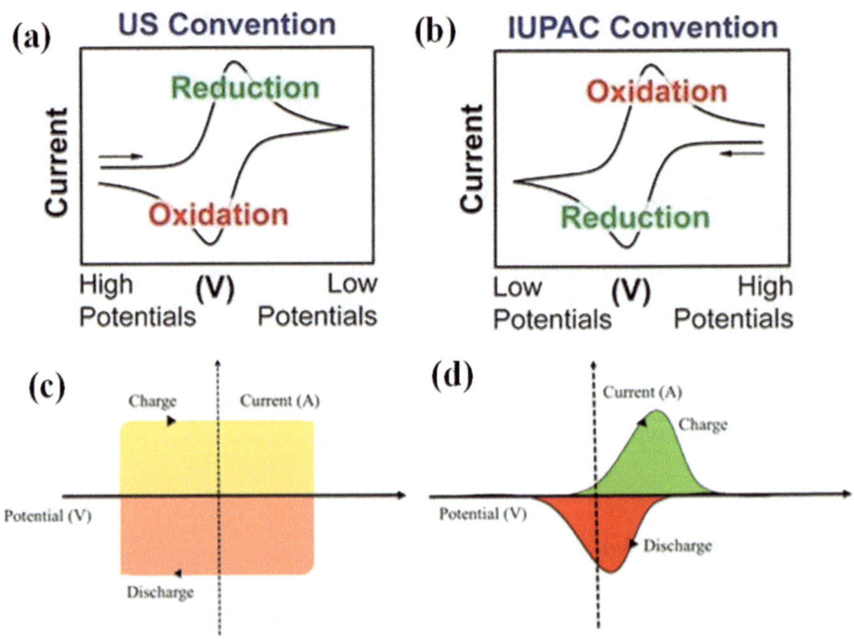

Fig. 3 **a, b** Visual representation of standard conventions in cyclic voltammetry tools [13]. Reproduced with permission @ copyright (2018) American Chemical Society; **c** CV profiles for capacitors/supercapacitors depicting ideal capacitive behavior and **d** CV curves for batteries showing distinct redox peaks related to specific electrochemical reactions [14]. Reproduced with permission @ copyright (2021) Royal Society of Chemistry

- **Peak Currents**: The peak currents observed in the CV correspond to the maximum rate of electron transfer during the oxidation or reduction reactions. These peak currents can be used to estimate the concentration of electroactive species in the solution.
- **Peak Potentials**: CV allows for the determination of peak potentials, which represent the potentials at which oxidation or reduction reactions occur. These peak potentials provide insights into the redox potentials of the species existent in the solution.
- **Electroactive Species**: Metal ions, organic molecules, and redox active complexes can be identified and characterized through CV.
- **Reversibility**: The peaks in CV may have a shape and symmetry that can help determine the reversibility of such redox reactions. Reversible reactions show well defined and symmetric peaks while irreversible processes cause broad and asymmetric peaks.
- **Electrode Surface Properties**: The information about modification of the electrode surface (for example, adsorption of species or surface morphology) can deliver by changes in shape or position of CV.
- **Reaction Kinetics**: CV curves are thoroughly studied to understand the charge storage kinetics of electrode based on the power law equations represented below,

  The power law is,

$$i = a\upsilon^b \tag{1}$$

  In order to quantify the impact of each charge storage process, the preceding equation is employed,

$$i = k_1\upsilon + k_2\sqrt{\upsilon} \tag{2}$$

- **Electrochemical Mechanisms**: CV can help elucidate the underlying electrochemical mechanisms governing the redox reactions, including details about electron transfer pathways, reaction intermediates, and electrode kinetics [15, 16].

(B) **Linear Sweep Voltammetry (LSV)**

The technique is where the potential is swept unidirectional from a start to end value and the current measured as shown in Fig. 2c, d. It helps determine onset potential and reaction kinetics [17].

- **Key Features**
  - **Onset Potential**: The onset potential, which is a potential for redox reaction to estimate the activation energy associated with nanomaterial's involvement in electrochemical reactions.

- **Tafel Analysis**: This provides insights into reaction kinetics in fuel cells and batteries.
- **Reaction Mechanisms**: LSV can be used to elucidate the reaction mechanisms by analysing the current response trends [18].

- **Applications in Energy Storage**

In energy storage systems like fuel cells, LSV is used to evaluate the catalytic activity of nanomaterials by measuring the onset potential and the current density during redox reactions.

LSV is also employed to study corrosion in ESDs, helping researchers to identify degradation mechanisms in nanomaterials used in harsh electrochemical environments [17].

(C) **Differential Pulse Voltammetry (DPV)**

DPV applies voltage pulses superimposed on a linear sweep, improving the sensitivity and resolution of electrochemical signals as shown in Fig. 2e, f. This is especially advantageous for the testing of low electrochemical activity materials [19].

- **Key Features**
  - **High Sensitivity**: Enables detection of small current changes, crucial for nanomaterials with low charge storage capacity.
  - **Peak Resolution**: Distinguishes closely spaced redox peaks in complex systems.

- **Applications in Energy Storage**

DPV is suitable for studying the redox processes of nanomaterials applied for batteries and SCs, especially for the case where redox peaks of the material are weak or tightly spaced. It also employed in electrocatalytic studies, helping to assess the activity of nanomaterials in fuel cells and other energy conversion devices.

Voltammetry techniques are essential for characterizing the electrochemical characteristics, making them indispensable for the development of advanced ESDs. As ESTs continue to evolve, voltammetry will remain a crucial tool for smart materials and improving the productivity, safety, and durability of ESD.

3 Galvanometric Techniques

The galvanometric techniques consist of measuring the current or voltage with controlled conditions to evaluate the electrochemical materials behaviour. However, galvanometric methods offer detailed insights into charge/discharge mechanisms, capacity retention, and energy efficiency of smart materials used in the application of energy storage, and are essential for assessing and optimizing smart material performance.

This section presents key galvanometric techniques whereby galvanostatic charge discharge (GCD), galvanostatic intermittent titration technique (GITT), and potentiostatic intermittent titration technique (PITT) characterize energy storage nanomaterials.

3.1 Overview of Galvanometric Techniques

In galvanostatic mode, a constant current is applied, and voltammetric response is recorded as a function of time. These methods give researchers the ability to assess different properties of nanomaterials, like energy storage capacity, charge/discharge behavior and electrochemical efficiency.

- **Galvanostatic Techniques**: Galvanostatic techniques apply a constant current, and the resulting voltage response is monitored as a function of time.
- **Potentiostatic Techniques**: In potentiostatic techniques, we apply a constant potential and measure the current response. Although usually applied in conjunction with galvanostatic methods, they are especially appropriate for studies of ion diffusion and reaction kinetics.

(A) **Galvanostatic Charge–Discharge (GCD)**

GCD is a plot of change in voltage verses time. In this test, a constant current is applied to the electrodes and electrode is charged/discharged between potential limits. The time required for charging/discharging in specific potential limits is recorded. This measurement is also known as 'Chronopotentiometry test' [20] (Fig. 4).

- **Key Features**
  - **Charge-Discharge Kinetics**: The GCD curves yield insights concerning the kinetic processes of charge and discharge in the nanomaterials. Linear

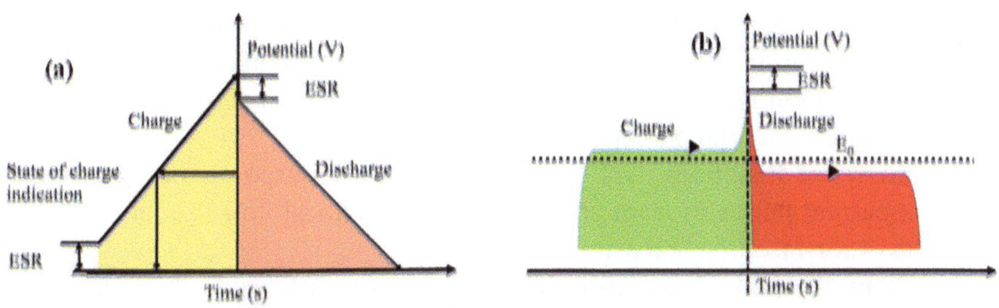

Fig. 4 Galvanostatic charge–discharge (GCD) profiles illustrating **a** nearly triangular curves for capacitors/supercapacitors, indicative of ideal capacitive behavior, and **b** charge–discharge plateaus for batteries, representing Faradaic processes [14]. Reproduced with permission @ copyright (2021) Royal Society of Chemistry

profiles indicate capacitive behavior; voltage plateaus indicate typical faradaic reactions of batteries.

- **Energy Capacity and Power Capability**: From GCD experiments, researchers can analyse the voltage and current data to calculate the energy capacity and power capability of the device. It is important information for assessing the performance of the device's performance in real world operations.
- **Coulombic Efficiency**: The reversibility of and efficiency of the material are evaluated based upon this relationship of the charge during discharge to that during charge. In other words, a high Coulombic efficiency indicates that the nanomaterial will store and release energy over many cycles.

• **Applications in Energy Storage**

- **Batteries**: GCD is applied to evaluate the specific capacity, rate ability, and lifetime in batteries of nanomaterials (such as anodes and cathodes of lithium ion batteries). The charge–discharge curves show these as voltage plateaus corresponding to the redox reactions and phase transitions occurring during the cycling process.
- **SCs**: GCD is useful for SCs to give information about the charge storage mechanism (electrical double layer capacitance or pseudocapacitance) and to calculate the specific capacitance of materials such as graphene or transition metal oxides.

(B) **Galvanostatic Intermittent Titration Technique (GITT)**

GITT is employed to study diffusion processes and the kinetics of electrochemical reactions in energy storage nanomaterial's. GITT operates by applying intermittent small current pulses, followed by rest periods where the current is set to zero, allowing the system to reach equilibrium. By measuring the voltage response during both the current pulses and rest periods, provides detailed information about the material's diffusion behaviour [21].

• **Key Features**

- **Diffusion Coefficients**: GITT is widely used to determine the diffusion coefficient of ions (such as lithium ions in batteries) in nanomaterials. The voltage relaxation during the rest periods provides information about the rate of ion diffusion within the material.
- **Kinetics of Reactions**: GITT analyzes the voltage response during and after the current pulses to learn about the electrochemical kinetics, including the rate of ion insertion and extraction processes.
- **Thermodynamic Properties**: GITT can also be used to obtain thermodynamic properties, such as open circuit voltage (OCV), Gibbs free energy change in the electrochemical reactions.

• **Applications in Energy Storage**

- **Batteries**: In most cases, GITT is employed to study ion transmission in battery batteries. For example, GITT can be used to measure how fast lithium ions penetrate the electrode materials in lithium based batteries, which is a key parameter for battery performance optimisation under the high charge/discharge rates.
- **Solid-State Batteries**: GITT is a useful tool to evaluate the ion diffusion characteristics in nanomaterials and solid state electrolytes for solid state batteries, where the ion transport could be slow because presence of liquid electrolytes is unavailable.

(C) **Potentiostatic Intermittent Titration Technique (PITT)**

GITT is analogous to PITT with constant current pulse; PITT is like GITT with constant potential steps. As the system approaches equilibrium, after each potential step, the current response is recorded. Studies of nanomaterial diffusion controlled processes and charge storage mechanisms are especially of interest using PITT [20].

- **Key Features**
  - **Current Transients**: PITT analyses the current decay after any potential step and gives information about ion diffusion and charge transport time constants in the material.
  - **Diffusion Coefficients**: As with GITT, PITT can be used to determine the diffusion coefficients of ions in nanomaterials for energy storage.
  - **Electrochemical Reactions**: Thermodynamics and kinetics of electrochemical reactions are provided by PITT as an aid to understand the energy storage processes taking place in batteries and SCs.
- **Applications in Energy Storage**
  - **Battery Systems**: Studying the diffusion of ions such as lithium or sodium in battery electrodes, especially in systems where the potential profile is critical to phase transitions, is made possible by PITT.
  - **Supercapacitors**: PITT trains charge–discharge tools in SCs, especially in materials in which capacitive and faradaic processes are present in energy storage.

(D) **Constant Current Cycling (CCC)**

CCC is another widely used galvanometric technique that involves cycling the battery or SCs at a controlled constant current can be measured by the voltage response of the battery during the charge and discharge phase [22].

- **Key Features**
  - **Capacity Retention**: CCC is applied to assess the ability of energy storage material to retain its charge storage capacity as it is cycled, which shows how effectively the material retains its storage capacity after many cycles.

- **Cycle Stability**: Cycling beneath a continuous current also provides researchers with insight on how the material is degrading, necessary for endurance applications.
- **Rate Capability**: CCC can be used to investigate the rate capability of nanomaterials, or their capacity to deliver capacity under different currents to determine if they are fit for fast charging or high power applications.

• **Applications in Energy Storage**

- **Lithium-ion Batteries**: CCC has been established in lithium-ion battery investigations to quantify the durable retentively and strength abandonment of nanomaterials such as silicon anode or nickel-rich cathode.
- **SCs**: Using CCC, it is possible to determine the practical capacity, specific capacitance, Ed and Pd for several charge–discharge cycles which is useful for selecting the electrode materials with better performance characteristics.

Due to their sensitivity, galvanometric methods including GCD, GITT, and PITT are crucial tools for gaining fundamental understanding of charge storage processes, reaction rates, ion transport and aging.

4 Impedance Measurement Techniques

Impedance measurement techniques are crucial for analyzing charge transfer, ion diffusion, and interfacial reactions in energy storage materials. These methods help optimize system performance, efficiency, and longevity. Among them, Electrochemical Impedance Spectroscopy (EIS) stands out as a powerful tool for evaluating dynamic electrochemical responses across a wide frequency range.

4.1 Overview of Impedance Measurement Techniques

Impedance measurement techniques use an alternating current (AC) signal to analyse the system's response across multiple frequencies. Researchers can decompose electrochemical processes into frequency-dependent components such as ion transport, charge transfer, and diffusion by calculating impedance as the ratio of AC voltage to AC current [23].

Impedance (Z) is the quantity expressed as $Z = Z' + jZ''$, where Z' is the real part (resistive component), and Z'' is the imaginary part (capacitive or inductive) component. Nyquist plots provide a graphical representation of impedance, with a semicircle in the high-frequency region reflecting charge transfer resistance and a straight line in the low-frequency region indicating ion diffusion. Meanwhile, Bode plots display impedance magnitude and phase angle as functions of frequency, aiding in the analysis of time constants and frequency-dependent behaviour.

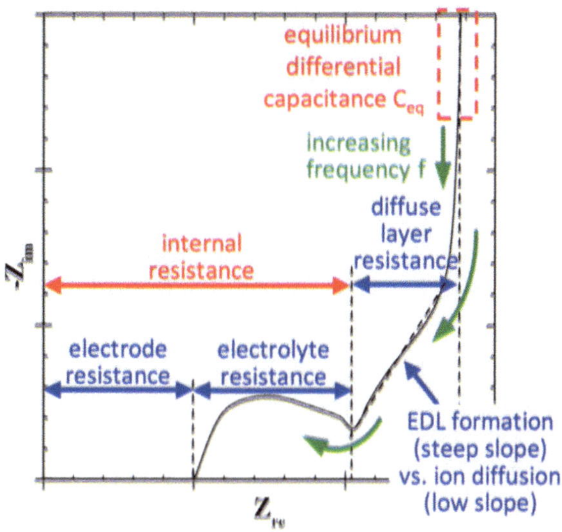

Fig. 5 Electrochemical impedance spectroscopy (EIS) spectrum highlighting key features such as solution resistance (R_s), charge transfer resistance (R_{ct}), and Warburg impedance related to ion diffusion [25]. Reproduced with permission @ copyright (2018) American Chemical Society

(A) **Electrochemical Impedance Spectroscopy (EIS)**

In EIS, a small AC signal is applied to the system, and the impedance is measured over a wide frequency range (mHz to MHz) to elucidate key electrochemical processes occurring within the system as depicted in Fig. 5, such as charge transfer reactions, ion diffusion, and electrode surface properties [24].

- **Key Features**
  - **Charge Transfer Resistance (R_{ct})**: Represents resistance to electron transfer at the electrode/electrolyte interface, critical factor in determining the efficiency of batteries and fuel cells. A low R_{ct} indicates faster electron transfer and improved performance.
  - **Solution Resistance (R_s)**: Reflects ion movement resistance through the electrolyte, affecting overall device impedance.
  - **Warburg Impedance (Z_w)**: Relates to ion diffusion in electrode materials, providing insights into charge/discharge kinetics.
  - **Double-Layer Capacitance (C_{DL})**: Corresponds to electrostatic energy storage at the electrode/electrolyte interface, crucial for SCs [25].

(B) **Nyquist and Bode Plot Analysis**

Nyquist and Bode plots provide detailed electrochemical insights:

- **Nyquist plot**

The Nyquist plot, as depicted in Fig. 6, displays a relationship between the imaginary impedance (Z'') and real impedance (Z'), with a semicircle indicating charge transfer resistance and a vertical line at low frequencies representing ion diffusion [27].

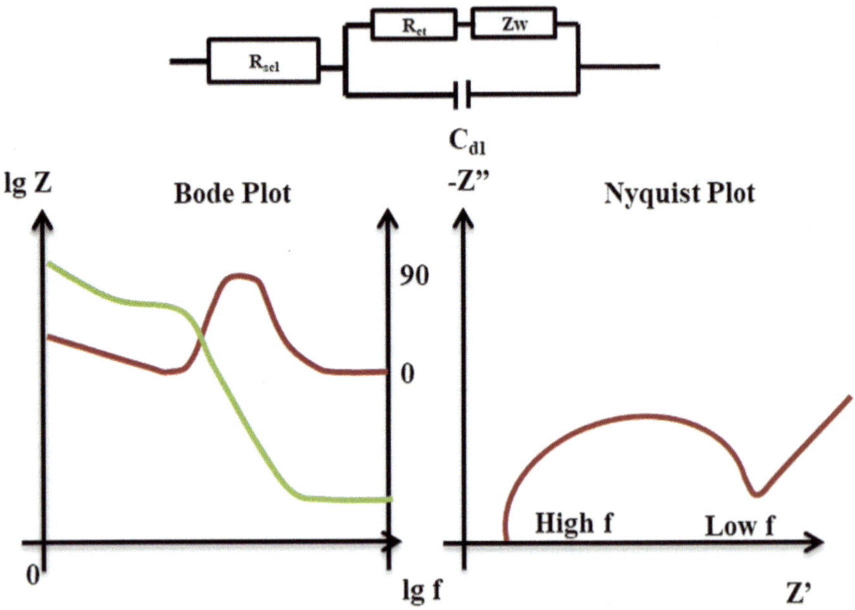

Fig. 6 Nyquist and Bode plots obtained from EIS measurements, showing frequency-dependent impedance characteristics that provide insights into the resistive and capacitive behaviour of materials [26]

- **Bode Plot**

Displays magnitude and phase response verses frequency, helping distinguish between capacitive and diffusive charge storage mechanisms.

- **Magnitude Plot**

This plot reveals system resistance and capacitance variations across frequencies.

- **Phase Angle Plot**

The phase angle plot indicates whether the system behaves more like a resistor (phase angle near 0°) or a capacitor (phase angle near −90°) [28].

(C) **Applications of EIS in Energy Storage Nanomaterials**

- **Material Optimization**: EIS identifies performance-limiting factors such as high R_{ct}, slow ion diffusion, and poor electrolyte conductivity, guiding nanomaterial enhancements.
- **Nanostructured Electrodes**: Evaluates improvements in charge transfer resistance and internal resistance through high-surface-area materials like graphene and carbon nanotubes.

- **Solid Electrolytes**: By analysing the impedance response, we can assess the ionic conductivity and interface resistance between the solid electrolyte and electrodes.
- **Cycle Life and Degradation Studies**: Monitors long-term performance by tracking R_{ct} changes, electrolyte degradation, and solid-electrolyte interphase layer formation.

(D) **Challenges and Limitations of EIS**

- **Complex Data Interpretation**: The fitting of impedance data to equivalent circuit models requires expertise and careful consideration of the physical meaning of each circuit element.
- **Measurement Time**: EIS measurements, particularly at low frequencies, can be time-consuming. For devices with long relaxation times (e.g., SCs with large double-layer capacitance).
- **Small Signal Approximation**: EIS is typically performed under small perturbations, which may not fully represent real-world operating conditions.

EIS play a vital role in understanding and optimizing smart nanomaterials for ESD applications. By providing detailed insights into R_{ct}, ion diffusion, and interface properties, EIS aids in designing high-performance materials. Through Nyquist and Bode plots, as well as equivalent circuit modelling, researchers gain precise insights into the resistive and capacitive components of ESDs, paving the way for the progress of advanced ESTs.

5 Other Electrochemical Characterization Techniques for Energy Storage Smart Nanomaterials

In addition to commonly used techniques such as EIS and Galvanometric Methods, several other electrochemical characterization methods are critical for sympathetic the behaviour and performance of smart nanomaterials in ESD. These techniques offer unique visions into charge storage mechanisms, ion transport, and material steadiness, providing valuable information for optimizing ESTs such as batteries, and SCs.

5.1 Chronoamperometry (CA) and Chronopotentiometry (CP)

CA and CP are two time-resolved electrochemical techniques employed to study charge–discharge behaviour, ion transport and electrode stability in EST. Data are

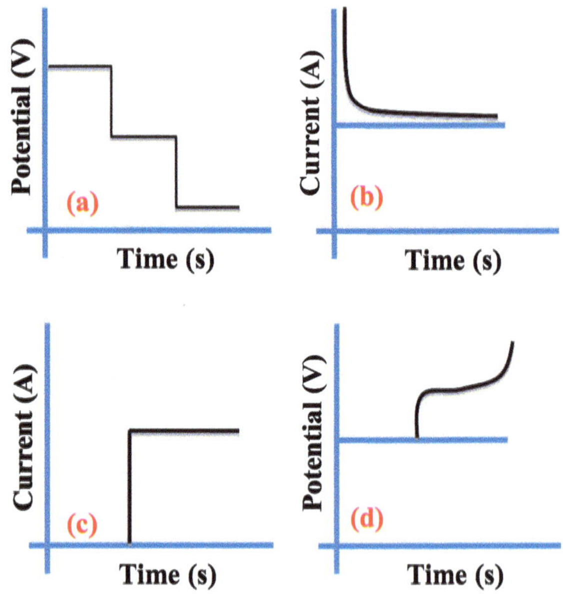

Fig. 7 Comparison of potential programs and techniques: **a**, **b** chronoamperometry curves showing current response to potential steps; **c**, **d** chronopotentiometry profiles illustrating potential response under constant current, used for stability and cycling analysis

obtained through applying a CA or a CP to the system and recording the resultant current (in CA) or potential (in CP) as a function of time [29].

- **Chronoamperometry** (CA)
  - **Current-Time Response**: CA is applied with a constant potential to the system and the resulting current is measured as a function of time represented in Fig. 7a, b. In particular, this technique is useful when studying diffusion controlled processes in which the current decays with time as the electroactive species concentration changes at the electrode surface [23].

Applications: In studying battery electrode materials, CA is used to evaluate ion diffusion rates, an essential part of characterizing the kinetics of charge/discharge processes. The process is also used in corrosion studies, and can be applied to the investigation of electrochemical deposition and dissolution processes.

- **Chronopotentiometry** (CP)
  - **Voltage-Time Response**: In CP a constant current is used and the corresponding potential is followed with time Fig. 7c, d. CP is frequently utilized to explore charge and discharge characteristics of the energy storage materials, like electrode of battery and SC.

Applications: The capacity retention and cycling stability of nanomaterials in batteries and SCs are provided by CP. In addition, it is used in electrochemical degradation studies in which potential profile with time informs about the stability and reversibility of the electrochemical reactions [30].

5.2 *Tafel Plot Analysis*

Tafel Plot Analysis is a technique used in an electrochemical system where the kinetics of electrochemical reactions is studied specifically when charge transfer processes are present for example in batteries and fuel cells. The Tafel equation is used to relate the overpotential (η) to the current density (j), and will prove useful in calculating the exchange current density and the empirical Tafel slope, these important kinetic parameters will provide information into the reaction mechanisms.

- **Key Features**
  - **Tafel Slope**: The slope of the Tafel plot (η vs. log(i)) gives some hints to the rate determining step of the reaction as well as the efficiency of the electrochemical reaction. Fast kinetics and better electron transfer is indicated by a smaller value of Tafel slope.
  - **Exchange Current Density**: The rate of an electrochemical reaction at equilibrium is given by this parameter and is important when assessing the performance of catalysts in fuel cells or reaction kinetics in batteries [31].
- **Applications in Energy Storage**
  - **Fuel Cells**: In fuel cell research, the ORR or HER performance of catalysts is commonly investigated via the use of Tafel analysis. Analysis of Tafel slope and exchange current density can enable researchers to determine efficiency and activity retention of various nanomaterials as electrocatalysts.
  - **Batteries**: In battery research, Tafel plots can be used to study the charge transfer kinetics at the electrode/electrolyte interface, helping to optimize electrode materials for faster charge/discharge rates and improved efficiency.

5.3 *Potentiostatic and Galvanostatic Cycling*

These methods are used to evaluate the long-term stability and cycling performance of energy storage devices. These techniques involve applying either a constant potential (potentiostatic) or constant current (galvanostatic) and monitoring the resulting charge/discharge cycles over an extended period.

- **Potentiostatic Cycling**: Potentiostatic cycling imposes a constant potential to the electrode, with measured current response as a function of charge–discharge cycles. This technique is also useful for studying the electrochemical stability of nanomaterials under applied high potentials leading to side reactions or degradation.
- **Galvanostatic Cycling**: In galvanostatic cycling the constant current is applied and the voltage profile during charging and discharging are recorded. The importance of this technique lies in the fact that it is used to determine the capacity retention, cycle life and rate capability of ESDs [32].

- **Applications**
  - **Cycle Life Testing**: This is routinely done by performing cycling tests of electrode materials under potentiostatic and galvanostatic conditions in batteries and supercapacitors. Long term nanomaterial performance is evaluated through metrics such as capacity degradation over time, Coulombic efficiency and voltage hysteresis.
  - **Rate Capability**: Secondly, these techniques enable an evaluation of the rate capability of nanomaterials, that is, their ability to deliver high powers at high rapid charge/discharge rates. For such applications such as electric vehicles where fast charging is essential, materials with good rate capability are a necessity.

5.4 Open Circuit Voltage (OCV) Measurements

OCV measurements involve monitoring the voltage of an electrochemical cell when no current is flowing (i.e., in an open-circuit state). This technique provides information about the thermodynamic equilibrium of the system and is particularly useful for studying battery materials and fuel cells [21].

- **Key Features**
  - **Equilibrium Potential**: In addition, the OCV represents the equilibrium potential of the electrochemical reactions in the system. The OCV gives information about the state of charge (SOC) and the energy storage capacity of the electrode materials in batteries.
  - **Voltage Relaxation**: After charge or discharge, the OCV tends to relax towards an equity value. This is from which it can deliver the rate of the relaxation and clues about the diffusion and redistribution of ions within the electrode materials.
- **Applications**
  - **Batteries**: Determination of battery state of charge and state of health is where OCV is widely used. And it also plays a central role in studies of the formation of the solid electrolyte interphase (SEI), which can impact battery performance and longevity.
  - **Fuel Cells**: OCV measurements are widely employed in fuel cell research to estimate the thermodynamic efficiency of the system and the quality of catalysts employed to facilitate electrochemical reactions in the system.

We complement EIS and galvanostatic cycling measurements with additional information on the electrochemical properties of energy storage nanomaterials, by means of other electrochemical characterization techniques: Chronoamperometry (CA), Chronopotentiometry (CP), Tafel plot analysis and OCV measurements. These methods, along with reaction kinetics, ion diffusion, and other phenomena pertinent to charge storage mechanisms in materials used in energy storage devices (batteries, supercapacitors, and fuel cells) are essential. Unless a combination of these techniques is employed, researchers are unlikely to achieve a full understanding of performance and limitations.

6 Applications of Characterization Techniques for Smart Nanomaterials in Supercapacitors and Batteries

SCs, also known as electrochemical capacitors, have gained significant attention as ESDs due to their high P_d, rapid charge–discharge capability, long cycle life, and stability. Structure and properties of materials used as electrodes for SCs, which can be prepared from advanced nanomaterials such as carbon based nanostructures, metal oxides, conducting polymers, etc., largely determine the performance of SCs. Tailored nanostructures and increased surface areas of smart nanomaterials have played a crucial role to improve SC performance [33].

The essential component of an ESD in a battery consists of two or more electrochemical cells capable of converting chemical energy into electrical energy. Even with a finite supply of fuel and oxidant, these devices serve both as energy converters and storage units, providing only limited periods of the electrical energy. Good electrical conductivity, a large surface area and compatibility with the electrolyte are the principal attributes of typical electrodes made of metal like zinc, lead or lithium.

Even with a finite supply of fuel and oxidant, these devices serve both as energy converters and storage units, providing only limited periods of the electrical energy. The electrodes, typically crafted from metals like zinc, lead, and lithium, play a pivotal role, requiring attributes such as good electrical conductivity, a high surface area to maximize capacitance, and compatibility with the electrolyte for optimal battery operation.

Batteries come into two types: primary and secondary. Primary batteries, exemplified by Lithium batteries, are non-rechargeable and rely on metallic components. Conversely, secondary batteries, such as Lithium-ion batteries, are rechargeable. Despite holding less energy, Li-ion batteries can supply higher load currents compared to their non-rechargeable counterparts. Both battery types include cathode, anode, separator, and electrolyte components. Nanostructured cathode materials, including $LiCoO_2$, $LiMnO_4$, and $LiNiO_2$, find application in Li-ion batteries. Lithium primary batteries are operated in devices for example heart pacemakers, instrumentation, and watches, while Li-ion batteries are employed in portable electronics, electric vehicles, and aerospace applications [34]. Characterization techniques play a vital role in optimizing these materials, understanding their behavior, and improving the design of SCs and batteries for various applications.

This section focuses on the key characterization techniques applied to smart nanomaterials used in SCs and batteries highlighting their role in understanding charge storage mechanisms, optimizing material performance, and enhancing device efficiency.

(A) **Cyclic Voltammetry (CV)**

CV is one of the most widely used techniques to evaluate the electrochemical behaviour of nanomaterials in SCs and batteries. It provides critical insights into the charge storage mechanisms (whether through electric double-layer capacitance (EDLC) or pseudocapacitance) and helps determine the specific capacitance and specific capacity of electrode materials illustrated in Fig. 8 [35].

The expressions for calculation of C_s and specific capacity C_{sp} from CV are as follow,

$$C_s = \frac{\int_{v1}^{v2} I(V)\mathrm{d}v}{m * v * \Delta v} \tag{3}$$

$$C_{sp} = \frac{\int_{v1}^{v2} I(V)\mathrm{d}v}{m * v * 3600} \tag{4}$$

The variation in the scan rates changes the capacitance of the supercapacitor devices. The lower capacitance value observed at higher scan rates is due to reduced efficiency in ion diffusion. In contrast, lower scan rates allow ample time for ionic diffusion into the inner pores of electrode materials, resulting in higher capacitance. By conducting CV at various scan rates, researchers can assess the rate capability of the nanomaterials. A stable capacitance over a wide range of scan rates indicates that the material can deliver high power densities, essential for energy storage applications [36]. CV curves can distinguish between electrochemical reactions, such as faradic or non-faradic, during charging and discharging. However, determining the average capacitance directly from CV analysis is challenging for battery-type and pseudocapacitive materials, as they exhibit redox peaks in CV profiles. Soon the careful selection of initial and final potential is crucial in CV measurements, as the operating potential window of SC electrodes depends on the redox and decomposition potential of the electrolyte. The area under the CV curve is directly related to the capacitance, and a more rectangular shape of the CV curve indicates ideal capacitive behaviour typical of EDLCs. For pseudocapacitive materials, the CV curve exhibits redox peaks, indicating faradaic reactions as shown in illustrated in Fig. 8a, b [37]. Carbon-based materials such as Graphene, carbon nanotubes (CNTs), and activated carbon typically exhibit EDLC behaviour, while transition metal oxides like MnO_2, NiO, and Co_3O_4 and conducting polymers for e.g. Polyaniline (PANI) and polypyrrole (PPy), etc. display pseudocapacitive behaviour due to reversible redox reactions.

(B) **Galvanostatic Charge–Discharge (GCD)**

GCD measurements are essential for evaluating the practical performance of SCs and batteries. This technique involves applying a constant current and recording the voltage response during the charge and discharge cycles, providing direct information

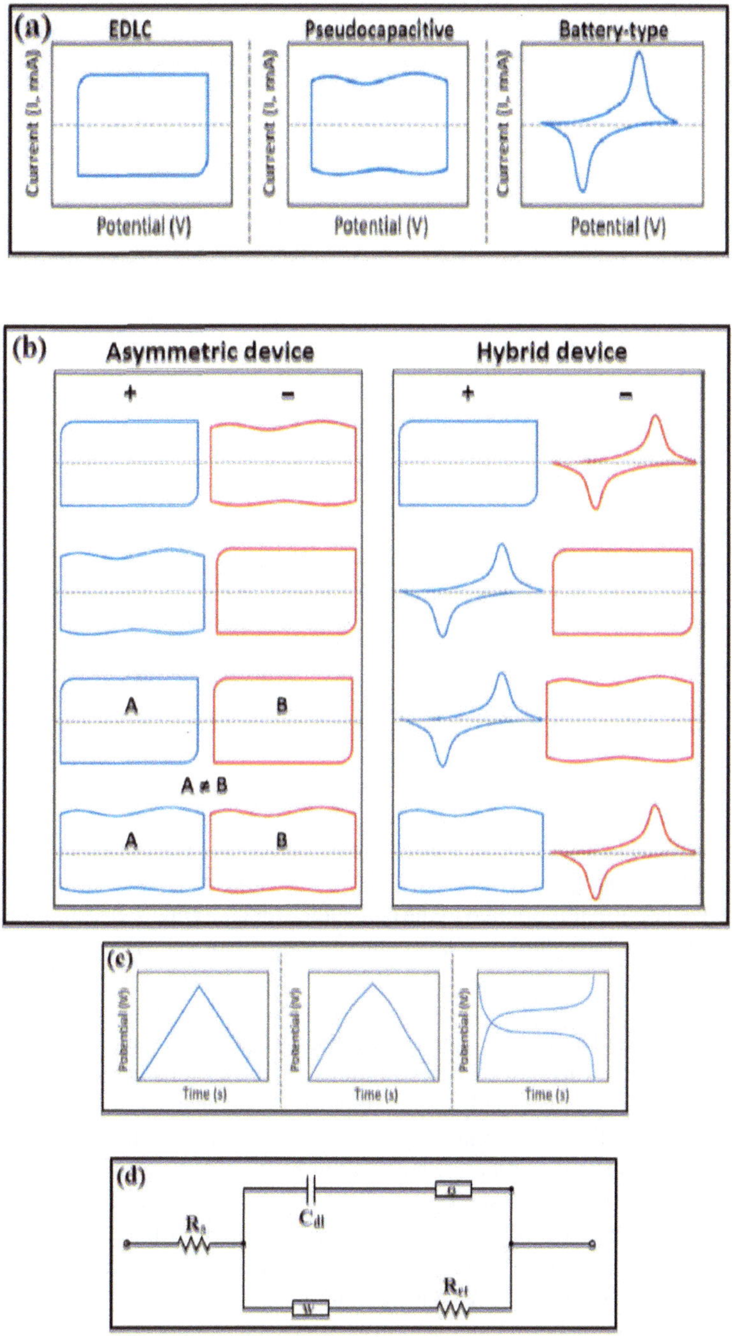
(a)
EDLC
Pseudocapacitive
Battery-type
Current (I, mA)
Potential (V)
(b)
Asymmetric device
Hybrid device
+
−
A
B
A ≠ B
(c)
Time (s)
(d)
Cdl
Rs
W
Rct

◂**Fig. 8** **a** Distinct shape of CV for EDLC, pseudocapacitive and battery type material. **b** Asymmetric and hybrid different configurations of distinct CV by combining materials A and B. **c** GCD profile to distinguish EDLC, pseudo capacitive and battery type material. **d** General circuit diagram for electrochemical impedance spectroscopy (EIS) [1]. Reproduced with permission @ copyright (2023) Elsevier

about the specific capacitance C_s, specific capacity C_{sp}, energy density E_d, power density P_d and cycle stability of the material [38]. For supercapacitive materials the GCD curves have a triangular shaped curve with linear charge/discharge profile. While for batteries a plateau is observed during the charging/discharge process. The expressions for the calculation of C_s, C_{sp}, E_d and P_d are as follows,

$$C = \frac{I_d \times t_d(s)}{m(g) \times \Delta V(V)} \tag{5}$$

$$C_{sp} = \frac{I_d(A) \times t_d(s)}{m(g) \times 3600} \tag{6}$$

$$\text{Energy density } (E_d)\left(\frac{\text{Wh}}{\text{kg}}\right) = \frac{C\left(\frac{F}{g}\right) \times V^2(V)}{2 \times 3.6} \tag{7}$$

$$\text{Power density } (P_d)\left(\frac{\text{W}}{\text{kg}}\right) = \frac{E_d\left(\frac{\text{Wh}}{\text{kg}}\right) \times 3600}{t_d(s)} \tag{8}$$

where I_d is the current density, t_d is the discharge time, m is the deposited mass and V is the working potential window.

GCD provides an accurate measurement of the specific capacitance by analysing the slope of the voltage–time curve during the discharge process. During the experiment, a steady current is applied to the cell, either for charging or discharging. Since the current flows through the cell, it induces electrochemical reactions at the electrodes, leading to changes in voltage across the cell over time. Plotting the recorded voltage data against time unveils characteristic charge–discharge curves as depicted in Fig. 8c. The GCD curve's shape reflects these materials' charge storage mechanism. Specifically, when GCD curves exhibit symmetrical and consistent triangular shapes across cell voltage and time, it indicates that charges are stored through the EDLC mechanism. The relatively low E_d values, particularly in terms of E_d, hinder the real-world implementation of carbon-based SCs. Conversely, if the curves are asymmetrical or vary in shape, it suggests the involvement of a pseudocapacitive mechanism in charge storage. The GCD curves often display voltage plateaus, reflecting the phase transition associated with ion intercalation/extraction, it recommends battery mechanism. The voltage–time data is carefully analysed to derive crucial parameters such as charge and discharge capacities, voltage characteristics, energy and power properties, and kinetic behaviour. These metrics are essential for evaluating the efficiency, cycling stability, rate capability, and capacity degradation of the ESD, providing valuable insights into its electrochemical performance. GCD also used to evaluate the Coulombic efficiency of electrode, which is the ratio of the discharge to charge capacity. High Coulombic efficiency (close to 100%) indicates that the energy storage and release processes are highly efficient, with minimal energy loss [1].

(C) **Electrochemical Impedance Spectroscopy (EIS)**

EIS is a crucial technique for understanding the charge transfer, ion diffusion, and internal resistances within SCs and batteries. By evaluating the impedance spectrum, researchers can optimize the electrode/electrolyte interface, like charge transfer kinetics, ion diffusion, and double-layer capacitance. This non-destructive technique widely employed in fields ranging from corrosion science to battery research, EIS is a fundamental tool for understanding and optimizing electrochemical systems for diverse applications [39].

Generally, in EIS measurements, the impedance data are obtained by giving an alternating voltage at a small amplitude at the open circuit potential over a vast range of frequencies [40]. At the higher frequencies, the electrode materials showing pure resistance resulting in electrolyte ions cannot penetrate the micropores of the materials. EIS can be demonstrated as a Nyquist plot, where it is a plot of the imaginary part of impedance versus the real part of the impedance. The Nyquist plot is composed of three regions—high frequency, high to the medium frequency and low frequency region marking a semicircle; R_{ct}; and nearly vertical line along the imaginary axis showing capacitive behaviour. The R_{ct} is calculated by the semicircle diameter of the plot as shown in Fig. 8. At low frequency, the Nyquist plot shows the pure capacitive behaviour which theoretically is parallel to the imaginary axis but normally there is an inclined angle of 45° or 90° between plotline and real axis that represents Warburg and ideal capacitive ionic diffusion [41].

The equivalent circuit is shown inset of Fig. 8d, which contains the low- and high-frequency components, where, C_{DL}—double-layer capacitor and W—Warburg impedance. The charge transfer may arise from faradic and non-faradic components and the faradic components originate from the redox reactions at the electrode surface while the non-faradic components arise from the C_{DL}. At the interface of electrode, charge transfer takes place which depends on the rate of electron transfer. Furthermore, the rate of electron transfer counts on the utilization of oxidants and the creation of reluctant near the electrode surface. Impedance is nothing but the ac resistance of the system, which consists of imaginary and real parts. As known, the electrochemical system exhibits resistive, capacitive, and inductive properties. The real part consists of resistive properties, while the imaginary part consists of capacitive and inductive properties.

While nanomaterials study, the highly porous carbon nanomaterials, such as activated carbon and mesoporous carbon, show improved ion diffusion properties, which is crucial for achieving fast charge–discharge cycles in SCs and combination of carbon material with metal oxides or conducting polymers enhances both the capacitance and the ion diffusion rate, improving the overall performance of the device.

(D) **Chronoamperometry (CA) and Chronopotentiometry (CP)**

To study the charge storage kinetics and stability of nanomaterials in the SCs and batteries CA and CP time-based techniques are used. In CA, a constant potential is applied, and the resulting current is measured over time, while in CP, a constant current is applied, and the potential response is recorded [22].

CA and CP provide insights into the charge storage mechanisms in supercapacitors, helping to differentiate between fast capacitive storage and slower diffusion-controlled processes. By applying constant current or voltage over time, these techniques help evaluate the stability of the electrode material during prolonged operation. Any degradation in performance, such as increased resistance or loss of capacitance, can be identified using these techniques.

In the smart material study, CA and CP are particularly useful for studying hybrid nanomaterials, such as carbon–metal oxide composites, to evaluate the balance between electric double-layer capacitance and pseudocapacitance contributions and in conducting polymer-based SCs, CA and CP help to assess the electrochemical stability of the polymer under repeated cycling, as these materials can suffer from swelling, shrinking, or degradation over time.

Along with the electrochemical characterization structural, morphological and thermal characterizations are necessary for understanding the phase purity and crystallinity, particle size, distribution, and surface characteristics, thermal stability and phase transitions of nanomaterials.

7 Future Directions in Smart Nanomaterials for Supercapacitors and Batteries

The continuous development of smart nanomaterials holds significant promise for the future of SCs and battery technology. Future research may focus on the following areas: The development of hybrid nanomaterials that combine different classes of materials (e.g., carbon-based materials with metal oxides or conducting polymers) could lead to synergistic effects, enhancing energy storage capacity and rate performance, the development of sustainable materials, incorporation of smart nanomaterials into solid-state battery designs could enhance safety and E_d.

Smart nanomaterials play a transformative role in the advancement of SCs and battery technology, contributing to enhanced energy storage capabilities, faster charge/discharge rates, and improved cycling stability. Through various characterization techniques, researchers can comprehensively understand the electrochemical behaviour and structural properties of these materials, facilitating the optimization of electrochemical performance. As the demand for efficient energy storage solutions continues to grow, the on-going development of innovative smart nanomaterials will be crucial in meeting the challenges of future energy storage applications, from portable electronics to electric vehicles and renewable energy systems [42].

8 Future Directions and Emerging Trends

The future of ESTs will heavily rely on the development and optimization of smart nanomaterials, coupled with advanced characterization techniques. As global demand for more efficient, sustainable, and high-performance EST grows, researchers are exploring innovative ways to enhance ESDs like batteries and SCs. This section delves into the potential future directions and emerging trends in the field, focusing on the evolution of characterization techniques and the next generation of smart nanomaterials.

8.1 Advanced Characterization Techniques for Nanomaterials

The continuous miniaturization and complexity of smart nanomaterials require cutting-edge characterization techniques that can probe at the nanoscale with unprecedented precision. Several emerging techniques are gaining prominence, as they offer deeper insights into the performance and behaviour of energy storage materials.

(A) **In-Situ and Operando Characterization Techniques**

In-situ and operando techniques allow real-time monitoring of electrochemical processes, structural changes, and phase transitions in nanomaterials during operation. These techniques provide a more comprehensive understanding of how nanomaterials behave under actual working conditions:

- **In-situ X-ray Diffraction (XRD)** and **In-situ Transmission Electron Microscopy (TEM)** allow researchers to observe structural changes in electrodes during charge–discharge cycles, leading to better material design.
- **Operando Raman Spectroscopy** helps monitor chemical changes and ion diffusion, offering insights into reaction mechanisms that occur within the nanomaterials.
- **In-situ Electrochemical Impedance Spectroscopy (EIS)** provides real-time data on the resistance and conductivity changes during battery and SCs operation.

These techniques will be crucial for optimizing smart nanomaterials and improving the overall performance, cycle life, and stability of ESDs.

New platforms are emerging that integrate multiple characterization techniques into a single experimental setup. For instance, integrating Atomic Force Microscopy (AFM) with electrochemical Analysis lets the researchers determine how mechanical characteristics are linked to the electrochemical behaviour. These platforms offer a more comprehensive picture of how a range of material properties, including mechanical strength, surface roughness, and electrical conductivity, cooperate and affect energy storage performance.

(B) **Challenges in Advanced Characterization Techniques**

While advanced characterization techniques provide deeper insights into nanomaterials and their energy storage performance, several challenges must be addressed to ensure their effective application:

- **Complexity in real-time characterization**: While in-situ and operando techniques (e.g., In-situ XRD, TEM, and EIS) offer real-time insights, challenges remain in achieving high-resolution data under operational conditions, especially for dynamic electrochemical processes.
- **Scalability of advanced characterization techniques**: Although techniques like Operando Raman Spectroscopy and AFM integration provide rich data, their high cost and technical complexity may limit large-scale application in industry.
- **Material integration challenges**: The integration of smart nanomaterials into solid-state and flexible energy storage devices (ESDs) requires addressing issues related to material compatibility, mechanical stability, and ion transport efficiency.
- **Standardization and reproducibility**: There is a need for standardized protocols in characterization to ensure consistent and reproducible results, which is crucial for commercialization and cross-laboratory comparisons.

9 Emerging Nanomaterials for Energy Storage

As the field of energy storage continues to evolve, several new classes of smart nanomaterials are emerging, each with distinct properties that could significantly impact future ESTs.

If we think of graphene as 2D material of choice, then other 2D materials such as transition metal dichalcogenides (TMDs), MXenes and phosphorene are currently gaining interest for their extraordinary electrical, mechanical and thermal properties. With high surface area and unusual electrochemical properties, these materials are ideal next generation energy storage materials. Metal organic frameworks (MOFs) and covalent organic frameworks (COFs) are porous materials with tunable structures consisting of large surface areas and tailored functionality. They are ideal for ESTs, due to their highly ordered and customizable frameworks. In addition, MOFs have exhibited great promise as high capacity electrodes and efficient catalysts for energy conversion reactions. When mixed with traditional battery and SC materials such as graphene or CNTs, these materials can greatly increase E_d and P_d performance [43].

Nanostructured conducting polymers and organic materials offer sustainable alternatives to traditional metal-based materials. Their flexibility, lightweight nature, and tunability make them ideal candidates for flexible energy storage devices, such as wearable electronics or foldable SCs.

New synthetic strategies are being developed to create highly porous, nanostructured polymers that exhibit excellent electrochemical properties, expanding the range of smart materials available for energy storage.

9.1 Literature Review

Recent studies have provided deeper insights into the role of emerging nanomaterials in energy storage technologies. For instance, Carbon-based composites, metal oxides, conducting polymers, and hybrid nano-architectonics have emerged as promising candidates for next-generation ESDs. Among these, carbon-based hybrid nanostructures stand out as some of the most promising materials for energy storage. By blending carbon materials with metal nanoparticles, metal oxides, or polymers, these composites harness synergistic effects that significantly boost their electrochemical performance. For instance, carbon-based composites have been shown to enhance rate capability, cycle stability, and energy storage capacity, thanks to their specially designed shapes, content, and surface chemistry. This makes them particularly effective in SCs where their high surface area and conductivity allow for rapid charge storage and release [44]. These materials are particularly effective in SCs, where their high surface area and conductivity enable rapid charge storage and release. Metal oxides and sulphides are widely studied for their pseudocapacitive properties, making them ideal for SC applications. Materials such as cobalt oxide, nickel oxide, and molybdenum sulfide have been integrated with conducting polymers like polyaniline (PANI) and polypyrrole (PPy) to enhance their conductivity and structural integrity [45]. These hybrid materials demonstrate improved electrochemical performance, including higher specific capacitance and better cycling stability. Additionally, metal oxides like $Co_{0.5}Ni_{0.5}MoO_4$ have been used in double-shelled hollow spheres, which exhibit excellent pseudocapacitive properties and high specific capacities in both SCs and lithium-ion batteries [46]. Conductive polymers such as PANI and PPy have gained attention for their high capacitance and electrochemical reversibility. When combined with inorganic materials like MoS_2, these polymers form hybrid electrodes that overcome their inherent limitations, such as low cyclic stability. For example, a MoS_2/PANI hybrid electrode has demonstrated a specific capacity of 1087.5 C/g, significantly higher than that of pure PANI. These materials are also being explored for their potential in hydrogen evolution reactions, where they exhibit low over potentials and high stability [47]. Conductive metal–organic frameworks (c-MOFs) are emerging as promising electrode materials due to their uniform pore sizes, customizable functional groups, and excellent electrical conductivity. By interweaving MOFs with conductive polymers like PANI, researchers have developed flexible, high-performance SCs. For instance, a PANI-interwoven MOF electrode has achieved an areal capacitance of 2146 mF cm^{-2}, making it suitable for flexible solid-state SCs [48]. Furthermore, Hybrid nanoarchitectonics, such as mesoporous C_{60}–BCN composites, has shown immense potential for energy storage applications. These materials combine the high surface area of mesoporous structures with the unique properties of fullerene and borocarbonitride, leading to enhanced ion charge transfer and electrochemical stability. Such composites have demonstrated exceptional specific capacitance stability of $\sim$100% after 6000th cycles and high storage capacities in lithium-ion batteries [49]. These studies, along with others, provide strong evidence for the continued advancement of smart nanomaterials in energy storage applications.

9.2 Practical Application and Case Studies

To enhance the relevance of these emerging nanomaterials, several practical applications and case studies are highlighted:

- Graphene-Based SCs in Wearable Electronics: Research demonstrates that graphene-enhanced SCs exhibit high flexibility and E_d, making them suitable for applications in wearable smart textiles and biomedical sensors [50].
- MXenes for High-Performance Batteries: Case studies on MXene-based electrodes show significant improvements in energy storage capacity and charge/discharge rates in lithium-ion and sodium-ion batteries [51].
- MOF-Derived Electrodes in Next-Generation Batteries: Studies indicate that MOF-derived nanostructured electrodes enhance charge storage mechanisms, leading to increased efficiency in zinc-air and lithium-sulphur batteries [43].
- Flexible Conducting Polymers in Stretchable Energy Devices: Research on nanostructured conducting polymers illustrates their role in the development of highly stretchable and foldable SCs for medical applications.
- Solid-State Electrolytes for Safer Energy Storage: Case studies on solid-state lithium batteries reveal the potential of integrating nanoporous solid-state electrolytes to improve safety and performance [52].

These case studies highlight the real-world impact of smart nanomaterials and demonstrate their potential for future energy storage advancements.

9.3 Sustainable and Green Nanomaterials

As sustainability becomes a key focus in material science, the demand for eco-friendly and biodegradable nanomaterials is growing. Future trends will likely focus on developing materials that have minimal environmental impact, both in their production and disposal. The uses of naturally derived carbon nanomaterials, such as those from biomass, and the development of non-toxic and recyclable electrode materials will be a major focus in future research.

Additionally, green chemistry approaches are being explored to reduce the environmental footprint of nanomaterial synthesis, ensuring that the next generation of energy storage devices is both efficient and sustainable [53].

9.4 Solid-State Energy Storage

One of the most exciting trends in energy storage is the development of solid-state batteries and solid-state SCs. The integration of smart nanomaterials into solid electrolytes could lead to safer, more efficient, and longer-lasting energy storage systems. Research is now focusing on solid electrolytes with high ionic conductivity and stable interfaces, essential for overcoming the limitations of current liquid electrolytes.

Smart nanomaterials are likely to play a key role in enhancing the performance of solid-state systems by providing improved electrode/electrolyte interfaces and higher energy densities. Nanostructured materials, such as nanoporous metals and 3D nanostructured scaffolds, are also being explored to enhance ion transport and mechanical stability in solid-state batteries.

9.5 Flexible and Wearable Energy Storage Devices

With the growing market for wearable electronics and the Internet of Things (IoT), flexible and stretchable energy storage devices are gaining attention. Future research will explore the integration of smart nanomaterials into flexible SCs and batteries that can maintain high performance under mechanical stress. Materials such as nanostructured polymers, flexible carbon materials, and 2D materials are essential for developing ESDs that can bend, fold, or stretch without compromising energy storage performance.

10 Conclusion and Outlook

The future of energy storage lies in the continuous advancement of smart nanomaterials and the refinement of characterization techniques that enable their full potential to be realized. As research moves toward real-time, in-situ characterization, integrated multifunctional platforms, and material discovery, the design and optimization of smart nanomaterials for energy storage will accelerate. The development of new, sustainable nanomaterials and their integration into solid-state and flexible devices will pave the way for more efficient, safer, and eco-friendly ESTs.

Emerging trends such as solid-state energy storage, flexible devices, and sustainable materials will drive the innovation needed to meet the growing global demand for clean, efficient, and durable energy storage solutions.

References

1. Sharma, S., Chand, P.: Supercapacitor and electrochemical techniques: a brief review. Results Chem. **5**, 100885 (2023). https://doi.org/10.1016/j.rechem.2023.100885
2. Anil Kumar, Y., Koyyada, G., Ramachandran, T., Kim, J.H., Sajid, S., Moniruzzaman, M., Alzahmi, S., Obaidat, I.M.: Carbon materials as a conductive skeleton for supercapacitor electrode applications: a review. Nanomaterials **13**(6), 1049 (2023). https://doi.org/10.3390/nano13061049
3. Shirsat, M.D., Sathe, B.R., Koinkar, P.M.: Editorial: smart materials for energy conversion and sensor based technologies. Front. Mater. **8**, 626397 (2021). https://doi.org/10.3389/fmats.2021.626397
4. Gurav, S.R., Waikar, M.R., Rasal, A.S., Sonker, R.K., Sonkawade, R.G.: Current development and challenges in textile-based flexible supercapacitors. In: Smart and Flexible Energy Devices, pp. 539–550. CRC Press (2022)
5. Kumar, S., Saeed, G., Zhu, L., Hui, K.N., Kim, N.H., Lee, J.H.: 0D to 3D carbon-based networks combined with pseudocapacitive electrode material for high energy density supercapacitor: a review. Chem. Eng. J. **403**, 126352 (2021). https://doi.org/10.1016/j.cej.2020.126352
6. Kumar, N., Kim, S.B., Lee, S.Y., Park, S.J.: Recent advanced supercapacitor: a review of storage mechanisms, electrode materials, modification, and perspectives. Nanomaterials **12**(20), 3708 (2022). https://doi.org/10.3390/nano12203708
7. Gohar, O., Khan, M.Z., Bibi, I., Bashir, N., Tariq, U., Abdul Karim, M.R., Ali, F., Hanif, M.B., Motola, M.: Nanomaterials for advanced energy applications: recent advancements and future trends. Mater. Des. **241**, 112930 (2024). https://doi.org/10.1016/j.matdes.2024.112930
8. Lukatskaya, M.R., Dunn, B., Gogotsi, Y.: Multidimensional materials and device architectures for future hybrid energy storage. Nat. Commun. **7**, 12647 (2016). https://doi.org/10.1038/ncomms12647
9. Rahul, Sonker, R.K., Shukla, P.K., Singh, P.K., Khan, Z.H.: Experimental and characterization techniques. In: Composite Materials, pp. 99–123. CRC Press (2021)
10. Chooto, P.: Cyclic voltammetry and its applications. In: Voltammetry. IntechOpen (2019). https://doi.org/10.5772/intechopen.83451
11. Forster, R.J., Walsh, D., Adamson, K., Spain, E.: Voltammetry | Overview, 3rd edn. Elsevier Inc. (2019). https://doi.org/10.1016/B978-0-12-409547-2.14562-7
12. Kosov, A.V., Grishenkova, O.V., Semerikova, O.L., Isaev, V.A., Zaikov, Y.P.: On the theory of cyclic voltammetry for multiple nucleation and growth: scan rate influence. J. Electroanal. Chem. **883**, 115056 (2021). https://doi.org/10.1016/j.jelechem.2021.115056
13. Shao, Y., El-Kady, M.F., Sun, J., Li, Y., Zhang, Q., Zhu, M., Wang, H., Dunn, B., Kaner, R.B.: Design and mechanisms of asymmetric supercapacitors. Chem. Rev. **118**, 9233–9280 (2018). https://doi.org/10.1021/acs.chemrev.8b00252
14. Namsheer, K., Rout, C.S.: Photo-powered integrated supercapacitors: a review on recent developments, challenges and future perspectives. J. Mater. Chem. A **9**, 8248–8278 (2021). https://doi.org/10.1039/d1ta00444a
15. Joshi, P.S., Sutrave, D.S.: A brief study of cyclic voltammetry and electrochemical analysis. Int. J. ChemTech Res. **11**, 77–88 (2018). https://doi.org/10.20902/ijctr.2018.110911
16. Puthongkham, P., Venton, B.J.: Recent advances in fast-scan cyclic voltammetry. Analyst **145**, 1087–1102 (2020). https://doi.org/10.1039/c9an01925a
17. Sweep, L., Stripping, A., Stripping, C., Compounds, I., Compounds, O.: Potential sweep and current responses. In: Elements, pp. 181–188 (2005)
18. O.O.F., Cyclic, I.T.S., Theoretical: Chapter-I: introduction and overview of cyclic voltammetry and its theoretical. Current (n.d.)
19. Batley, G.E., Florence, T.M.: An evaluation and comparison of some techniques of anodic stripping voltammetry. J. Electroanal. Chem. **55**, 23–43 (1974). https://doi.org/10.1016/S0022-0728(74)80469-9
20. Application, A., Bat, N.: Potentiostatic intermittent titration technique. Metrohm Autolab Appl. Note. **2**, 1–3 (2014)

21. Kim, J., Park, S., Hwang, S., Yoon, W.S.: Principles and applications of galvanostatic intermittent titration technique for lithium-ion batteries. J. Electrochem. Sci. Technol. **13**, 19–31 (2022). https://doi.org/10.33961/jecst.2021.00836
22. Licht, F., Davis, M.A., Andreas, H.A.: Charge redistribution and electrode history impact galvanostatic charging/discharging and associated figures of merit. J. Power Sources **446**, 227354 (2020). https://doi.org/10.1016/j.jpowsour.2019.227354
23. Choudhary, Y.S., Jothi, L., Nageswaran, G.: Electrochemical characterization. Elsevier Inc. (2017). https://doi.org/10.1016/B978-0-323-46140-5.00002-9
24. Magar, H.S., Hassan, R.Y.A., Mulchandani, A.: Electrochemical impedance spectroscopy (EIS): Principles, construction, and biosensing applications. Sensors **21**, 6578 (2021). https://doi.org/10.3390/s21196578
25. Mei, B.A., Munteshari, O., Lau, J., Dunn, B., Pilon, L.: Physical interpretations of Nyquist plots for EDLC electrodes and devices. J. Phys. Chem. C **122**, 194–206 (2018). https://doi.org/10.1021/acs.jpcc.7b10582
26. Tigris, Bode and Nyquist Plot, pp. 1–8. Palmsens (2021)
27. Wagner, N.: Basics of electrochemical impedance spectroscopy introduction. In PEM Fuel Cell Diagnostic Tools, pp. 37–70 (2011). https://doi.org/10.1201/b11100-5
28. Patil, S.S., Patil, P.S.: 3D bode analysis of nickel pyrophosphate electrode: a key to understanding the charge storage dynamics. Electrochim. Acta **451**, 142278 (2023). https://doi.org/10.1016/j.electacta.2023.142278
29. Simões, F.R., Xavier, M.G.: Chapter 6—electrochemical sensors. In Electrochemical Sensors. Elsevier Inc. (2017). https://doi.org/10.1016/B978-0-323-49780-0/00006-5
30. Radhika, M.G., Gopalakrishna, B., Chaitra, K., Bhatta, L.K.G., Venkatesh, K., Kamath, M.K.S., Kathyayini, N.: Electrochemical studies on Ni, Co and Ni/Co-MOFs for high-performance hybrid supercapacitors. Mater. Res. Express **7** (2020). https://doi.org/10.1088/2053-1591/ab8d5d
31. Bian, W., Huang, Y., Xu, X., Ud Din, M.A., Xie, G., Wang, X.: Iron hydroxide-modified nickel hydroxylphosphate single-wall nanotubes as efficient electrocatalysts for oxygen evolution reactions. ACS Appl. Mater. Interfaces **10**, 9407–9414 (2018). https://doi.org/10.1021/acsami.7b18875
32. Wu, Q., He, T., Zhang, Y., Zhang, J., Wang, Z., Liu, Y., Zhao, L., Wu, Y., Ran, F.: Cyclic stability of supercapacitors: materials, energy storage mechanism, test methods, and device. J. Mater. Chem. A **9**, 24094–24147 (2021). https://doi.org/10.1039/d1ta06815f
33. Venkata, K., Raghavendra, G., Vinoth, R., Zeb, K., Obaidat, I.M., Je, H.: An intuitive review of supercapacitors with recent progress and novel device applications. J. Energy Storage **31**, 101652 (2020). https://doi.org/10.1016/j.est.2020.101652
34. Conway, B.E.: Transition from "Supercapacitor" to "Battery" behavior in electrochemical energy storage. Proc. Int. Power Sources Symp. **138**, 319–327 (1991). https://doi.org/10.1109/ipss.1990.145856
35. Mendoza, S., Bustos, E., Manríquez, J., Godínez, L.A.: Voltammetric techniques. In: Agricultural and Food Electroanalysis, pp. 21–48 (2015). https://doi.org/10.1002/9781118684030.ch2
36. Sawant, S.A., Waikar, M.R., Rasal, A.S., Chodankar, G.R., Dhas, S.D., Moholkar, A.V., Shirsat, M.D., Chakarvarti, S.K., Sonkawade, R.G.: Chemical synthesis and supercapacitive evaluation of polyaniline nanofibers (PANINFs). J. Mater. Sci. Mater. Electron. **32**, 11865–11876 (2021). https://doi.org/10.1007/s10854-021-05816-7
37. Wang, Y., Song, Y., Xia, Y.: Electrochemical capacitors: mechanism, materials, systems, characterization and applications. Chem. Soc. Rev. **45**, 5925–5950 (2016). https://doi.org/10.1039/c5cs00580a
38. Abid, M.A.A.M., Radzi, M.I., Mupit, M., Osman, H., Munawar, R.F., Samat, K.F., Suan, M.S.M., Isomura, K., Islam, M.R.: Cyclic voltammetry and galvanostatic charge-discharge analyses of polyaniline/graphene oxide nanocomposite-based supercapacitor. Malays. J. Compos. Sci. Manuf. **3**, 14–26 (2020). https://doi.org/10.37934/mjcsm.3.1.1426

39. Lazanas, A.C., Prodromidis, M.I.: Electrochemical impedance spectroscopy—a tutorial. ACS Meas. Sci. Au **3**, 162–193 (2023). https://doi.org/10.1021/acsmeasuresciau.2c00070
40. Zhang, J., Zhao, X.S.: On the configuration of supercapacitors for maximizing electrochemical performance. ChemSusChem **5**, 818–841 (2012). https://doi.org/10.1002/cssc.201100571
41. Waikar, M.R., Shaikh, A.A., Sonkawade, R.G.: PANINFs synthesized electrochemically as an electrode material for energy storage application. Polym. Bull. **76**, 4703–4718 (2019). https://doi.org/10.1007/s00289-018-2634-1
42. Banerjee, S., De, B., Sinha, P., Cherusseri, J., Kar, K.K.: Applications of supercapacitors. In Springer Series in Materials Science, pp. 341–350. Springer (2020). https://doi.org/10.1007/978-3-030-43009-2_13
43. Raje, P.G., Gurav, S.R., Waikar, M.R., Rasal, A.S., Chang, J.Y., Sonkawade, R.G.: The review of different dimensionalities based pristine metal organic frameworks for supercapacitor application. J. Energy Storage **56**, 105700 (2022). https://doi.org/10.1016/j.est.2022.105700
44. Zhang, L., Zhao, X.S.: Carbon-based materials as supercapacitor electrodes. Chem. Soc. Rev. **38**, 2520–2531 (2009). https://doi.org/10.1039/b813846j
45. Annu, Park, S.S., Alam, M.N., Yewale, M., Shin, D.K.: Unraveling the electrochemical insights of cobalt oxide/conducting polymer hybrid materials for supercapacitor, battery, and supercapattery applications. Polymers (Basel) **16**, 2907 (2024). https://doi.org/10.3390/polym16202907
46. Li, L., Chen, W., Luo, W., Xiao, Z., Zhao, Y., Owusu, K.A., Liu, Z., Zhou, L., Mai, L.: $Co_{0.5}Ni_{0.5}MoO_4$ double-shelled hollow spheres with enhanced electrochemical performance for supercapacitors and lithium-ion batteries. Energy Technol. **7** (2019). https://doi.org/10.1002/ente.201801160
47. Alsaiari, M., Imran, M., Afzal, A.M., Mumtaz, S., Algethami, J.S., Harraz, F.A.: Designing high-performance polyaniline@MoS_2@AC hybrid electrode for electrochemical-based next-generation battery—supercapacitor hybrid energy storage device and hydrogen evolution reaction. Phys. Scr. **99** (2024). https://doi.org/10.1088/1402-4896/ad25d6
48. Wang, L., Feng, X., Ren, L., Piao, Q., Zhong, J., Wang, Y., Li, H., Chen, Y., Wang, B.: Flexible solid-state supercapacitor based on a metal-organic framework interwoven by electrochemically-deposited PANI. J. Am. Chem. Soc. **137**, 4920–4923 (2015). https://doi.org/10.1021/jacs.5b01613
49. Bahadur, R., Singh, G., Li, Z., Singh, B., Srivastava, R., Sakamoto, Y., Chang, S., Murugavel, R., Vinu, A.: Hybrid nanoarchitectonics of ordered mesoporous C_{60}–BCN with high surface area for supercapacitors and lithium-ion batteries. Carbon **216**, 118568 (2024). https://doi.org/10.1016/j.carbon.2023.118568
50. Sawant, S.A., Patil, A.V., Waikar, M.R., Rasal, A.S., Dhas, S.D., Moholkar, A.V., Vhatkar, R.S., Sonkawade, R.G.: Advances in chemical and biomass-derived graphene/graphene-like nanomaterials for supercapacitors. J. Energy Storage **51**, 104445 (2022). https://doi.org/10.1016/j.est.2022.104445
51. Dubal, D.P., Chodankar, N.R., Kim, D.H., Gomez-Romero, P.: Towards flexible solid-state supercapacitors for smart and wearable electronics. Chem. Soc. Rev. **47**, 2065–2129 (2018). https://doi.org/10.1039/c7cs00505a
52. Nasrin, K., Gokulnath, S., Karnan, M., Subramani, K., Sathish, M.: Redox-additives in aqueous, non-aqueous, and all-solid-state electrolytes for carbon-based supercapacitor: a mini-review. Energy Fuels **35**, 6465–6482 (2021). https://doi.org/10.1021/acs.energyfuels.1c00341
53. Wang, Y., Hu, Y.J., Hao, X., Peng, P., Shi, J.Y., Peng, F., Sun, R.C.: Hydrothermal synthesis and applications of advanced carbonaceous materials from biomass: a review. Adv. Compos. Hybrid Mater. **3**, 267–284 (2020). https://doi.org/10.1007/s42114-020-00158-0

Low Dimensional Nanomaterials for Hydrogen Storage

Yogesh Kumar Yadav, Abhishek Kumar, Mohammad Abu Shaz, and Thakur Prasad Yadav

Abstract Economic expansion and population growth are the main causes of the growing demand for energy on a worldwide scale. Nowadays, fossil fuels, which are limited and non-renewable, provide 90% of the energy used worldwide. Furthermore, the main cause of greenhouse gas emissions, which have detrimental effects on the environment and human health, is the burning of fossil fuels. Because of this, switching to clean alternative energy sources is crucial. Because hydrogen fuel is sustainable, non-toxic, and emits no pollutants, it has drawn attention from all over the world. Hydrogen has an energy density of 142 MJ/kg, which is more than any other element and almost three times that of fossil fuels. In contrast to the damaging and depleting fossil fuels, hydrogen is plentiful and can be generated responsibly from water and other renewable resources. Effective hydrogen storage is still a significant obstacle, though. Storage procedures are made more difficult by its low gravimetric density (0.08988 g/L). Although hydrogen can be kept as a gas, liquid, or solid, solid-state storage is the most effective and safest option. This technique enables high-capacity storage at low pressure and close to room temperature. While materials such as carbon nanotubes and metal–organic frameworks store hydrogen by surface adsorption, metal hydrides absorb it internally. Recent developments in metal hydrides based on high-entropy alloys, which may be synthesized a variety of ways, exhibit encouraging potential for use in hydrogen production and storage.

Keywords Greenhouse gas · Hydrogen storage · Solid-state storage · Metal hydride · Carbon nanotubes · High entropy alloys

Y. K. Yadav · A. Kumar · M. A. Shaz · T. P. Yadav (✉)
Department of Physics, Institute of Science, Banaras Hindu University, Varanasi, India
e-mail: tpyadav@allduniv.ac.in; yadavtp@gmail.com

T. P. Yadav
Department of Physics, Faculty of Science, University of Allahabad, Prayagraj, India

R. K. Sonker et al. (eds.), *Low-Dimensional Multifunctional Hybrid Nanostructured Materials*, Engineering Materials, https://doi.org/10.1007/978-981-96-7952-2_3

1 Introduction

Energy is the basic demand of the living and non-living things. This demand can be fulfilled by using energy derived from fossil fuel, solar energy, wind energy, geothermal energy, hydrogen energy, tidal energy, hydroelectric energy, etc. [1–3]. Generally, we are bound with fossil fuels for producing energy. But on consumption of fossil fuels for the production of energy, they emit harmful gases particularly the greenhouse gas CO_2. An increase in CO_2 concentration by continued CO_2 emission leads to climate change/global warming effect, which is the most dreaded effect of the use of fossil fuels. Besides climate change, burning of fossil fuels also produces urban air pollution. Clean and renewable fuel sources are desperately needed to address climate change, cut pollution, and fulfil the world's expanding energy needs [4–7]. The world's energy supply is currently mostly dependent on finite fossil fuels, which deteriorates the environment. Hydrogen, on the other hand, is a renewable and sustainable substitute that emits no carbon dioxide. Hydrogen is one of the cleanest and most promising energy carriers among the several fuel choices [8–11]. It is widely available in nature and has the largest energy content per unit mass (142 MJ/kg), almost three times that of fossil fuels. However, effective techniques for production, storage, and use must be developed in order to realize its full potential. It can be used in an electrochemical cell, internal combustion engine to power vehicles, electric devices. It produces water after use in these devices. There are several ways of producing hydrogen, and each has a different sustainability and environmental impact. Almost 70% of the hydrogen produced worldwide comes from the most popular process, steam methane reforming (SMR). Despite being less expensive, SMR is less sustainable because it uses fossil fuels and releases a lot of CO_2. Likewise, another industrial process that produces considerably more carbon emissions is coal gasification. On the other hand, electrolysis is the process of utilizing electricity to divide water into hydrogen and oxygen offers a cleaner substitute, especially when it is driven by renewable energy. Although high manufacturing costs are still an issue, green hydrogen which is produced by electrolysis fueled by renewable energy sources is thought to be the most environmentally friendly choice. Furthermore, photocatalytic water splitting and biomass gasification are showing promise as environmentally benign hydrogen production techniques. Reducing reliance on fossil fuels and minimising environmental damage require a shift to sustainable hydrogen production. Making hydrogen a practical long-term energy source will require both more investment in renewable energy sources and advancements in electrolysis technology. In fact, 60 million tons of hydrogen are being produced worldwide [12]. The major issue related to the use of hydrogen as a fuel is its storage. Due to its very low gravimetric density (0.08988 g/L), hydrogen storage presents a pivotal challenge. Liquid hydrogen has a far higher density of about 70 g/L than gaseous hydrogen, which has a low gravimetric density of 0.08988 g/L. Metal hydrides, on the other hand, are a more effective and portable choice for storing hydrogen since they can reach even higher storage densities, frequently surpassing 100 g/L. The benefits of solid-state storage over gaseous and liquid versions are shown by this comparison.

For mobile applications, it is well inferred that as compared to the storage of hydrogen as a compressed gas or in liquid form, solid-state storage of hydrogen in the form of hydrides is the safest and efficient mode [13, 14]. The storage method of solid-state brings better performance than both gaseous and liquid hydrogen storage systems. The storage method enables higher density distribution and affords safety benefits and releases hydrogen at moderate pressures efficiently.

The storage of hydrogen as solids requires operational conditions at room temperature and provides enhanced safety benefits providing an effective hydrogen transport system compared to gaseous and liquid methods. Table 1 displays the differences between hydrogen storage methods. Hydrides, unlike gaseous storage, do not require very high pressure (>50 atm) nor do they require very low temperature (~−253 °C) as needed for liquid state storage of hydrogen. The U.S. Department of Energy (US DOE) has set important performance goals for onboard hydrogen storage in light-duty fuel cell cars in order to guarantee effective, realistic, and affordable hydrogen storage options. In order to make hydrogen a competitive alternative to traditional fuels, especially in transportation applications, these goals center on gravimetric and volumetric capacity, cost reduction, and overall system performance. The system gravimetric capacity, which calculates the useable specific energy from hydrogen per unit mass of the storage system, is an important parameter. 1.5 kWh/kg (0.045 kg H_2/kg system) was the 2020 aim; by 2025, it would have increased to 1.8 kWh/kg (0.055 kg H_2/kg system). Space-constrained applications also heavily rely on the system volumetric capacity, which indicates the useful energy density from hydrogen in relation to system volume. 1.0 kWh/L (0.030 kg H_2/L system) was the target for 2020; by 2025, it is anticipated that this will have increased to 1.3 kWh/L (0.040 kg H_2/L system), and then to 1.7 kWh/L (0.050 kg H_2/L system). These goals are to maximize energy density while creating small storage options that improve vehicle design. Cost reduction is still a significant obstacle, even after capacity enhancements. The DOE seeks to retain safety, durability, and operational efficiency while drastically reducing storage system costs to make hydrogen fuel cell vehicles commercially feasible. For broad acceptance and competitiveness with conventional fossil fuel-based technology, these financial obstacles must be removed. Continuous improvements in material science, engineering, and system integration are necessary to meet these challenging objectives. Several hydrides show promise in approaching DOE targets: MgH_2 (7.6 wt% H_2, improved with catalysts), $LiBH_4$ (18.5 wt% H_2, enhanced via destabilization), $NaAlH_4$ (5.6 wt% H_2, Ti-doped for lower desorption temperature), High-Entropy Hydrides (HEHs) (tunable properties for better sorption), and MOF-hydride composites (improving kinetics and reversibility). While challenges remain, these materials are advancing toward practical hydrogen storage. Intermetallic based high entropy materials are one of the most promising candidates for solid-state hydrogen storage material owing to their high hydrogen storage capacity. This is one of the other factors, which makes multicomponent high entropy materials as front running candidates for hydrogen storage applications. Hydrogen storage benefits from the special properties of HEAs which provide superior storage capabilities. For example, their high configurational entropy and diverse composition lead to increased structural stability

Table 1 The following table presents a comparative overview of different hydrogen storage methods, highlighting their key characteristics, and advantages

Storage form	Energy density (MJ/kg)	Density (g/L)	Storage pressure/ temperature	Key advantages
Gaseous hydrogen	142	0.08988	350–700 bar (high pressure)	Simple, scalable, low-cost infrastructure
Liquid hydrogen	141	70.85	Cryogenic temperatures (~−253 °C)	High density, efficient for large-scale storage
Solid-state storage	Varies (up to 1.5–2.0 g/L)	1.5–2.0	Ambient temperature, low pressure	Safe, compact, high density, reduces safety concerns

and resistance to degradation under extreme conditions. Results from Ti–V–Zr–Cr–Ni HEA research have indicated beneficial storage abilities for hydrogen molecules according to study by Kumar et al. [15]. The mechanical properties of HEAs include high strength along with ductility which makes them particularly suitable for use in hydrogen storage systems requiring pressure and temperature stability. Current research indicates that HEAs are established as top candidates for future hydrogen storage applications because of their excellent properties which include corrosion resistance and high thermal stability.

2 Concept of Multicomponent Smart Nanomaterials

Multicomponent smart nanomaterials, which offer improved efficiency and versatility across a range of areas, represent a significant advancement in materials research [16]. Metallurgy has historically concentrated on individual elements or compounds, however alloying numerous elements enhance qualities in a similar way to mixing ingredients to make a healthy juice [17]. Franz Karl Archard studied multicomponent alloys in the eighteenth century, but there were few real-world uses for them [18]. Most of the current research ignores the potential of multicomponent alloys, especially high-entropy alloys, which show promise for novel structures and uses in materials science, in favor of binary or ternary systems.

Promising study topic, HEAs first proposed by Prof. Jien-Wei Yeh in 2004 because of their remarkable strength, thermal stability, and resistance to wear and corrosion at room temperature and extreme temperatures [19, 20]. HEAs, which are made up of five or more primary elements in concentrations ranging from 5 to 35%, provide special chemical and physical characteristics that make them useful for applications such as wear-resistant coatings, hydrogen storage, and anticorrosion

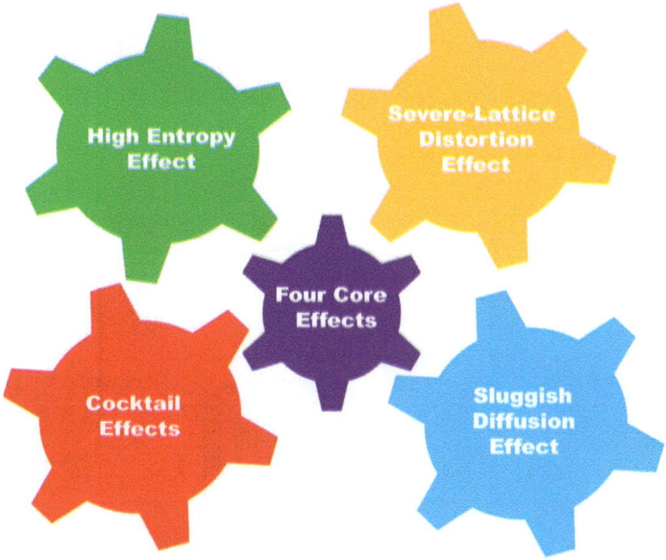

Fig. 1 A graphical representation of the four core effects of HEAs. Each effect highlights the unique properties that emerge from the combination of multiple principal elements, contributing to the exceptional performance of HEAs

[21]. HEAs with multi-principal elements have special qualities that set them apart from other traditional alloys. These include high entropy from multiple components, and lattice distortion that improves mechanical and functional properties, slow diffusion because of complex atomic arrangements affecting performance, cocktail effects from synergistic interactions that improve properties (Fig. 1) [22].

Their unique properties from conventional alloys are a result of their high configurational entropy, which is caused by the random arrangement of elements and promotes the production of stable solid solutions over intermetallic compounds [23]. HEAs are divided into three groups according to the number of primary elements: low, medium, and high entropy alloys shown in Fig. 2 [24].

Under constant pressure and temperature, a system's goal in thermodynamics is to minimize its Gibbs free energy (G) [17]. When G is minimized, equilibrium is obtained, as shown by the equation:

$$G = H - TS$$

where, T is the temperature, S is the entropy, and H is the enthalpy. Thus, a system entropy and enthalpy are important factors in figuring out its equilibrium state at a particular temperature [25]. The free energy transitions between the elemental and mixed states are compared to determine which state has the lowest mixing free energy, which helps predict the equilibrium state of an alloy. It is possible to express this relationship as:

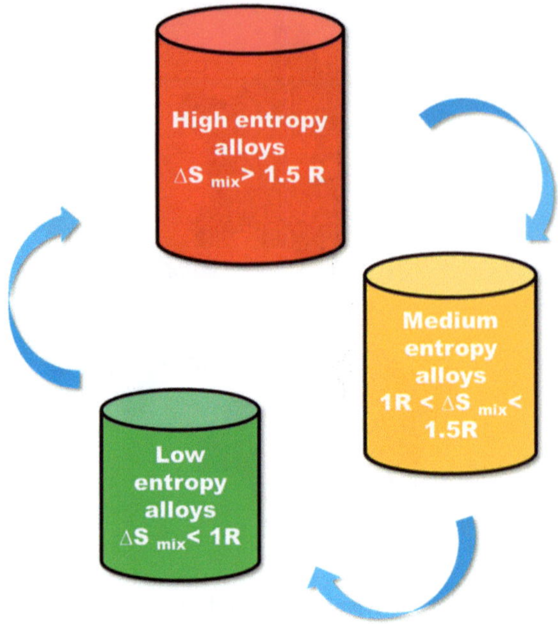

Fig. 2 The relationship between composition and entropy in HEAs is illustrated in this figure by showing how these materials are categorized according to the ranges of their elemental composition and the accompanying configurational entropy

$$\Delta G_{\text{mix}} = \Delta H_{\text{mix}} - T\Delta S_{\text{mix}}$$

The configurational entropy of an equimolar alloy of n elements that is changing from an elemental state to a random solution state (ideal or regular solution), as per Boltzmann's theory, can be computed using:

$$\Delta S_{\text{mix}} = -R\ln(n)$$

where, n is the number of distinct elements in the alloy and $R = 8.31$ J/K mol is the gas constant. Figure 3 illustrates the configurational entropy (expressed in terms of R for equiatomic alloy) as a function of the number of constituent elements, shown for up to 13 elements. An essential feature of HEAs is their entropy, which increases with the number of elements [26].

The mixing enthalpy

$$\Delta H_{\text{mix}} = \sum_{i=1\ i\neq j}^{n} \Omega_{ij} C_i C_j$$

$$\Omega_{ij} = 4\Delta H_{\text{mix}}^{\text{AB}}$$

In this context, the mixing enthalpy $\Delta H_{\text{mix}}^{\text{AB}}$ for binary liquid AB alloys is closely associated with the phase stability of HEAs [27]. It usually takes between -15 kJ/mol and 5 kJ/mol of needed mixing enthalpy to generate a stable solid solution

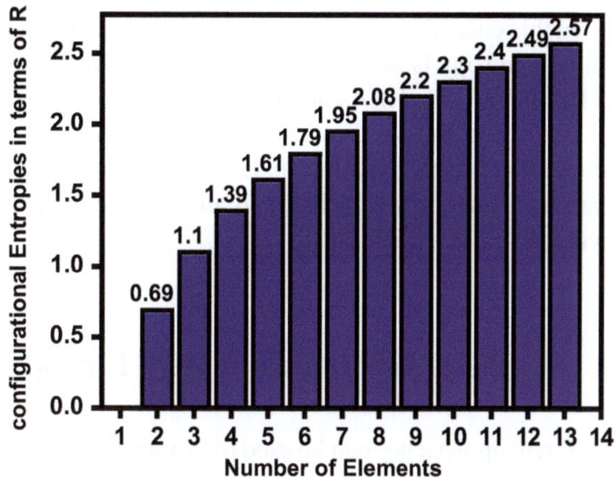

Fig. 3 This figure presents the calculated entropy of mixing, also known as configurational entropy, as a function of the number of constituent elements in a HEA system. The graph illustrates how entropy increases progressively with the number of elements, extending up to a maximum of 13 components

in HEAs. The atomic size difference between the constituent elements, which is a critical parameter in defining the evolution of the solid solution phase, is another important aspect that impacts the formation of solid solutions in HEAs [26].

$$\delta = 100\sqrt{\sum_{i=1}^{n} C_i\left(1 - \frac{r_i}{\bar{r}}\right)^2}, \ \bar{r} = \sum_{i=1}^{n} C_i r_i,$$

where C_i and r_i are atomic percentage and atomic radius.

The development of HEAs requires an atomic size difference of between 1 and 6%, according to the Hume-Rothery criteria [28]. Furthermore, the valence electron concentration (VEC), which is defined as [29]: plays a critical role in forecasting the phase structure of HEAs.

$$\text{VEC} = \sum_{i=1}^{n} C_i(\text{VEC})_i,$$

where, C_i is the concentration and $(\text{VEC})_i$ is the valence electron concentration of the ith element. Based on the VEC value, HEAs tend to form different crystal structures:

- A BCC structure when VEC < 6.87
- A mixture of BCC and FCC structures when VEC is between 6.87 and 8
- An FCC structure when VEC > 8.

Another important parameter for predicting the formation of solid solution HEAs is Ω, which is calculated as [30]:

$$\Omega = T_{\text{m}}\Delta S_{\text{mix}}/\Delta H_{\text{mix}}$$

In this case, the melting temperature is T_m, the mixing enthalpy is ΔH_{mix}, and the mixing entropy is ΔS_{mix}. When Ω is more than 1.1 and the atomic size difference δ is less than 6.6%, HEAs are expected to form a stable solid solution [31].

Predicting the synthesis of HEAs with solid solution phases is made easier by computing thermodynamic characteristics. Essential for predicting phase structures, parameters such as Ω, VEC, and atomic size difference are useful in directing the synthesis of new materials.

3 Synthesis Techniques of High Entropy Alloys

Different approaches can be used to synthesize HEAs, and depending on the required material qualities and uses, each approach has advantages of its own (Fig. 4). The following are thorough steps to synthesize HEAs.

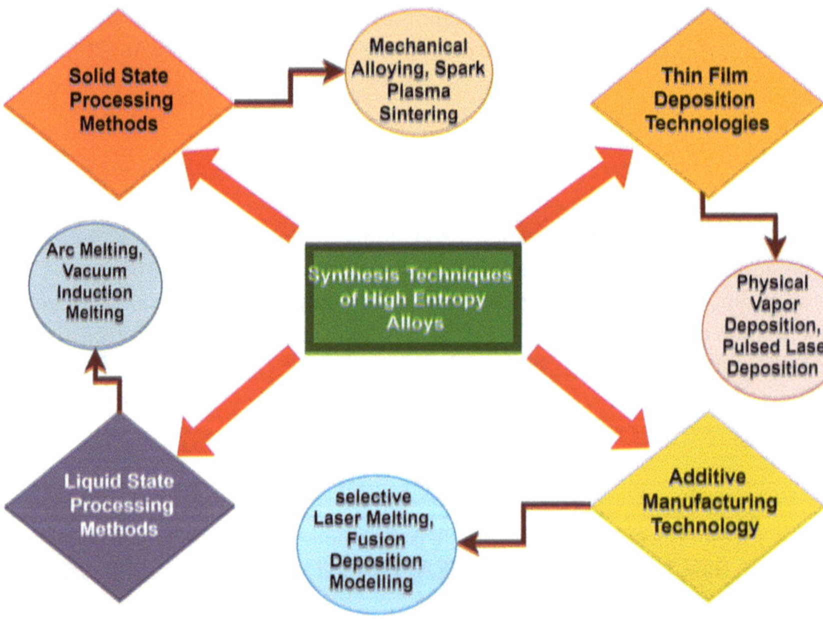

Fig. 4 This schematic figure illustrates the multiple synthesis routes for HEAs, highlighting the diverse techniques and procedures employed in their fabrication. Each pathway represents a distinct approach used to tailor the material properties of HEAs, showcasing variations in processing methods such as mechanical alloying, arc melting, and spark plasma sintering

3.1 *Solid-State Processing of HEAs*

High-energy ball milling (HEBM) and powder atomization are two techniques used in the solid-state processing of HEAs [32]. Powder atomization, which is essential for additional processing, quickly cools molten metals into homogeneous particles using gas or water [33]. By continuously welding and fracturing alloy powders, HEBM, a form of mechanical alloying, creates fine-grained or nanostructured HEAs while improving their mechanical characteristics [34]. In order to create a green body for further procedures like hot pressing or sintering, cold uniaxial pressing compacts HEA particles without using heat [35]. Through solid-state diffusion, sintering improves the mechanical characteristics and density of HEA. In order to further densify and increase mechanical strength and density, hot pressing combines pressure and heat. By employing pressure and electric current to consolidate powders at lower temperatures, spark plasma sintering (SPS) preserves fine microstructure and improves mechanical properties [36].

3.2 *Liquid-State Processing Methods*

Additionally, HEAs can be synthesized by melting and solidifying the alloy constituent parts in a liquid state [37]. The arc melting process, in which raw materials are melted using an electric arc, is one popular method for this [38]. Another method is vacuum induction melting (VIM), which involves utilizing an induction furnace to melt the raw materials in a regulated environment or under vacuum [39]. VIM is particularly well-suited for the synthesis of HEAs because it reduces contamination and offers perfect control over the alloy composition [40]. Another method, electromagnetic stirring, involves applying an electromagnetic field to the molten alloy in order to promote uniform mixing and lessen element segregation [41]. Increased uniformity and better microstructural control are guaranteed in the final HEAs by this technique.

3.3 *Thin Film Deposition Technologies*

HEA thin films are produced for cutting-edge uses such as protective coatings and microelectronics [42]. Physical vapor deposition techniques like magnetron sputtering are popular because of their consistent film formation and ability to precisely control composition [43]. A solid HEA target is vaporized using a high-energy laser in pulsed laser deposition (PLD), which provides superior control over film thickness and stoichiometry and is appropriate for ultra-thin films [44]. Melting HEA powders in a high-temperature plasma jet creates films with exceptional wear and corrosion resistance, making plasma spray deposition perfect for thicker HEA coatings [45].

Because it can efficiently coat complex and large-scale surfaces, this technique is ideal for industrial, automotive, and aerospace applications.

3.4 Additive Manufacturing Technology

Layer-by-layer fabrication of HEA components is made possible by additive manufacturing technologies (AMT), which provide reduced waste, design freedom, and customization [46]. AMT encompasses a number of techniques, each with unique potential. Although the technique is mainly utilized for polymers, fusion deposition modelling (FDM), which builds components by extruding material layer by layer, can be modified for metals and alloys such as HEAs, especially for basic geometries [47]. Layer by layer, liquid resin is solidified using lasers in stereolithography (SLA) [48]. It was first developed for polymers but may be adjusted to provide smooth and accurate finishes for metal alloys. For complicated HEA components, direct ink writing (3D plotting) is perfect since it puts material precisely onto a substrate [49]. By using a laser to sinter HEA powder without totally melting it, selective laser sintering (SLS) preserves microstructures and works with intricate designs [50]. Strong, thick HEA components used in the automotive, biomedical, and aerospace industries are produced using selective laser melting (SLM), which uses a laser to completely melt powder layers [51]. For refractory HEAs that require high-temperature stability and robustness, selective electron beam melting (SEBM) uses an electron beam to melt powder layers [52]. The HEA features, component complexity, and application needs all influence the AMT technique selection.

4 Application of Multicomponent Alloys for Hydrogen Production and Storage

Due to a special set of features, multicomponent alloys such as HEAs in particular are receiving a lot of attention for their potential uses in the production and storage of hydrogen. These alloys have a high degree of configurational entropy due to the near-equiatomic proportions of five or more primary elements, which improves their structural stability, resistance to corrosion, and tunable characteristics. An extensive examination of the use of HEA in the production and storage of hydrogen is provided below.

4.1 High-Entropy Alloys in the Hydrogen Production

Because of their distinct structural and catalytic characteristics, HEAs are becoming more and more significant in the hydrogen energy cycle. Performance of fuel cells and hydrogen production is improved by their multicomponent design, especially in important reactions like hydrogen evolution (HER) and oxidation (HOR). Superior catalytic activity, stability, and durability are attributes of HEAs that make them perfect for enhancing processes such as water electrolysis.

4.1.1 HEAs as Electrocatalysts for Water Splitting

As electrocatalysts for the hydrogen evolution reaction (HER), HEAs are very effective due to a number of beneficial features that they possess [53]. Their unique multi-element compositions contribute to higher catalytic activity, greater stability, and superior corrosion resistance, which are vital for their employment in electrolyzers. Protons (H^+) are reduced in the HER at the cathode during electrolysis to form hydrogen gas (H_2) [54]. The following sums up the general response to HER:

$$2H^+ + 2e^- \rightarrow H_2 \quad \text{(in acidic media)}$$

In alkaline media, the reaction involves water molecules:

$$2H_2O + 2e^- \rightarrow H_2 + 2OH^-$$

As electrocatalysts in both scenarios, HEAs maximize H_2 or H_2O adsorption and facilitate H_2 desorption, hence raising the HER efficiency.

Because of their adjustable multi-element compositions, which promote hydrogen gas, desorption and water adsorption during HER, HEAs provide increased catalytic activity [55]. Platinum-based HEAs lower platinum use while maintaining good performance, and transition metal-based HEAs, like those with nickel, cobalt, or iron, offer a cost-effective option for large-scale HER applications [56]. Recent research shows that NiCoFeMoMn HEA performs exceptionally well in alkaline electrolytes [57]. With a Tafel slope of 29 mV dec^{-1} and an exceptionally low HER overpotential of 150 mV at 1,000 mA cm^{-2}, this HEA shows remarkable reaction kinetics. Additionally, at the same current density, the overpotential for the oxygen evolution reaction (OER) is just 350 mV. This HEA exhibits great efficiency as a bifunctional electrocatalyst for overall water splitting, requiring just 1.47 V to maintain a stable current density of 10 mA cm^{-2} when utilized as both anode and cathode in an alkaline electrolyzer. Furthermore, the electrocatalyst offers a more affordable option to platinum-based catalysts by remaining stable for more than 375 h of continuous hydrogen production. By providing scalable and long-lasting solutions for effective hydrogen generation, these findings support the promise of HEAs to advance hydrogen energy technology.

4.1.2 Hydrogen Oxidation Reaction (HOR) and HEAs

The hydrogen oxidation reaction, or HOR, is essential to fuel cell energy production. Fuel cell performance is improved by HEAs with optimized compositions because they catalyze the HOR effectively. HEAs are perfect for the harsh conditions in fuel cell operations because of their innate resistance to wear and corrosion. HEA compositions are scalable for use in commercial fuel cell applications because they can be customized.

The HOR is essential to the process of producing energy in fuel cells. Protons (H^+) and electrons (e^-) are released when hydrogen (H_2) molecules are oxidized at the anode, which is where the HOR occurs [58]. The following can be used to express the total HOR reactivity in acidic media:

$$H_2 \rightarrow 2H^+ + 2e^-$$

In alkaline media, the reaction proceeds slightly differently:

$$H_2 + 2OH^- \rightarrow 2H_2O + 2e^-$$

In the HOR, protons travel to the cathode and combine with oxygen to form water, while electrons are transferred to create electricity. Because of their adjustable compositions, which maximize hydrogen adsorption and dissociation and increase fuel cell performance, HEAs perform exceptionally well as electrocatalysts for HOR [59]. They outperform traditional catalysts like platinum in severe environments due to their resistance to corrosion and wear. By adding elements like Ni, Co, and Fe, HEAs can also lessen their dependency on pricey metals, which makes them a scalable and affordable solution for commercial fuel cell applications.

4.1.3 HEAs in Other Hydrogen Generation Processes

Methane (CH_4), the main component of natural gas, is converted into hydrogen (H_2) and carbon monoxide (CO) by a reaction with steam (H_2O) in the commonly used industrial process known as steam methane reforming (SMR) [60]. The following can be used to illustrate the entire response:

$$CH_4 + H_2O \rightarrow CO + 3H_2$$

SMR is essential for producing hydrogen, which is used in sectors like fuel cells and ammonia synthesis. Conventional nickel catalysts work well, but they have problems with sintering and carbon accumulation that cause them to become inactive. HEAs with components like Ni, Co, and Fe provide better SMR performance [61]. The multi-element design of HEAs increases stability and reaction efficiency, making them a promising method for producing hydrogen sustainably.

4.1.4 Hydrogen Production from Ammonia Catalyzed by HEAs

Because of its high hydrogen content and ease of liquefaction at relatively low pressures, ammonia (NH_3) is becoming more and more considered as a promising hydrogen carrier [62]. Because of this, it is a desirable method of storing and moving hydrogen, particularly in the framework of a sustainable energy economy. Efficiently splitting ammonia into H_2 and N_2 at the site of application is one of the major difficulties in using it as a hydrogen carrier. Catalysts aid in this breakdown process, and HEAs have proven to be particularly useful in this regard [63]. Ammonia is broken down by HEAs into hydrogen and nitrogen, a process that can be shown as follows:

$$2NH_3 \rightarrow 3H_2 + N_2$$

Since ammonia is catalyzed by HEAs to break down into H_2 and N_2, it can be used as a viable hydrogen carrier in renewable energy systems. Because of its multi-element design, catalytic activity may be tuned, resulting in decreased activation energy and improved reaction speeds. In high-temperature settings, HEAs function better than conventional catalysts and are resistant to deterioration, providing long-term stability. Because of their scalability, durability, and efficiency, HEAs are essential for producing hydrogen on a wide scale and facilitating the shift to an energy economy based on hydrogen.

4.2 Hydrogen Storage in Multicomponent Alloys

In order to efficiently store and release hydrogen for use in fuel cells and transportation, among other applications, hydrogen storage is an essential part of hydrogen energy systems [1]. Multicomponent alloys, in particular HEAs, have special microstructural and thermodynamic features that make them viable candidates for solid-state hydrogen storage [64]. The kinetics of hydrogen absorption and storage capacity can be controlled by adjusting their compositions. For high-density storage applications like fuel cells and hydrogen-powered cars, HEAs synthesize stable metal hydrides and are perfect. Excellent hydrogen storage capacities, enhanced absorption/desorption kinetics, and greater cycling stability are provided by alloys based on titanium, zinc, and magnesium [10]. Because of its high capacity and lightweight design, hydrogen extraction adapters HEAs can be used for both stationary and mobile hydrogen storage applications [13].

The implementation of HEAs serves as fundamental material for hydrogen production alongside storage operations (Fig. 5). HEAs serve as powerful electrocatalysts for water electrolysis because they transform the efficiency of HER during hydrogen production specifically using renewable energy. The hydrogen storage systems benefit from HEAs because they provide both enhanced absorption features and stability characteristics along with improved resistance to degradation. The integration of HEAs in these processes appears through a visual representation which

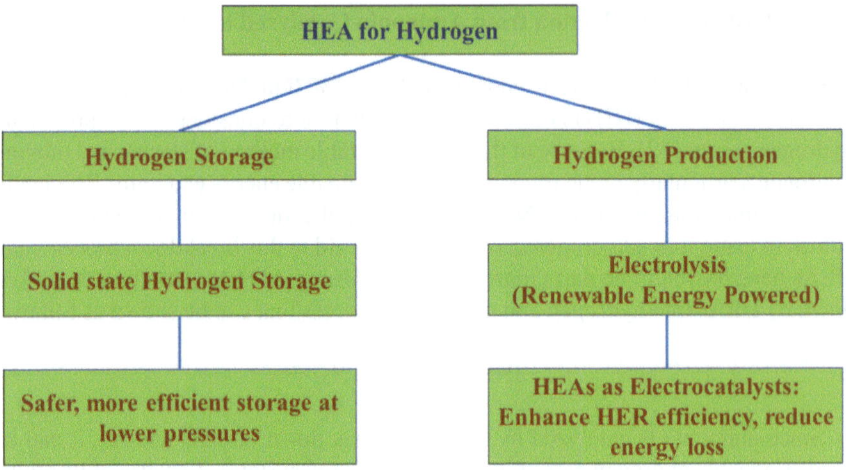

Fig. 5 Flowchart illustrating the role of HEAs as essential materials for both hydrogen production and storage

shows their utilization from production through storage. The environmental impact of hydrogen production methods requires assessment and it becomes essential to understand how renewable-powered electrolysis minimizes carbon emissions above traditional steam methane reforming. HEAs are essential for developing the hydrogen economy by enhancing catalyst performance, hydrogen production, and storage due to their stability and reversible hydrogen cycling across numerous cycles.

5 Utility of Hydrogen Storage Materials

A number of technological challenges must be addressed in order to make the shift to a hydrogen economy. One of the main ways to produce hydrogen is by natural gas reforming, although this process releases carbon dioxide [54]. In addition to developments in fuel cells and hydrogen infrastructure, carbon-neutral production methods utilizing renewable energy sources such as wind, solar, and hydro are crucial to achieving the full potential of hydrogen [65]. Efficient hydrogen storage is still a major problem, especially for mobile applications. Because of their low volumetric density, compressed gas and liquid hydrogen cannot meet the U.S. Department of Energy requirements for onboard storage in automobiles [66]. A good substitute with higher energy densities is solid-state hydrogen storage, in which materials absorb and release hydrogen [1, 2]. Fuel cells and hydrogen-powered vehicles depend on safe, portable, and effective hydrogen storage made possible by high-capacity materials like metal hydrides and HEAs [4, 10]. These components improve performance and fuel efficiency, making hydrogen a viable substitute for fossil fuels in transportation and aviation. During periods of high energy production, hydrogen storage materials in renewable energy systems store excess energy as hydrogen, which can subsequently be transformed back into electricity to guarantee a consistent energy supply. As energy systems transition to intermittent renewable sources, this skill is essential for grid resilience. Large-scale energy storage for grid systems and industrial applications is another benefit of hydrogen storage materials, which guarantee long-term energy storage with little degradation. In order to help global initiatives to cut emissions and make the switch to sustainable energy technologies such as grid storage, energy integration, mobility, and industrial operations they tackle the problems of effective, secure, and lightweight storage.

6 Hydrogen Storage in Nanomaterials

Cutting-edge research in nanomaterials is addressing the challenge of hydrogen storage for applications like fuel cells, hydrogen-powered vehicles, and renewable energy systems (Fig. 6) [67]. Nanomaterials have high surface area-to-volume ratios, enhancing hydrogen adsorption [68]. Transition metal nanoparticles, such as palladium, platinum, and nickel, absorb hydrogen and form metal hydrides, improving storage capacity [69]. Metal–organic frameworks (MOFs) and carbon-based nanomaterials like graphene and carbon nanotubes (CNTs) also show promise for hydrogen storage [70, 71]. Nanomaterials offer better kinetics and lower desorption temperatures than bulk materials, making them suitable for hydrogen storage [68]. However, challenges like long-term durability, scalability, and production costs remain. Despite these obstacles, nanomaterials hold potential for advancing hydrogen storage technologies and supporting a sustainable energy future.

7 Conclusions

The materials science challenge of hydrogen storage is to understand the interaction of hydrogen with other elements better, especially metals. The analysis of the current trends in smart materials is agreeable that hydrogen economy energy sustainability can be boosted in the future. Other approaches like advanced composites materials, MOFs, and nanomaterials and HEAs could become more effective, lighter and safer for the solution of hydrogen storage. These materials present higher energy density, increase and decrease rates and higher cyclability as well. Since storage of hydrogen is becoming better and better while solid stage is much safer, it can help hydrogen to be widely used and used as clean source energy, thus helping to cut on the use of fossil-based fuels and therefore minimize on the effects on the environment. In the long run, these advancements are vital for bringing about the energy transition process. Multicomponent smart materials like high entropy alloys-based metal hydrides for applications involving solid-state hydrogen storage have a promising material for hydrogen storage and new multicomponent compounds of intermetallic have to be

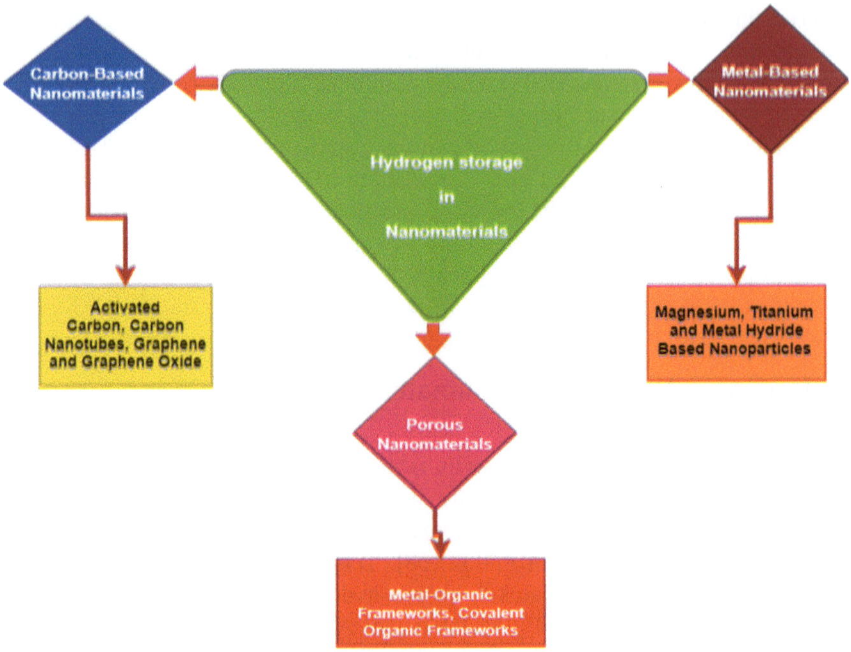

Fig. 6 This schematic picture, which uses metal-based, carbon-based, and porous nanoparticles, shows how hydrogen can be stored in nanomaterials. It draws attention to the unique capacities and storage methods of each kind, highlighting how they might improve hydrogen storage options

investigated discovered. Hydrogen production, storage, and conversion has reached a technological level, although plenty of improvements and new discoveries are still possible.

References

1. Srivastava, O.N., Yadav, T.P., Shahi, R.R., Pandey, S.K., Shaz, M.A., Bhatnagar, A.S.H.I.S.H.: Hydrogen energy in India: storage to application. Proc. Indian Natl. Sci. Acad. **81**(4), 915–937 (2015). https://doi.org/10.16943/ptinsa/2015/v81i4/48303
2. Shukla, V., Yadav, T.P.: Complex hydrides: lightweight, high gravimetric hydrogen storage materials. In: Enabling Methodologies for Renewable and Sustainable Energy, pp. 163–179. CRC Press (2023). https://doi.org/10.1201/9781003272717-10
3. Singh, D., Shahi, R.R., Yadav, T.P., Mandal, R.K., Tiwari, R.S., Srivastava, O.N.: Hydrogenation of $(Zr_{69.5}Al_{7.5}Cu_{12}Ni_{11})_{100-x}Ti_x$ quasicrystalline alloys and its effect on their structural and microhardness behavior. J. Non-Cryst. Solids **380**, 11–16 (2013). https://doi.org/10.1016/j.jnoncrysol.2013.06.019
4. Shahi, R.R., Bhatnagar, A., Pandey, S.K., Shukla, V., Yadav, T.P., Shaz, M.A., Srivastava, O.N.: MgH_2–$ZrFe_2Hx$ nanocomposites for improved hydrogen storage characteristics of MgH_2. Int. J. Hydrogen Energy **40**(35), 11506–11513 (2015). https://doi.org/10.1016/j.ijhydene.2015.06.056

5. Shukla, V., Bhatnagar, A., Pandey, S.K., Shahi, R.R., Yadav, T.P., Shaz, M.A., Srivastava, O.N.: On the synthesis, characterization and hydrogen storage behavior of $ZrFe_2$ catalyzed Li–Mg–N–H hydrogen storage material. Int. J. Hydrogen Energy **40**(36), 12294–12302 (2015). https://doi.org/10.1016/j.ijhydene.2015.07.063
6. Gurav, S.R., Waikar, M.R., Rasal, A.S., Sonker, R.K., Sonkawade, R.G.: Current development and challenges in textile-based flexible supercapacitors. In: Smart and Flexible Energy Devices, pp. 539–550. CRC Press (2022)
7. Shahi, R.R., Yadav, T.P., Shaz, M.A., Srivastava, O.N.: Synthesis characterization and hydrogenation behaviour of As quenched $Ti_{41.5+x}Zr_{41.5-x}Ni_{17}$ (x = 0, 3.5, 11.5 and 13.5) nano quasicrystalline ribbons. J. Phys.: Conf. Ser. **809**(1), 012011 (2017). https://doi.org/10.1088/1742-6596/809/1/012011
8. Pandey, S.K., Bhatnagar, A., Mishra, S.S., Yadav, T.P., Shaz, M.A., Srivastava, O.N.: Curious catalytic characteristics of Al–Cu–Fe quasicrystal for de/rehydrogenation of MgH_2. J. Phys. Chem. C **121**(45), 24936–24944 (2017). https://doi.org/10.1021/acs.jpcc.7b08624
9. Shukla, V., Bhatnagar, A., Singh, S., Soni, P.K., Verma, S.K., Yadav, T.P., Srivastava, O.N.: A dual borohydride (Li and Na borohydride) catalyst/additive together with intermetallic FeTi for the optimization of the hydrogen sorption characteristics of $Mg(NH_2)_2/2LiH$. Dalton Trans. **48**(30), 11391–11403 (2019). https://doi.org/10.1039/C9DT01782A
10. Yadav, T.P., Kumar, A., Verma, S.K., Mukhopadhyay, N.K.: High-entropy alloys for solid hydrogen storage: potentials and prospects. Trans. Indian Natl. Acad. Eng. **7**(1), 147–156 (2022). https://doi.org/10.1007/s41403-021-00242-7
11. Shukla, V., Shaz, M.A., Yadav, T.P.: Improved hydrogen sorption characteristics of $Mg(NH_2)_2$–$2LiH/Li_4BH_4(NH_2)_3$ using SrH_2 and MgF_2 auto-catalyst. Int. J. Hydrogen Energy **57**, 1244–1252 (2024)
12. Verma, S.K., Shaz, M.A., Yadav, T.P.: Introducing 2D layered WS_2 and MoS_2 as an active catalyst to enhance the hydrogen storage properties of MgH_2. Int. J. Hydrogen Energy **87**, 1035–1046 (2024)
13. Yadav, Y.K., Shaz, M.A., Yadav, T.P.: Al–Cu–Fe–Ni–Ti high entropy alloy nanoparticles as new catalyst for hydrogen sorption in MgH_2. Int. J. Hydrogen Energy (2024), in press
14. Yadav, Y.K., Shaz, M.A., Yadav, T.P.: Storage of hydrogen in solid-state materials. In: Hydrogen Energy, pp. 181–208. CRC Press (2024). https://doi.org/10.1201/9781003310123-11
15. Kumar, A., Yadav, T.P., Shaz, M.A., Mukhopadhyay, N.K.: Hydrogen storage properties in rapidly solidified TiZrVCrNi high-entropy alloys. Energy Storage **6**(1), e532 (2024). https://doi.org/10.1002/est2.532
16. Verma, S.K., Mishra, S.S., Mukhopadhyay, N.K., Yadav, T.P.: Superior catalytic action of high-entropy alloy on hydrogen sorption properties of MgH_2. Int. J. Hydrogen Energy **50**, 749–762 (2024). https://doi.org/10.1016/j.ijhydene.2023.10.107
17. Gao, M.C., Yeh, J.W., Liaw, P.K., Zhang, Y. (eds.): High-Entropy Alloys: Fundamentals and Applications. Springer (2016). https://doi.org/10.1007/978-3-319-27013-5
18. Sarma, P.P., Barman, K., Baruah, P.K.: A brief history of high entropy materials. In: High Entropy Materials, pp. 14–25. CRC Press (2024). https://doi.org/10.1201/9781003333634-2
19. Yeh, J.W., Chen, S.K., Lin, S.J., Gan, J.Y., Chin, T.S., Shun, T.T., Tsau, C.H., Chang, S.Y.: Nanostructured high-entropy alloys with multiple principal elements: novel alloy design concepts and outcomes. Adv. Eng. Mater. **6**(5), 299–303 (2004). https://doi.org/10.1002/adem.200300567
20. Cantor, B., Chang, I.T.H., Knight, P., Vincent, A.J.B.: Microstructural development in equiatomic multicomponent alloys. Mater. Sci. Eng. A **375**, 213–218 (2004). https://doi.org/10.1016/j.msea.2003.10.257
21. Zhang, Y., Zuo, T.T., Tang, Z., Gao, M.C., Dahmen, K.A., Liaw, P.K., Lu, Z.P.: Microstructures and properties of high-entropy alloys. Prog. Mater. Sci. **61**, 1–93 (2014). https://doi.org/10.1016/j.pmatsci.2013.10.001
22. Yeh, J.W.: Recent progress in high entropy alloys. Ann. Chim. Sci. Mat. **31**(6), 633–648 (2006)
23. Ibrahim, P.A., Özkul, İ, Canbay, C.A.: An overview of high-entropy alloys. Emergent Mater. **5**(6), 1779–1796 (2022). https://doi.org/10.1007/s42247-022-00388-w

24. George, E.P., Raabe, D., Ritchie, R.O.: High-entropy alloys. Nat. Rev. Mater. **4**(8), 515–534 (2019). https://doi.org/10.1038/s41578-019-0121-4
25. Qian, H., Hopfield, J.J.: Entropy-enthalpy compensation: perturbation and relaxation in thermodynamic systems. J. Chem. Phys. **105**(20), 9292–9298 (1996). https://doi.org/10.1063/1.472890
26. Yadav, Y.K., Shaz, M.A., Mukhopadhyay, N.K., Yadav, T.P.: Formation of B2 phase and its stability in equiatomic Al–Cu–Fe–Ni–Ti high entropy alloy. J. Alloys Metall. Syst. **8**, 100137 (2024). https://doi.org/10.1016/j.jams.2024.100137
27. Mishra, S.S., Bajpai, A., Biswas, K.: $TiVCrNiZrFe_x$ high entropy alloy: phase evolution, magnetic and mechanical properties. J. Alloys Compd. **871**, 159572 (2021). https://doi.org/10.1016/j.jallcom.2021.159572
28. Troparevsky, M.C., Morris, J.R., Daene, M., Wang, Y., Lupini, A.R., Stocks, G.M.: Beyond atomic sizes and Hume-Rothery rules: understanding and predicting high-entropy alloys. JOM **67**, 2350–2363 (2015). https://doi.org/10.1007/s11837-015-1594-5
29. Mishra, S.S., Mukhopadhyay, S., Yadav, T.P., Mukhopadhyay, N.K., Srivastava, O.N.: Synthesis and characterization of hexanary Ti–Zr–V–Cr–Ni–Fe high-entropy laves phase. J. Mater. Res. **34**(5), 807–818 (2019). https://doi.org/10.1557/jmr.2018.449
30. Yadav, Y.K., Yadav, S., Shaz, M.A., Yadav, T.P.: A facile synthesis of high entropy alloy nanoparticles and notable catalytic activity for methylene blue degradation. Mater. Lett. **397**, 138854 (2025). https://doi.org/10.1016/j.matlet.2025.138854
31. Yang, X., Zhang, Y.: Prediction of high-entropy stabilized solid-solution in multi-component alloys. Mater. Chem. Phys. **132**(2–3), 233–238 (2012). https://doi.org/10.1016/j.matchemphys.2011.11.021
32. Sun, X., Sun, Y.: Synthesis of metallic high-entropy alloy nanoparticles. Chem. Soc. Rev. (2024), in press. https://doi.org/10.1039/D3CS00899K
33. Alshataif, Y.A., Sivasankaran, S., Al-Mufadi, F.A., Alaboodi, A.S., Ammar, H.R.: Manufacturing methods, microstructural and mechanical properties evolutions of high-entropy alloys: a review. Met. Mater. Int. **26**, 1099–1133 (2020). https://doi.org/10.1007/s12540-019-00419-w
34. Yadav, Y.K., Shaz, M.A., Mukhopadhyay, N.K., Yadav, T.P.: High entropy alloys synthesized by mechanical alloying: a review. J. Alloy. Metall. Syst. **9**, 100170 (2025)
35. Attia, U.M.: Cold-isostatic pressing of metal powders: a review of the technology and recent developments. Crit. Rev. Solid State Mater. Sci. **46**(6), 587–610 (2021). https://doi.org/10.1080/10408436.2020.1787170
36. Riva, S., Brown, S.G., Lavery, N.P., Tudball, A., Yusenko, K.V.: Spark plasma sintering of high entropy alloys. In: Spark Plasma Sintering of Materials: Advances in Processing and Applications, 1st edn., pp. 517–538. Woodhead Publishing (2019). https://doi.org/10.1016/B978-0-12-813820-5.00017-0
37. Onawale, O.T., Cobbinah, P.V., Nzeukou, R.A., Matizamhuka, W.R.: Synthesis route, microstructural evolution, and mechanical property relationship of high-entropy alloys (HEAs): a review. Materials **14**(11), 3065 (2021). https://doi.org/10.3390/ma14113065
38. Skalska, M.: Properties and application of raw materials produced by electric arc melting. Mater. Ceram./Ceram. Mater. **72**(1), 49–55 (2020). https://doi.org/10.2298/MCM2001049S
39. Hinte, C., Fantin, A., Barienti, K., Herbst, S., Maier, H.J.: A comparative study on vacuum arc- and vacuum induction-melting for high entropy shape memory alloy production. SSRN Electron. J. (2024). https://doi.org/10.2139/ssrn.4858201
40. Kang, M., Won, J.W., Lim, K.R., Park, S.H., Seo, S.M., Na, Y.S.: Microstructure and mechanical properties of As-cast CoCrFeMnNi high entropy alloy. Korean J. Met. Mater. **55**(10), 732–738 (2017). https://doi.org/10.3365/KJMM.2017.55.10.732
41. Guo, Y., Liu, L., Zhang, W., Yao, K.D., Chen, W., Ren, J., Xiang, J.: A new method for preparing high entropy alloys: electromagnetic pulse treatment and its effects on mechanical and corrosion properties. Mater. Sci. Eng. A **774**, 138916 (2020). https://doi.org/10.1016/j.msea.2019.138916
42. Li, W., Liu, P., Liaw, P.K.: Microstructures and properties of high-entropy alloy films and coatings: a review. Mater. Res. Lett. **6**(4), 199–229 (2018). https://doi.org/10.1080/21663831.2018.1502330

43. Deng, Y., Chen, W., Li, B., Wang, C., Kuang, T., Li, Y.: Physical vapor deposition technology for coated cutting tools: a review. Ceram. Int. **46**(11), 18373–18390 (2020). https://doi.org/10.1016/j.ceramint.2020.03.160
44. Cropper, M.D.: Thin films of AlCrFeCoNiCu high-entropy alloy by pulsed laser deposition. Appl. Surf. Sci. **455**, 153–159 (2018). https://doi.org/10.1016/j.apsusc.2018.05.110
45. Anupam, A., Kottada, R.S., Kashyap, S., Meghwal, A., Murty, B.S., Berndt, C.C., Ang, A.S.M.: Understanding the microstructural evolution of high entropy alloy coatings manufactured by atmospheric plasma spray processing. Appl. Surf. Sci. **505**, 144117 (2020). https://doi.org/10.1016/j.apsusc.2019.144117
46. Najmon, J.C., Raeisi, S., Tovar, A.: Review of additive manufacturing technologies and applications in the aerospace industry. In: Additive Manufacturing for the Aerospace Industry; 1st edn., pp. 7–31. Elsevier (2019). https://doi.org/10.1016/B978-0-12-814062-8.00002-9
47. Vyavahare, S., Teraiya, S., Panghal, D., Kumar, S.: Fused deposition modelling: a review. Rapid Prototyping J. **26**(1), 176–201 (2020). https://doi.org/10.1108/RPJ-04-2018-0106
48. Huang, J., Qin, Q., Wang, J.: A review of stereolithography: processes and systems. Processes **8**(9), 1138 (2020). https://doi.org/10.3390/pr8091138
49. Peng, S., Mooraj, S., Feng, R., Liu, L., Ren, J., Liu, Y., Chen, W.: Additive manufacturing of three-dimensional (3D)-architected CoCrFeNiMn high-entropy alloy with great energy absorption. Scr. Mater. **190**, 46–51 (2021). https://doi.org/10.1016/j.scriptamat.2020.08.028
50. Tiwari, S.K., Pande, S., Agrawal, S., Bobade, S.M.: Selection of selective laser sintering materials for different applications. Rapid Prototyping J. **21**(6), 630–648 (2015). https://doi.org/10.1108/RPJ-06-2014-0072
51. Zhang, C., Zhu, J., Zheng, H., Li, H., Liu, S., Cheng, G.J.: A review on microstructures and properties of high entropy alloys manufactured by selective laser melting. Int. J. Extreme Manuf. **2**(3), 032003 (2020). https://doi.org/10.1088/2631-7990/ab98e4
52. Popov, V.V., Katz-Demyanetz, A., Koptyug, A., Bamberger, M.: Selective electron beam melting of $Al_{0.5}CrMoNbTa_{0.5}$ high entropy alloys using elemental powder blend. Heliyon **5**(2), e01234 (2019). https://doi.org/10.1016/j.heliyon.2019.e01234
53. Zhang, G., Ming, K., Kang, J., Huang, Q., Zhang, Z., Zheng, X., Bi, X.: High entropy alloy as a highly active and stable electrocatalyst for hydrogen evolution reaction. Electrochim. Acta **279**, 19–23 (2018). https://doi.org/10.1016/j.electacta.2018.05.130
54. Huo, X., Yu, H., Xing, B., Zuo, X., Zhang, N.: Review of high entropy alloys electrocatalysts for hydrogen evolution, oxygen evolution, and oxygen reduction reaction. Chem. Rec. **22**(12), e202200175 (2022). https://doi.org/10.1002/tcr.202200175
55. Wang, B., Yao, Y., Yu, X., Wang, C., Wu, C., Zou, Z.: Understanding the enhanced catalytic activity of high entropy alloys: from theory to experiment. J. Mater. Chem. A **9**(35), 19410–19438 (2021). https://doi.org/10.1039/D1TA02817F
56. Kumar, A.: Synthesis, characterisation, and properties of powder metallurgy transition metal-based high entropy alloys for electrocatalytic application. Doctoral dissertation, The University of Waikato (2023)
57. Liu, H., Qin, H., Kang, J., Ma, L., Chen, G., Huang, Q., Zhao, N.: A freestanding nanoporous NiCoFeMoMn high-entropy alloy as an efficient electrocatalyst for rapid water splitting. Chem. Eng. J. **435**, 134898 (2022). https://doi.org/10.1016/j.cej.2022.134898
58. Lu, S., Zhuang, Z.: Electrocatalysts for hydrogen oxidation and evolution reactions. Sci. China Mater. **59**(3), 217–238 (2016). https://doi.org/10.1007/s40843-016-0124-4
59. Yan, X., Zhou, Y., Wang, S.: Nano-high entropy materials in electrocatalysis. Adv. Funct. Mater. 241311 (2024). https://doi.org/10.1002/adfm.202413311
60. Mishra, A., Kim, D., Altahtamouni, T., Kasak, P., Popelka, A., Park, H., Han, D.S.: A comparative study on carbon neutral hydrogen carrier production: formic acid from CO_2 vs. ammonia. J. CO_2 Util. **82**, 102756 (2024). https://doi.org/10.1016/j.jcou.2024.102756
61. Gonçalves, J.M., Ruiz-Montoya, J.G.: Emerging high-entropy coordination compounds and their derivatives for energy application. J. Mater. Chem. A **11**(39), 20872–20885 (2023). https://doi.org/10.1039/D3TA03500J

62. Chatterjee, S., Parsapur, R.K., Huang, K.W.: Limitations of ammonia as a hydrogen energy carrier for the transportation sector. ACS Energy Lett. **6**(12), 4390–4394 (2021). https://doi.org/10.1021/acsenergylett.1c02105
63. Xie, P., Yao, Y., Huang, Z., Liu, Z., Zhang, J., Li, T., Wang, C.: Highly efficient decomposition of ammonia using high-entropy alloy catalysts. Nat. Commun. **10**, 4011 (2019). https://doi.org/10.1038/s41467-019-11900-5
64. Yadav, Y.K., Shaz, M.A., Yadav, T.P.: Solid-state hydrogen storage properties of Al–Cu–Fe–Ni–Ti high entropy alloy. Int. J. Hydrogen Energy **99**, 985–995 (2025)
65. Evro, S., Oni, B.A., Tomomewo, O.S.: Carbon neutrality and hydrogen energy systems. Int. J. Hydrogen Energy **78**, 1449–1467 (2024)
66. Rusman, N.A.A., Dahari, M.: A review on the current progress of metal hydrides material for solid-state hydrogen storage applications. Int. J. Hydrogen Energy **41**(28), 12108–12126 (2016). https://doi.org/10.1016/j.ijhydene.2016.05.252
67. Kumar, A., Mukhopadhyay, N.K., Yadav, T.P.: Hydrogen storage in high entropy alloys. In: Towards Hydrogen Infrastructure, pp. 133–164. Elsevier (2024)
68. Pandey, S.K., Bhatnagar, A., Shukla, V., Kesarwani, R., Deshpandey, U., Yadav, T.P.: Catalytic mechanism of TiO_2 quantum dots on the de/re-hydrogenation characteristics of magnesium hydride. Int. J. Hydrogen Energy **46**(75), 37340–37350 (2021). https://doi.org/10.1016/j.ijhydene.2021.08.097
69. Pal, P., Ting, J.M., Agarwal, S., Ichikawa, T., Jain, A.: The catalytic role of D-block elements and their compounds for improving sorption kinetics of hydride materials: a review. Reactions **2**(3), 333–364 (2021). https://doi.org/10.3390/reactions2030020
70. Collins, D.J., Zhou, H.C.: Hydrogen storage in metal-organic frameworks. J. Mater. Chem. **17**(30), 3154–3160 (2007). https://doi.org/10.1039/B618320A
71. Rather, S.U.: Preparation, characterization and hydrogen storage studies of carbon nanotubes and their composites: a review. Int. J. Hydrogen Energy **45**(7), 4653–4672 (2020). https://doi.org/10.1016/j.ijhydene.2019.12.082

Li-Ion Batteries for Energy Storage

Aman Joshi, Sunaina Saini, and Prakash Chand

Abstract In the present scenario of fast-paced world, Lithium-ion batteries (LIBs) are perhaps the most widely supported forms of energy storage in the world because of their easy accessibility and reliability. At present, LIBs are the dominant battery technology and are extensively utilised in the sector of transportable electronics automotive, and hybrid electric vehicles due to their desirable characteristics for instance high efficiencies, an elongated life cycle, elevated power and energy densities. Therefore, to satisfy the growing requirements for the storage of energy, especially from the growing popularity of electric vehicles, massive research and development efforts are essential for manufacturing the LIB of the foreseeable future with significantly enhanced outcomes. These performances include better stability, safety, charging rate, cyclability, specific energy, and volumetric energy density. However, there continue to be a number of significant obstacles to overcome in the process of developing next-generation LIBs. To go beyond the capabilities of LIB in the upcoming years, new battery concepts will need to be further developed. Regarding LIB, the primary goals of the current book chapter are to first present an overview of the fundamental ideas involved, and embrace the most recent advancements in the related field that have been made and also challenges in LIBs included. A concise discussion on the latest developments in active anode and cathode and inactive electrolyte, separator, binder, and current collector materials presently utilized in commercialized LIBs is also included in this chapter.

Keywords Active anode and cathode materials · Electrolytes · Energy storage · Lithium-ion battery (LIB)

A. Joshi
Department of Physics, J.C. Bose University of Science and Technology, YMCA, Faridabad, Haryana, India

S. Saini · P. Chand (✉)
Department of Physics, National Institute of Technology, Kurukshetra, India
e-mail: Prakash@nitkkr.ac.in

R. K. Sonker et al. (eds.), *Low-Dimensional Multifunctional Hybrid Nanostructured Materials*, Engineering Materials, https://doi.org/10.1007/978-981-96-7952-2_4

1 Introduction

The energy storage device market has completely changed due to lithium-ion or Li-ion batteries. Today, they serve as the foundation of numerous applications, including the incorporation of handheld devices, electric cars and renewable energy sources [1]. Lithium-ion batteries (LIBs), which are employed in numerous applications, are among the highest in energy power sources available today. Because of their effective energy storage and release capabilities, LIBs have emerged as one of the most significant energy storage technologies. From electric cars and handheld electronics to massive energy storage systems that stabilize renewable energy-dependent grids, these batteries power a vast array of gadgets. However, the advancements in their underlying science and the historical events that influenced their evolution are just as important to their success as their current performance. Electric vehicles (EVs), laptops, cell phones, and large-scale renewable energy systems are all powered by lithium-ion (Li-ion) batteries, one of the most popular and cutting-edge energy storage technologies available today. These batteries are preferred because of their effective electrical energy storage and release capabilities, which include a high energy density, extended cycle life, and comparatively low rates of self-discharge. As the need for energy storage increases in a variety of industries, from renewable energy applications to portable electronics, lithium-ion batteries are essential to solving today's urgent energy storage problems. There has never been a greater demand for better energy storage options. Efficient storage systems are essential to maintaining a balance between the supply and demand for energy as the world's energy consumption rises as it moves toward cleaner, more sustainable energy sources. Intermittent renewable energy sources, such as wind and solar, only produce electricity when the sun is shining or the wind is blowing. Lithium-ion batteries are thought to be one of the most important technologies to close this gap because of their high efficiency and scalability, which allow excess renewable energy to be stored for use during periods of high demand or low generation. Furthermore, as EVs mostly depend on these batteries for durability, long range, and fast charging, the market for sophisticated, dependable, and reasonably priced battery systems is growing rapidly. Lithium-ion batteries are at the center of the changing energy landscape because of their capacity to supply energy solutions for both small and large-scale applications, from portable devices to grid-scale energy storage. They also play a significant role in the battle against climate change and the movement toward a decentralized, more resilient energy grid, and as the technology develops, advancements in battery life, cost reduction, charging speeds, and sustainability are set to further boost their contribution. According to recent research, the global electric vehicle market is expected to develop at a compound annual growth rate (CAGR) of more than 20% between 2023 and 2030, which would significantly boost the requirement for LIBs. Additionally, LIBs are essential to the integration of renewable energy sources since they may be used to store energy produced by wind and solar power, which helps to mitigate the intermittent character of these energy sources. LIBs are playing an increasingly important role as the globe moves toward more sustainable energy sources, and continuous technological developments are crucial to supplying future energy storage demands. In conclusion, lithium-ion batteries represent a key component of the shift to cleaner, more effective energy systems and are not merely a technological advancement. Key issues in energy storage, grid stability, and the uptake of renewable energy sources are addressed by their rapid and effective energy storage and discharge capabilities. Lithium-ion batteries are still essential to creating a sustainable and energy-efficient future as we investigate new uses and advancements in this area. Although it has a range of additional applications, on account of its high energy density, it is commonly seen in vehicles and electrical equipment. The separator, electrolyte, anode, cathode, and current collector are all present in LIB. Lithium is a soft metal that has a silvery-white colour and an atomic number of 3. Arfwedson and Berzelius investigated petalite ore ($LiAlSi_4O_{10}$) and found lithium in 1817 [1]. Lewis noted that Li-ion may be utilised as the anode in a battery due to its exceptional electrochemical and physical properties, which include low reactiveness, strong polarizing power, poor density, and low specific capacity. Early in 1958, Harris carried out an experimental investigation on the disintegration of lithium using a range of various electrolytes and other compounds [2]. He discovered that the passivation layer in the LIB allowed ionic transport to occur while preventing the electrolyte from interacting with the lithium. This is why the battery remained stable. This was crucial to improving the LIBs. Adam Heller built the Lithium-thionyl chloride battery in 1973. Many health-care, military, and other vehicle uses are made possible by its high power density, extended shelf life, and additional advanced features. Energy is still obtained from this battery [3]. An apparatus consisting of positive and negative electrodes composed of Lithium-Cobalt-Oxide and Lithium metal respectively, was demonstrated in 1979 by a team under the direction of Godshall et al. In this investigation, the acceptable range of voltages was discovered to be 4 V. This system has cathode material is an essential ingredient of lithium ions, and uniform anode material may tolerate a range of cathode materials in addition to lithium metal [4–6]. Godshall established a new benchmark in the industry in 1982 when he demonstrated the usage of cathode ($LiCoO_2$) within LIBs for the first time [7]. LIB technology had substantial advancement between 1983 and 1990. For example, in 1983, Manganese Spinel (Mn_2O_4) was developed by Michael M. Thackeray et al. as a cathode material for LIBs [8]. Akira Yoshino created a carbonaceous model cell in 1985 using lithium ions and cobalt oxide ($LiCoO_2$) as electrodes, which demonstrated excellent LIB safety. $LiCoO_2$ was also said to have a bright future in commercial and industrial applications, which would spur more advancements [9]. Arumugam and Goodenough [10] made a polyanion class cathode material containing the Lithium-Iron-Phosphate compound, this was discovered in 1989. Jeff Dahn used ethylene carbonate as a solvent in 1990 to help create the most promising contemporary LIB [11]. He demonstrated how to utilize the ethylene carbonate as a solvent might reverse the process of lithium-ion intercalation into graphite. After then, the commercial LIBs were introduced in 1991 by Asahi Kasei and Sony Corporation [12]. Padhi and Goodenough subsequently introduced $LiFePO_4$ as a positive electrode in 1996 [13]. A cathodic material having a high voltage and excellent efficacy, notably a nickel-manganese-cobalt compound rich in lithium, was found in 1998 by Johnson et al. [14]. There might be possible breakthrough in 2002. By doping the materials with elements like zirconium, aluminum, and niobium, Nevertheless, Chiang succeeded to further improve the thermal conductivity of each material and boost improve only lithium batteries' performance [15]. A ground-breaking modification to LIB was introduced in 2004 by Chiang. He employed fewer than 100 nm-diameter nano phosphate particles to boost circumference of the positive electrodes and battery capability. Therefore the electrode surface and the electrolyte have more contact [16]. In 2005, vanadium phosphate came into existence, establishing the idea of two electrons [17]. Over time, technological developments have resulted in the creation of scalable techniques, such as the usage of sub micrometre-sized $LiCoO_2$ particles. Ranging from solid-state batteries to lithium–sulfur batteries and beyond, our review covers the possible pros and cons of advanced battery technology. In conclusion, lithium-ion batteries represent a key component of the shift to cleaner, more effective energy systems and are not merely a technological advancement. Key issues in energy storage, grid stability, and the uptake of renewable energy sources are addressed by their rapid and effective energy storage and discharge capabilities. Lithium-ion batteries are still essential to creating a sustainable and energy-efficient future as we investigate new uses and advancements in this area. To provide readers with a comprehensive understanding of LIBs for energy storage, in this chapter, a recognised variety of research paper is cited with sources, including industry articles, scientific literature, and industrial journals.

2 Historical Development and Innovations in LIBs

Scientists started working on lithium-ion batteries in the twentieth century in an effort to create more effective, larger-capacity substitutes for the then-dominant lead-acid and nickel–cadmium batteries. There have been numerous significant turning points in the development of lithium-ion batteries:

(i) **Exploration in the 1970s and 1980s**: When scientists started looking at lithium as a possible battery technology option in the 1970s, the first indications of lithium-based batteries appeared. Compared to traditional battery materials, lithium provided the potential for better energy densities because it is the lightest metal with a high electrochemical potential.

(ii) 1970s: Exxon Corporation (now ExxonMobil) researchers conducted the first lithium battery experiments. Lithium metal was utilized as the anode in these early batteries, but because lithium is reactive with moisture, there were safety concerns.

(iii) 1980s: During this time, John B. Goodenough, a pivotal figure in the creation of lithium-ion batteries, made important contributions. The discovery of the lithium cobalt oxide ($LiCoO_2$) cathode material by him and his colleagues at the University of Oxford in 1980 made rechargeable batteries more stable.

(iv) Commercialization (1990s): Early in the 1990s, a discovery was made that led to the development of lithium-ion batteries as we know them today. Under the direction of Dr. Akira Yoshino, Sony Corporation created the first commercial lithium-ion battery using graphite as the anode material rather than lithium metal. This invention ushered in the contemporary Li-ion battery by making it safer, more effective, and rechargeable.

(v) The first Li-ion battery to be sold commercially was introduced by Sony in 1991, and it soon gained popularity for usage in portable devices like laptops and cell phones. Due in large part to advancements in energy density and affordability, lithium-ion technology has spread beyond consumer electronics to encompass electric vehicles (EVs) and renewable energy storage applications as the need for batteries that are lighter, more compact, and have a longer lifespan increased.

(vi) Recent Developments (2000s–Present): The widespread use of electric cars and the growing integration of renewable energy sources have propelled the exponential growth of lithium-ion battery technology in the twenty-first century. Higher energy densities, quicker charging times, and lower production costs have been the results of recent advancements in battery chemistry, new anode and cathode materials, and enhanced manufacturing techniques. The use of substitute materials to increase the sustainability of Li-ion batteries is still being investigated; this includes initiatives to lessen dependency on cobalt, which has sparked ethical and environmental questions.

(vii) Today: Grid-scale energy storage, electric buses, and stationary storage for solar and wind energy systems are just a few of the large-scale uses for lithium-ion batteries.

The latest advancements in LIBs are evidence of the consistent exploration for modern propels in vitality capacity innovation [1]. Since LIBs were presented to showcase within the early 1990s, they have come a long way much appreciated for steady inquiries about, innovative progressions, and buyer requests. This area gives an outline of the key revelations and turning focuses that impelled the advancement of LIBs and turned them into the flexible and broadly utilized vitality capacity gadgets they are nowadays [7]. Goodenough and his colleagues made the imperative revelation of chemical intercalation within the 1980s, which revolutionized LIB improvement. This ground-breaking investigation has brought about in a rechargeable LIB that has tall vitality density and a moderately more self-discharge rate compared to customary chemical batteries. The marketing of LIBs within the early 1990s stamped a critical turning point when Sony Enterprise presented the primary LIBs to the showcase in 1991. In recognition of their long life expectancy and tall vitality thickness compared to Nickel–cadmium and Nickel-metal Hydride batteries, these lightweight and compact batteries have rapidly become prevalent for utilisation in convenient electronic gadgets such as advanced cameras and portable workstation computers and versatile phones [6]. Subsequently, broad inquiries have been conducted in order to progress the effectiveness and security of LIBs, driving a number of striking revelations.

Advancement of new cathode materials, such as lithium manganese oxide ($LiMn_2O_4$), lithium iron phosphate ($LiFePO_4$), and lithium cobalt oxide ($LiCoO_2$) has essentially moved forward the vitality thickness and steadiness of LIBs. Graphite has customarily been the fabric of choice for the anode due to its electrical conductivity and strength. In any case, propels in silicon-based anodes and high-capacity materials may empower Li-ion batteries to realise even higher vitality densities. The primary objectives of inquiring about electrolyte composition are to move forward ionic conductivity, increment security, and increment the working temperature extension of Li-ion batteries [1]. The improvement of strong electrolytes speaks of potential progress in addressing safety issues related to fluid electrolytes. Progresses in battery fabricating strategies, counting roll-to-roll and added substance fabricating, have empowered the mass generation of high-quality, low-cost Li-ion batteries at decreased costs. With the presentation of security highlights such as warm administration frameworks and battery administration frameworks, which diminish the chance of disastrous disappointment and warm runaway, the security and unwavering quality of LIBs progressively progressed. LIB innovation is continually advancing due to the growing demand for wearable gadgets, renewable energy capacity, and electric vehicles (EVs). In order to meet the constantly changing requirements of numerous applications, investigative endeavours are focused on ceaselessly progressing vitality thickness, lifetime, charging speed and taking a toll proficiency. In spite of the fact that there have been noteworthy propels in Li-ion battery innovation, challenges remain. These incorporate issues related to asset accessibility, natural effects and reusing [7]. Proceeded inquiry about endeavours and intrigue collaboration will be required to overcome these challenges and create feasible and versatile arrangements. In an outline, Li-ion battery development has seen mind-blowing development over the past few decades, changing vitality capacity and utilisation in a progressively zapped world. The need for more feasible, proficient and naturally neighbourly vitality capacity arrangements is what drives Li-ion battery technology in its current direction of development and advancement [3].

3 Fundamentals of LIBs

It is required to understand the basic ideas behind the operation of LIBs to optimize their performance and make full use of them, as shown in Fig. 1. One way to think of a LIB is as an electrochemical cell which may generate a great deal of energy. The key elements of the LIBs are the anodes, the cathodes, an electrolyte, separators, and current collector. Throughout the process of transition of lithium ions, the electrolyte functions as a medium between the anode and cathode, which are referred to as the negative and positive electrodes, respectively [18]. During the charge process phase, Li-ions flow from the cathode material to the anode one. On the other hand, the positive electrode, also known as the cathode, absorbs those lithium ions during the discharging phase [19, 20]. Multi-cell LIBs recharging has three stages: the constant

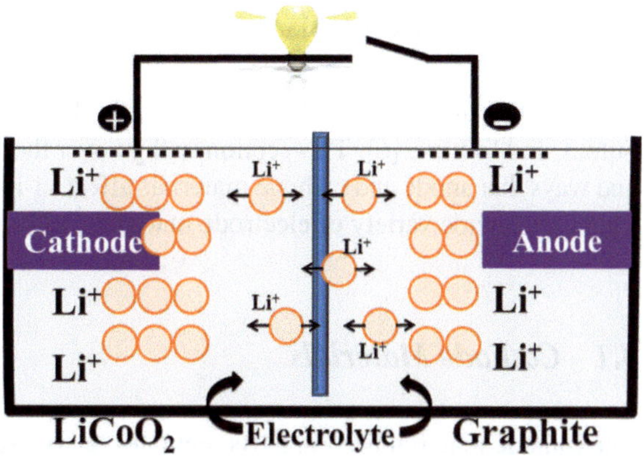

Fig. 1 Basic principle of charging and discharging Li-ion battery

voltage, balance, and the constant current, whereas single-cell LIBs charging consists of 2 phases: constant current or constant voltage [21].

This is a general approach to rearrange the partial response that occurs on the anode and cathode sides of a lithium ion battery [22];

$$LiCoO_2 \leftrightarrow Li_{1-y}CoO_2 + yLi^+ + ye^- \quad \text{(cathodic half)}$$

$$yLi^+ + ye^- + 6C \leftrightarrow Li_yC_6 \quad \text{(anodic half)}$$

A LIB's separator essentially maintains sufficient distance among the anode as well as cathode in order to avoiding short circuit, and this features a thinner membrane with a porous structure that allows ions to move during charging and discharging [23]. Al is often utilised as a current collector on the cathodic part of LIB, whilst Copper is employed on the anodic part. As a load during the discharge phase, these current collector's materials that conduct current and maintain contact with each electrode assisting in supplying the electrical current generated the in the battery to a variety of external applications, including cars, cell phones, lap-tops, cameras, and so forth [24]. A number of important elements are frequently taken into account when assessing LIB performance. The term "cyclability" describes how many cycles of charging and discharging a battery can withstand before experiencing a noticeable decline in capacity. For both consumer gadgets and electric cars, a battery with high cyclability can be utilized for longer periods of time. Volumetric energy density, or how much energy a battery can store per unit of volume, is another crucial consideration. For applications like electric vehicles, where weight and size are crucial factors, a battery with a higher volumetric energy density can store more energy in a smaller, more compact space.

4 Electrode Materials

The material used for the electrodes in LIBs determines its energy density, power output, and lifetime [6]. This section will go over the characteristics, advancements, and ways that anode and cathode materials affect Li-ion battery performance overall. There are a wide variety of electrode materials, as Fig. 2 shows.

4.1 Cathode Materials

The cathode material used in LIBs regulates factors such as voltage, energy density, and stability; this material acts an important role in the battery's performance. Numerous cathode materials have been developed and improved over time, each with specific advantages and disadvantages.

(a) **Lithium Cobalt Oxide ($LiCoO_2$)**

$LiCoO_2$ was one was among the earliest commercially available cathode materials for LIBs because of its stable cyclic behavior and high specific capacity [25]. However, concerns about the cost, availability, and environmental impact of cobalt have sparked a search for alternative cathode materials.

(b) **Lithium Iron Phosphate ($LiFePO_4$)**

$LiFePO_4$ has garnered interest because of its long cycle life, high thermal stability, and safety. Despite having a lower specific capacity than $LiCoO_2$, $LiFePO_4$ is safer and more environmentally friendly [26]. As such, it's a fantastic solution for many different applications.

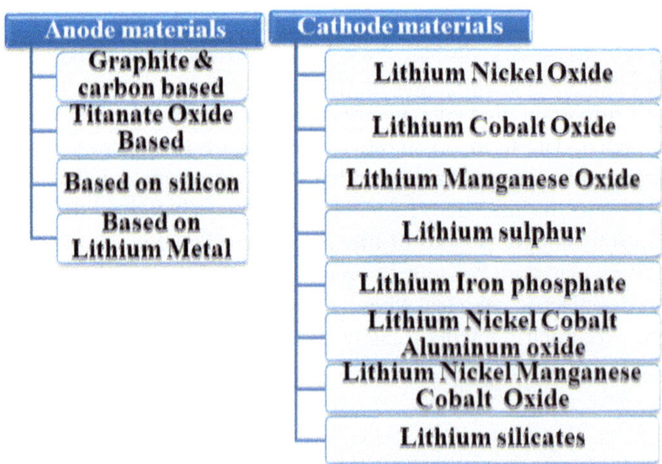

Fig. 2 Types of electrode materials

(c) **Lithium Nickel Manganese Cobalt Oxide** (NMC)

High specific capacitance, voltage stability, and thermal stability are all balanced in the NMC cathode [27]. Because of their greater energy density and longer lifespan as compared to other cathode materials, they are extensively utilized in wearable technology and electric vehicles (EVs).

(d) **Lithium Nickel Cobalt Aluminium Oxide** (NCA)

Because of their exceptional power retention and high energy density, NCA cathodes are ideal for high-performance uses like electric vehicles. Nonetheless, concerns over cobalt's impact on the environment and morality have spurred further research into cobalt alternatives.

NCA cathodes have a high energy density and remarkable power retention, which makes them perfect for high-performance applications like electric cars. However, more study into cobalt substitutes has been prompted by worries about cobalt's effects on the environment and ethical standards [28].

4.2 Anode Materials

Anode materials are essential for the operation of LIBs because they contain lithium ions during the charge and release cycle [5]. Several anode materials have been researched with the aim of improving specific power, safety and stability when cycling:

(a) **Graphite**: Graphite is the primary material for anodes utilized in commercial LIBs due to its stability, low cost, and reasonable specific capacity (~372 mAh/g). However, due to limited capacity, research into high-capacity alternatives has been promoted [29].
(b) **Silicon**: Anodes made from silicon have a significantly higher specific capacity (>4000 mAh/g) than graphite anodes. Yet, because of the volume augmentation that takes place throughout the adhesion process, they experience reduced capacity and electrode damage [30]. At 4 A g^{-1}, diatomite-derived Si@C has a reversible storage capacity of 1359.7 mAh g^{-1}, and beyond 500 cycles, it maintains 764.6 mAh g^{-1}. Highly efficient Si@C anode materials for LIBs can be synthesized in a scalable manner using the method described [31].
(c) **Lithium Metal**: Lithium metal anodes have significant potential for high-density batteries due to their low voltage and theoretical specific capacity greater than about 3860 mAh/g. However, challenges such as dendritic formation and security concerns make implementation difficult.
(d) **Metal Oxides and Alloy-Based Anodes**: Since LIBs have a significant specific capacitance and ability to overcome the limitations of graphite, some metal oxides, including titanium dioxide and titanium tin oxide, as well as materials made from alloys, especially silicon. Tin alloy has been researched for use as an anode material.

(e) Improving electrode–electrolyte interface compatibility, minimizing volume changes, and improving cycling stability are key goals of anode materials research. Strategies used to overcome these challenges include composite electrode design, surface coatings, and nanostructures [3].

In summary, cathode and anode materials are essential to Li-ion battery technology and ongoing research aims to optimize their properties to maximize the utilization of LIBs possible for battery energy storage applications [1].

5 Electrolyte

The electrolyte is an essential component of LIBs as it facilitates less difficulty for lithium ions to migrate among the cathode and anode through the charge and discharge phases [32]. Figure 3 shows several kinds of electrolytes. This section discusses the function, make-up, conductivity, and effects of electrolytes on batteries in terms of performance and security about LIB technology [8]. Lithium salts frequently produce an electrolyte when they dissolve in a solvent. In a LIB, an electrolyte is a chemical that serves as an ion transition medium. Numerous highly particular characteristics, including sufficient ionic conductivity, climate adaptability, non-toxicity, thermal stability, electrochemical stability, and others, are often required for the perfect electrolyte in order to support lithium-ion batteries [33]. When various forms of lithium salts, such as $LiPF_6$, $LiAsF_6$, and $LiClO_4$, are combined with organic carbonates, such as carbonate or diethyl carbonate, ethyl methyl carbonate, and others [34].

An electrolyte can be any combination of liquid and/or solid. The electrolyte utilized in rechargeable LIBs usually comes in two varieties. Whereas the second electrolyte is based on polymers, the initial is a liquid electrolyte. Because of the conductive properties of these polymer electrolytes, composites including ceramic nanoparticles develop, giving rise to high voltage resistance. Because of these electrolytes' increased viscosity and quasi-solid characteristics—which also lessen the likelihood of lithium dendrite formation—they may be used in conjunction with lithium metal anodes.

The polymer in the electrolyte primarily determines its electrochemical long-term viability which is required for the polymer-based electrolyte. One major issue with these polymer-based electrolytes is the lack of sufficient ion conduction. By choosing a liquid electrolyte with a range of solvents, and suitable dielectric and viscosity characteristics that also permit the big rate of ion conduction, one may get over this restriction, which is insufficient ion conduction. When examining the case of liquid electrolyte, the field-trial technique is governed by the dielectric constant and viscosity constants in the ion conduction system [35, 36].

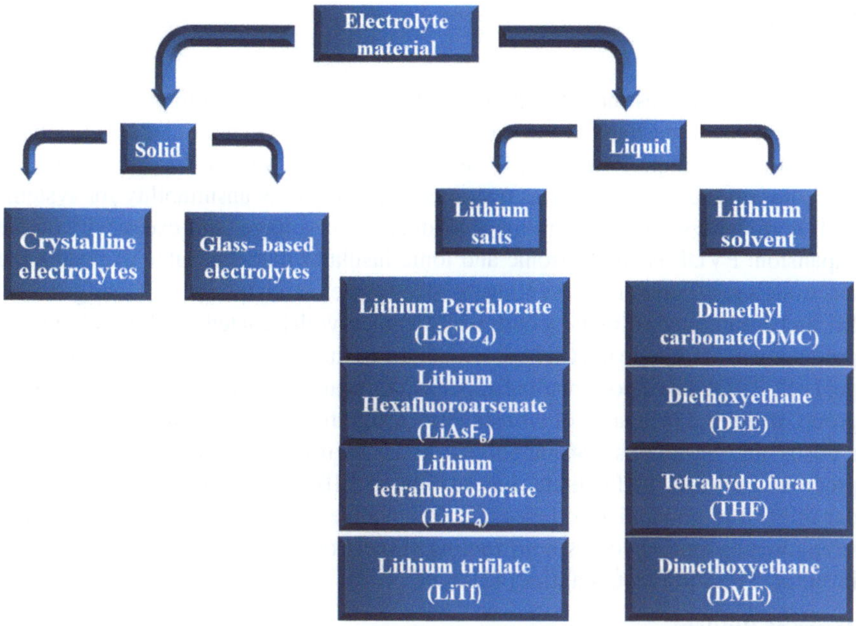

Fig. 3 Different types of electrolytes

6 Current Collector

Another essential component of LIBs is the current collector, which is usually connected to both the anode and cathode of an electrode and moves electrons from the inside to the outside. Aluminum foil serves as a current collector on the cathode side of LIBs, while copper foil is used on the anode side [37]. The current collector has a major influence on a number of LIB performance characteristics, such as electrical conductivity, cycle stability, capacity rate, contact resistance, corrosion resistance, and sustainability. In conclusion, this current collector should be at least a little thick, but it's a great rule to enhance battery performance. Then, it will raise the energy density and decrease the weight of the batteries [37]. According to recent technology, the thickness of 10 μm should be adequate to ensure that the current collectors for Al and Cu in LIBs have the right energy density. The materials used as LIB collectors today are frequently divided into a number of categories, including carbonaceous materials, aluminium, nickel, copper, stainless steel, tin, and so forth. Every one of these groups has distinct traits and attributes [38].

7 Binder

Strong binding forces should be produced by the binder in order to provide the electrode with better mechanical and stable qualities, which will minimize side reactions and result in great performance [39, 40]. With its weak Vander Waals force, PVDF is a conventional binder that has restricted uses due to its unsuitability for systems with high energy density, high mass loading, high voltage, and excessive volume expansion. PVDF is an electronic and ionic insulator that is capable of raising the electrode's resistance and preventing rapid ion and electron transport. It may generate LiF and unsaturated >C=CF– bonds while it reacts with lithiated graphite or Li metal, producing a lot of joule heat at high temperatures that cause the electrode material to peel off. PVDF has a poor binding capacity, a linear structure, and a C–F functional group, making it incapable of withstanding a significant volume change [41]. NMP, a costly and hazardous organic solvent, must be used throughout the preparation process of PVDF. To facilitate the creation of LIBs with high power and energy densities, it is imperative to investigate innovative binders from the viewpoints of strengthening binding forces, constructing 3D networks, improving conductivity, and inventing specific functionalities [42].

Types of Binders

(a) **Polyvinylidene Fluoride (PVDF)**

Due to its outstanding adhesive properties, chemical resistance, and versatility with a wide range of solvents and electrolytes, PVDF is a commonly used adhesive in Li-ion battery electrodes.

(b) **Polyvinyl Alcohol (PVA)**

PVA adhesives have excellent adhesive and film-forming properties, making them suitable for use in ecologically friendly electrode manufacturing processes and water treatment.

(c) **Carboxymethyl Cellulose (CMC)**

CMC glue has remarkable flexibility, binding strength, and water retention due to its natural cellulose composition. To improve electrode performance, they are commonly utilized with PVDF or PVA adhesives.

(d) **Styrene Butadiene Rubber (SBR)**

SBR adhesives are elastomeric polymers that provide remarkable mechanical strength and flexibility that is advantageous to Li-ion battery electrodes. They are frequently used in high-power electrode and flexible battery applications [2].

8 Separators

A crucial component of a LIB is the separator, which serves as a protective barrier between the anode and cathode electrodes and maintains the battery's thermal stability by maintaining a proper distance between them [43]. LIB performance is influenced by a number of factors, including separator behavior and structure. These factors include energy density, cycle life, power density, safety, and others. Ion transport while charging and discharging is only facilitated by a porous structure, despite the fact that there is no discernible chemical interaction with the separator. In order to prevent thermal runaway during a short circuit, the separator should impart a blocking interface function [44].

Separator categorization in LIBs is based on both chemical and physical behavior. These can be paper-based, microporous, bonded, woven, moulded, nonwoven, or laminated. These days, LIBs separators are made from nonwoven textiles or microporous polymeric films. A separator is typically not needed for batteries that use solid-state and polymer-based electrolytes. Microporous separators, which are often used in LIB systems with liquid electrolyte, are composed of polyolefin (PE, PP, PP/PE/PP) [45].

The current collector, that's usually attached to the anode and cathode of the electrode and transfers electrons from the inside to the outside, is an additional vital part of lithium-ion batteries. Cu foil is utilized on the anode side of LIBs, whereas Al foil is employed on the cathode side as a current collector. The current collector has a significant impact on a number of lithium-ion battery performance characteristics, including capacity rate, electrical conductivity, contact resistance, sustainability corrosion resistance, and cycle stability. To put it briefly, it's a fantastic rule to improve battery performance, however, there should be a minimum thickness for this current collector.

Then, it will raise the energy density and decrease the weight of the batteries [37]. According to freshly invented technology, an average thickness of 10 μm ought to be enough to guarantee an adequate energy density for Al and Cu current collectors in LIBs. Current collector materials for LIBs are often classified into many groups, such as carbonaceous materials, nickel, aluminium, stainless steel, copper, tin, and so on. Each of these categories has unique qualities and characteristics [24]. Figure 4 displays several types of separators. This section discusses separator composition, materials, impacts on battery performance, and their significance in Li-ion battery technology.

(a) **Materials for Separators**

(a) **Polyethylene (PE)**

Polyethylene separators are ideal for a range of Li battery applications in spite of their remarkable mechanical strength, chemical resistance, and thermal stability [17].

(b) **Polypropylene (PP)**

Fig. 4 Types of separators

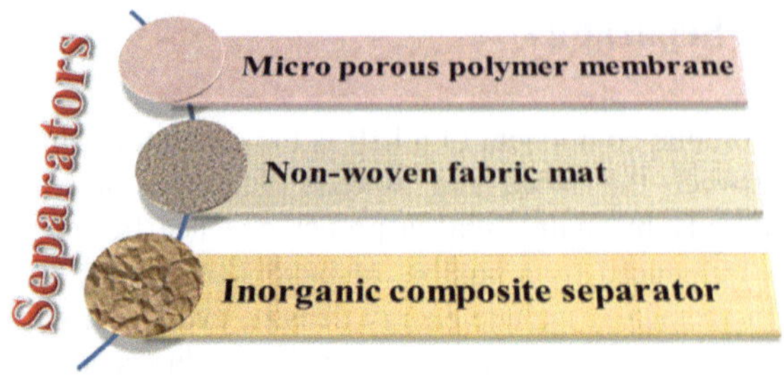

Polypropylene separators, which are similar to polyethylene separators in that they provide increased electrolyte wettability and improved puncture resistance, improve battery efficiency and safety [18].

(c) **Composite Materials**

Composite separators use polymers with inorganic or ceramic fillers to maintain excellent ionic conductivity while improving thermal stability, mechanical strength, and electrolyte retention [19].

9 Limitations in LIBs

(a) **Challenges in LIB Energy**

Due to its higher energy density and specific energy, LIB technology has supplanted competing battery chemistries in almost every industry and application. LIBs may maintain cell voltage levels that are approximately three times higher than those of the more well-known aqueous rechargeable chemistries including Nickel-Metal-Hydride (NiMH) and Nickel-Cadmium (NiCd) because of their non-aqueous electrolyte system. The fundamental electrochemical pair of high-energy designs ($Co^{3+/4+}$ and $Ni^{3+/4+}$ vs. graphite) hasn't altered since their inception, though, hence the nominal voltages of LIBs have only slightly improved [46].

(b) **Challenges in LIB Life**

The capacity of batteries to withstand damage is essential to EV production. For EV battery packs, several Original Equipment Manufacturers (OEMs) provide warranties that provide battery replacement in the event that capacity declines below a predetermined threshold. Because warranty periods are long—several decades and thousands of miles—and batteries need to be validated within months in order to implement cutting-edge power and energy density, besides other attributes on time, this presents technical obstacles during the battery development process. Furthermore, to prevent spreading discrepancies among derived quantities like battery power and energy, sufficiently exact test apparatus is needed [47].

(c) **Challenges in LIB Cost**

The high expense of battery electric vehicles (BEVs) compared to internal combustion engine vehicles (ICEVs) is one of the primary barriers to their market acceptance. Because of the high cost of materials and the complexity of the engineering involved, the battery pack is the most costly part of an EV. Nonetheless, an approx. Due to significant investments in battery production capacity and the elimination of cobalt from cathodes, an 80% price reduction has been targeted in the last ten years. If this trend continues, the total cost of ownership parity between ICEVs and BEVs will be closer than before. Additional cost reduction is necessary to improve market acceptance because purchase price parity is still elusive [48].

(d) **Solid-State Electrolytes**

It may provide merits in safety and energy density, but low ionic conductivity and low interface resistance hinder their commercialization. Research focuses on developing new solid-state electrolyte materials and fabrication techniques to solve these problems [21].

To overcome these difficulties, academic, business and government groups will need to collaborate across sectors and continue their research efforts. By overcoming these challenges, Li-ion batteries will be able to evolve into a flexible and sustainable energy storage technology, accelerating the transition to a cleaner, more electric future.

10 Beyond LIBs

Although LIBs have transformed the energy storage industry, research is currently underway to find alternative energy storage solutions and battery chemistries to meet the changing needs and problems of different applications. This section looks at some attractive Li-ion battery alternatives and their potential impact on future energy storage.

(a) **Solid-State Batteries**: The batteries have the potential to increase safety, energy density, and cycle life by replacing the liquid electrolyte in conventional LIBs with a solid electrolyte. The upcoming research focuses on introducing new solid electrolyte materials and interfaces to create high-performance solid-state batteries [21].
(b) **Lithium-Sulfur Batteries**: The lithium-sulfur battery provides a higher theoretical energy density than a Li-ion battery due to the large specific capacity of the sulfur cathode. Polysulfide dissolution and poor cycling stability are two issues that are being addressed using new sulfur hosts, modified electrolytes, and advanced electrode designs [49].
(c) **Metal-Air Batteries**: Metal-air batteries, potentially very energy-rich and affordable, use oxygen from the air as the cathode material. Research aims to overcome challenges such as slow kinetics, limited cycle life, and electrolyte stability to create practical metal-air battery systems for a variety of applications [50].

(d) **Series Batteries**: Series batteries store energy in an electrolyte solution held in external tanks, giving them scalability, flexibility, and long life. The goal of the research is to improve the energy density, efficiency, and affordability of flow battery systems for grid-scale energy storage and renewable energy integration [51].
(e) **Supercapacitors and Hybrid Capacitors**: Supercapacitors and hybrid capacitors combine the energy storage features of capacitors with their high energy density of batteries to deliver fast charging, long life and excellent performance. The main goals of the research are to improve energy density, refine electrode materials, and expand device integration for various applications [52].
(f) **Beyond Lithium Chemistry**: Research into alternative battery chemistries, such as polyvalent, potassium, and sodium-ion batteries, has two main goals: reduce costs and reduce supply dependence Limited lithium supply. The potential richness, affordability and environmental sustainability of these alternatives are among their advantages [53].

To fulfill the numerous demands of emerging applications, such wearable technology, electric cars, and incorporating renewable energy sources, it is imperative to investigate these alternative strategies for energy storage. Sustainable energy systems of the future, will be shaped by the status of battery chemical composition, materials science, research, and technological innovation. These fields will be the driving forces behind the establishment of energy storage technologies of the coming years.

11 Applications of Lithium-Ion Batteries (LIBS)

Because of its special blend of high energy density, extended lifespan, and relatively small weight, lithium-ion (Li-ion) batteries have become indispensable in a variety of industries. These characteristics make them perfect for a wide range of uses in large-scale energy storage systems as well as consumer gadgets. The main uses and the distinct functions that Li-ion batteries perform in each are covered in detail below.

(a) **Consumer Electronics**: Lithium-ion batteries are most commonly used in portable consumer electronics. Because of their small size and effective energy storage characteristics, Li-ion batteries are widely employed in devices ranging from computers and smartphones to tablets and smartwatches. These gadgets need batteries that are both thin and capable of delivering enough power for extended periods of time. Li-ion batteries' comparatively high energy density enables producers to create durable, lightweight electronics without appreciably enlarging their dimensions.

    Smartphones: Lithium-ion batteries are used in most modern smartphones because they provide a small, dependable power source that can frequently run for a whole day between charges.

Similar to laptops and tablets, these gadgets rely on lithium-ion batteries to last longer, allowing users to work or watch media without constantly plugging in.

(b) **Electric Vehicles (EVs)**: Electric vehicles are among the most revolutionary applications for lithium-ion batteries. Li-ion technology has been quickly embraced by the automotive sector to power electric trucks, buses, and cars. Long driving ranges, quick charging times, and overall vehicle performance are all made possible by these batteries. For the electric vehicle market to grow and compete with conventional internal combustion engine vehicles, it must be able to store a significant amount of energy in a comparatively small and light packaging.

Lithium-ion batteries are used in electric cars by top automakers like Tesla and Nissan because of their exceptional energy density, which enables faster acceleration and greater driving range.

Public Transportation: Li-ion batteries are also being used more and more to power electric buses and trucks, which offer cleaner, quieter, and more effective substitutes for diesel-powered vehicles in metropolitan areas.

(c) **Renewable Energy Storage**: Effective energy storage is more important than ever as the world moves toward renewable energy sources like solar and wind. In order to address the intermittent nature of renewable energy sources, lithium-ion batteries are essential for enabling their storage. For instance, wind energy varies according to weather patterns, and solar energy is only available during the day. When energy demand exceeds supply, excess energy produced during periods of peak production can be stored and released by Li-ion batteries.

  (i) **Grid Storage**: Lithium-ion batteries are widely utilized to stabilize the grid. Large battery systems assist guarantee a consistent supply of electricity even during periods when generation from renewable sources is low. These systems are frequently installed in conjunction with renewable energy farms.
  (ii) **Home Energy Storage**: On a smaller scale, homeowners can store energy from rooftop solar panels using lithium-ion batteries. Energy security can be improved by using this stored energy at night or during a power outage.

Medical Devices

(i) Additionally, lithium-ion batteries are used in vital medical equipment where durability and dependability are crucial. These batteries are found in portable medical equipment like defibrillators, hearing aids, and insulin pumps. They are perfect for applications where failure is not an option because of their long cycle life and reliable power delivery.
(ii) Wearable Medical equipment: Wearable health monitoring equipment, like glucose monitors, that must function continuously for extended periods of time without needing to be charged frequently are powered by lithium-ion batteries.
(iii) Implantable Devices: Small lithium-ion batteries are sometimes used to power medical implants, such pacemakers, so they can last for a long time before needing to be replaced.

12 Limitations of Lithium-Ion Batteries (LIBs)

Despite their widespread use and clear advantages, lithium-ion batteries have certain disadvantages. These restrictions reduce their efficacy in particular applications and highlight areas that need more study and improvement.

(a) **Cost**: Although economies of scale and technical developments have led to a dramatic drop in the price of lithium-ion batteries over the last ten years, they are still somewhat pricey when compared to alternative energy storage options. The price of these batteries may increase due to the cost of raw materials, especially nickel, cobalt, and lithium. This continues to be a major obstacle to broad adoption for sectors such as electric vehicles, since the battery makes up a sizable amount of the overall vehicle cost.

(b) **Electric Vehicles**: Although continuous advancements in battery technology may eventually help lower this cost, the high cost of lithium-ion batteries is one of the key reasons electric vehicles continue to be more expensive than their internal combustion engine equivalents.

(c) **Energy Density and Weight**: The amount of energy that lithium-ion batteries can store for a given weight is still limited, even with their comparatively high energy density. They are therefore less suitable for some high-energy applications. Even while improvements have resulted in greater ranges, there is still a trade-off between battery size and vehicle weight in the case of electric vehicles. Although larger batteries can extend an electric car's range, they also add weight to the vehicle, which may reduce performance and efficiency.

(d) **Electric Aviation**: Due to their low energy density, lithium-ion batteries are currently unsuitable for powering commercial aircraft because they cannot deliver the necessary range and power for big aircraft.

(e) **Space Exploration**: Due to their limited energy storage capacity and propensity to deteriorate in harsh environments, Li-ion batteries are not the best choice for extended space trips, despite their utility in some applications.

(f) **Degradation Over Time**: The capacity and performance of lithium-ion batteries decline with time as a result of chemical alterations in the battery cells. A decrease in the number of charge-discharge cycles the battery can withstand before needing to be replaced is a common indicator of this degradation. Therefore, as their batteries age, users of lithium-ion-powered devices—especially owners of electric vehicles—may notice a decrease in battery performance.

Electric Vehicles: Over time, battery deterioration in EVs may result in a shorter driving range, which may be problematic for long-term owners. This problem is being lessened by advancements in battery management systems and material enhancements, but it is still a restriction.

Safety Concerns

Thermal runaway, a hazardous situation in which the battery overheats and may cause a fire or explosion, is known to occur with lithium-ion batteries. Usually, things like overcharging, physical damage, or poor manufacturing are to blame for this. Even though contemporary battery management systems are made to reduce these hazards, safety is still an issue, particularly in high-density applications like large-scale energy storage or electric cars.

(a) **Electric Vehicles**: One of the main concerns with EVs is battery safety, especially after collisions or accidents. Even though manufacturers use a range of safety procedures, incidents involving electric car fires or explosions have brought attention to the importance of stringent safety rules.
(b) **Consumer Electronics**: Similar incidents of computers or cell phones catching fire due to battery problems have been reported, underscoring the importance of proper design and safety measures.

Environmental Impact: Even though lithium-ion batteries are better for the environment than conventional lead-acid batteries, there are still major environmental issues with their manufacture and disposal. Concerns regarding environmental harm and violations of human rights are raised by the mining of lithium, cobalt, and other rare earth metals, particularly in areas with lax regulations. Furthermore, recycling lithium-ion batteries is still a difficult and costly procedure, raising questions regarding the battery lifecycle's long-term viability.

(a) **Battery Disposal**: As lithium-ion batteries become more widely used, it becomes more difficult to recycle or get rid of them safely. Because some battery components might seep into the environment, improper disposal can contaminate the ecosystem.
(b) **Scarcity of Resources**: Concerns regarding the long-term supply of vital raw materials, especially lithium and cobalt, which are necessary for battery production, have been raised by the rising demand for lithium-ion batteries.

In conclusion, although lithium-ion batteries have transformed energy storage in many industries by providing small, dependable, and effective solutions for anything from electric cars and renewable energy storage to personal gadgets, they are not without drawbacks. As the world's reliance on these batteries grows, there are a number of important issues that need to be resolved, including high costs, energy density limitations, deterioration with time, safety hazards, and environmental problems.

Many of these problems will probably be resolved by ongoing research and technical developments, leading to advances in energy density, safety, cost, and sustainability. Lithium-ion batteries will continue to play a key role in supplying energy in the future as the need for greener and more effective solutions increases, but their drawbacks must be carefully controlled by legislation, innovation, and better recycling techniques.

13 Sustainability Focus

As the world moves toward cleaner energy sources, the sustainability of lithium-ion batteries (LIBs) has become a major concern. While these batteries are critical to the development of electric vehicles (EVs) and renewable energy storage, their environmental impact—particularly in terms of resource extraction, production, and disposal—is becoming more scrutinized. As a result, improving the sustainability of LIB technology has become a key focus of continuous research and innovation.

(a) **Recycling Initiatives**: One of the most severe sustainability challenges with LIBs is a lack of effective recycling mechanisms. Currently, the recycling rate for LIBs is minimal, with just a small percentage of materials recovered and reused. This inefficiency generates enormous environmental loss, as important metals such as lithium, cobalt, and nickel are frequently lost. In response, academics and firms are attempting to create closed-loop recycling systems in which battery components can be entirely collected, refined, and reused in new batteries. One notable example is Tesla's battery recycling project, which seeks to reclaim a major portion of the materials used in its batteries while reducing dependency on mining. Furthermore, firms such as Li-Cycle and Umicore are developing novel recycling processes that promise to enhance the output of valuable materials from wasted batteries while reducing environmental impact.

(b) **Sustainable Sourcing of Materials**: The sourcing of resources for LIBs, particularly cobalt, nickel, and lithium, poses sustainability problems because these materials are frequently mined in areas with weak labor rights and environmental laws. In response, attempts are being made to implement more sustainable and ethical sourcing processes. For example, BMW and Volkswagen are increasingly looking for conflict-free cobalt and ensuring that their supply chains are transparent and accountable. Furthermore, there is a rising emphasis on investigating alternate materials that may lessen reliance on vital resources. Lithium iron phosphate (LFP), which eliminates the requirement for cobalt and nickel, is becoming popular as an alternative cathode material. This move may assist to alleviate the environmental and ethical difficulties connected with traditional battery chemistries.

(c) **Second-Life Applications**: Another possible route for increasing sustainability is the usage of second-hand batteries. As LIBs in electric vehicles and renewable energy systems approach the end of their useful life in high-demand applications, they can still maintain a large amount of capacity. Researchers are looking into the possibility of reusing these "retired" batteries for less demanding uses, like stationary energy storage. This not only improves the life of the batteries, but it also minimizes the environmental impact of their production. For example, Nissan has established a second-life battery project in which EV batteries are repurposed to store energy from solar panels for homes, so contributing to the sustainability of both the transportation and energy sectors.

(d) **Emerging Technologies**: Innovative ways are also being investigated to improve battery performance while also reducing environmental impact. Solid-state batteries, for example, have the potential for higher energy densities and greater safety than typical lithium-ion batteries. Furthermore, they have the potential to reduce the use of hazardous materials such as liquid electrolytes while also improving recycling. Researchers are also investigating sodium-ion batteries as a possible alternative that, due to sodium's availability, may be more sustainable than lithium batteries.

14 Future Perspectives

Advances in technology and on-going research targeted at resolving significant issues and establishing new development avenues bode well for LIBs in the future. This section discusses the challenges and future directions in the development and use of Li-ion battery technology.

Prospects

(a) **Performance Improvement**: Enhancing the functioning of LIBs on the basis of lifetime, safety, power density, and power density is the goal of continuous research. Technological advances in electrolytes, cell design and electrode materials can significantly improve battery performance, leading to new applications for batteries in electric vehicles, portable electronics devices and renewable energy storage.

(b) **Application diversification**: The Li-ion battery is well positioned to become an important part of the transition to a clean energy future because its ability to power a variety of applications beyond consumer electronics. Automotive electrification, integration of renewable energy sources, and grid-scale energy storage are some of the innovative ways Li-ion batteries can contribute to decarbonisation and sustainable energy.

(c) **Innovation in Battery Chemistries**: Research on battery chemistries beyond Li-ion, such as solid-state batteries, metal-air batteries, and lithium-sulfur batteries, shows promise for density-increased energy, lower costs and safer operations. Research into these other chemicals could lead to new developments in energy storage technology.

(d) **Sustainable Battery Technologies**: Concerns about resource scarcity and environmental impact are increasingly focusing on the development of sustainable battery technology. Li-ion battery production must be sustainable, and resource recovery, recycling, and the use of environmentally friendly components are essential to managing the battery's shelf life.

15 Summary

Li-ion batteries have completely changed the way energy is stored and will be key to enabling the use of sustainable energy sources in the future. The advancement of Li-ion battery technology depends on research, development and sustained collaboration to overcome obstacles, take advantage of opportunities and meet society's growing needs because they can power everything from electric cars to grid-scale energy storage. From systems to wearable devices, LIBs have become the industry standard for energy storage technology. This chapter has provided an overview of the main components, advancements, difficulties, and possible uses of Li-ion battery technology. This chapter has addressed the fundamental parts of Li-ion batteries, including the separator, electrode materials, current collector, binder, and electrolyte. Every part is crucial to the performance, safety, and dependability of batteries.

Li-ion battery technology has advanced significantly through innovative research and advances in chemistry, engineering and materials science. Improvements in energy density and cycling stability can be attributed to advances in electrode materials, including silicon-based anodes and high-power cathodes. At the same time, advances in electrolyte composition, separator structures and adhesive materials have improved battery life, safety and reliability. Li-ion batteries have been successful but there are still some issues that need to be resolved if they are to reach their full potential. There are many obstacles to overcome, including safety concerns, energy density limitations, financial constraints and material sustainability challenges. Cross-industry collaboration, innovative research and calculated investment in battery technology are needed to solve these problems. Li-ion batteries have great potential to accelerate the transition to a cleaner, more electric future. It is anticipated that developments in battery chemistry, manufacturing techniques and system integration will further improve the performance of Li-ion batteries, reduce cost and increase the diversity of uses. Additionally, research into sustainable materials, new energy storage techniques and other battery chemistries will advance the field of energy storage technology.

References

1. Arfwedson, A.: Untersuchung einiger bei der Eisen-Grube von Utö vorkommenden Fossilien und von einem darin gefundenen neuen feuerfesten alkali. J. Chem. Phys. **22**, 93–117 (1818)
2. Harris, W.S.: Electrochemical studies in cyclic esters. Ph.D. thesis, University of California, Berkeley, CA (1958). https://doi.org/10.2172/4305596
3. Heller, A.: Hydrogen-evolving solar cells. Science **223**, 1141–1148 (1984). https://doi.org/10.1126/science.223.4641.1141
4. Godshall, N.A., Raistrick, I.D., Huggins, R.A.: Thermodynamic investigations of ternary lithium-transition metal-oxygen cathode materials. Mater. Res. Bull. **15**, 561–571 (1980). https://doi.org/10.1016/0025-5408(80)90107-4
5. Mizushima, K., Jones, P.C., Wiseman, P.J., Goodenough, J.B.: $LixCoO_2$ ($0 < x < 1$): a new cathode material for batteries of high energy density. Mater. Res. Bull. **15**, 783–789 (1980). https://doi.org/10.1016/0025-5408(80)90012-4

6. Poizot, P., Laruelle, S., Grugeon, S., Dupont, L., Tarascon, J.M.: Nano-sized transition-metal oxides as negative-electrode materials for lithium-ion batteries. Nature **407**, 496–499 (2000). https://doi.org/10.1038/35035045
7. Godshall, N.A., Huggins, R.A., Raistrick, I.D.: Ternary compound electrode for lithium cells. J. Electrochem. Soc. **129**, 262–266 (1982). https://doi.org/10.1149/1.2123865
8. Thackeray, M.M., David, W.I.F., Bruce, P.G., Goodenough, J.B.: Lithium insertion into manganese spinels. Mater. Res. Bull. **18**, 461–472 (1983). https://doi.org/10.1016/0025-5408(83)90087-0
9. Ogawa, S.: Profile of Akira Yoshino, Dr. Eng., and overview of his invention of the lithium-ion battery. Electrochemistry **87**, 1–5 (2019). https://doi.org/10.5796/electrochemistry.87.1
10. Manthiram, A., Goodenough, J.B.: Lithium insertion into $Fe_2(SO_4)_3$ frameworks. J. Power. Sources **26**, 403–408 (1989). https://doi.org/10.1016/0378-7753(89)80034-0
11. Fong, R., Von Sacken, U., Dahn, J.R.: Studies of lithium intercalation into carbons using nonaqueous electrochemical cells. J. Electrochem. Soc. **137**, 2009–2013 (1990). https://doi.org/10.1149/1.2086194
12. Abuaqilah, M.K., Saeed, R.A.A.: Ion lithium battery. Int. J. Electrochem. Sci. **14**, 12345–12356 (2019). https://doi.org/10.20964/2019.12.45
13. Padhi, A.K., Nanjundaswamy, K.S., Goodenough, J.B.: $LiFePO_4$: a novel cathode material for rechargeable batteries. J. Electrochem. Soc. **144**, 1188–1194 (1997). https://doi.org/10.1149/1.1837571
14. Johnson, C.S., Li, N., Vaughey, J.T., Hackney, S.A., Thackeray, M.M.: Lithium–manganese oxide electrodes with layered-spinel composite structures $xLi_2MnO_{3(1-x)}Li_{1+y}Mn_{2-y}O_4$ ($0 < x < 1, 0 \leq y \leq 0.33$) for lithium batteries. Electrochem. Commun. **7**, 528–532 (2005). https://doi.org/10.1016/j.elecom.2005.03.005
15. Chung, S.-Y., Bloking, J.T., Chiang, Y.-M.: Electronically conductive phospho-olivines as lithium storage electrodes. Nat. Mater. **1**, 123–128 (2002). https://doi.org/10.1038/nmat732
16. Yacoub, F.: Geological Report on the Lithium Potential of the Canadian Lithium and Augustus Prospects. Ministry of Natural Resources of Quebec (2016)
17. Lim, S.C., Vaughey, J.T., Harrison, W.T.A., Dussack, L.L., Jacobson, A.J., Johnson, J.W.: Redox transformations of simple vanadium phosphates: the synthesis of ε-$VOPO_4$. Solid State Ion. **84**, 219–226 (1996). https://doi.org/10.1016/0167-2738(96)00007-0
18. Mahmud, S., Rahman, M., Kamruzzaman, M., Ali, M.O., Emon, M.S.A., Khatun, H., Ali, M.R.: Recent advances in lithium-ion battery materials for improved electrochemical performance: a review. Results Eng. **15**, 100472 (2022). https://doi.org/10.1016/j.rineng.2022.100472
19. Roy, P., Srivastava, S.K.: Nanostructured anode materials for lithium-ion batteries. J. Mater. Chem. A **3**, 2454–2484 (2015). https://doi.org/10.1039/C4TA04980F
20. Li, X., Wang, C.: Engineering nanostructured anodes via electrostatic spray deposition for high performance lithium ion battery application. J. Mater. Chem. A **1**, 165–182 (2013). https://doi.org/10.1039/C2TA00074K
21. Zhang, S.S.: The effect of the charging protocol on the cycle life of a Li-ion battery. J. Power. Sources **161**, 1385–1391 (2006). https://doi.org/10.1016/j.jpowsour.2006.04.016
22. Zhang, J., Zhang, L., Liu, H., Sun, A., Liu, R.-S.: Electrochemical Technologies for Energy Storage and Conversion, 2 Volume Set, vol. 1. John Wiley & Sons (2011)
23. Orendorff, C.J.: The role of separators in lithium-ion cell safety. Electrochem. Soc. Interface **21**, 61–65 (2012)
24. Zhu, P., Gastol, D., Marshall, J., Sommerville, R., Goodship, V., Kendrick, E.: A review of current collectors for lithium-ion batteries. J. Power. Sources **485**, 229321 (2021). https://doi.org/10.1016/j.jpowsour.2020.229321
25. Lin, F.P.C., Li, J., Yin, Z.-W., Huang, W., Zhao, Q., Weng, Q., Liu, Q., Sun, J., Chen, G.: Structural understanding for high-voltage stabilization of lithium cobalt oxide. Adv. Mater. **36**, 2308805 (2024). https://doi.org/10.1002/adma.202308805
26. Zhao, T., et al.: An overview on the life cycle of lithium iron phosphate: synthesis, modification, application, and recycling. Chem. Eng. J. **485**, 149923 (2024). https://doi.org/10.1016/j.cej.2023.149923

27. Tancin, R.J., Özdoğru, B., Dutta, N.S., Finegan, D.P., Tremolet de Villers, B.J.: Direct reuse of graphite and lithium nickel manganese cobalt oxide (NMC) recovered from ultrafast-laser ablation debris in Li-ion battery electrodes. J. Power Sources **596**, 234027 (2024). https://doi.org/10.1016/j.jpowsour.2023.234027
28. Javaid, A., Khalid, H.A., Kazmi, S.A.A., Ali, G.: Analysis of lithium diffusion and overpotential in lithium nickel cobalt aluminum oxide based lithium-ion cells. J. Electroanal. Chem. **952**, 117991 (2024). https://doi.org/10.1016/j.jelechem.2023.117991
29. Zhao, W., Zhao, C., Wu, H., Li, L., Zhang, C.: Progress, challenge and perspective of graphite-based anode materials for lithium batteries: a review. J. Energy Storage **81**, 110409 (2024). https://doi.org/10.1016/j.est.2023.110409
30. Wang, Y., Yang, F., Wu, T., Huang, G.: Roundly exploring the synthesis, structural design, performance modification, and practical applications of silicon–carbon composite anodes for lithium-ion batteries. J. Energy Storage **101**, 113794 (2024). https://doi.org/10.1016/j.est.2024.113794
31. Huang, X., Ding, Y., Li, K., Guo, X., Zhu, Y., Zhang, Y., Bao, Z.: Spontaneous formation of the conformal carbon nanolayer coated Si nanostructures as the stable anode for lithium-ion batteries from silica nanomaterials. J. Power Sources **496**, 229833 (2021). https://doi.org/10.1016/j.jpowsour.2021.229833
32. Kainat, S., Anwer, J., Hamid, A., Gull, N., Khan, S.M.: Electrolytes in lithium-ion batteries: advancements in the era of twenties (2020's). Mater. Chem. Phys. **313**, 128796 (2024). https://doi.org/10.1016/j.matchemphys.2023.128796
33. Li, M., Lu, J., Chen, Z., Amine, K.: 30 years of lithium-ion batteries. Adv. Mater. **30**(33), 1800561 (2018). https://doi.org/10.1002/adma.201800561
34. Bhatt, M.D., O'Dwyer, C.: Density functional theory calculations for ethylene carbonate-based binary electrolyte mixtures in lithium ion batteries. Curr. Appl. Phys. **14**(2), 349–361 (2014). https://doi.org/10.1016/j.cap.2013.11.022
35. Chanson, C., Wiaux, J.P.: Safety of lithium-ion batteries. Eur. Assoc. Adv. Recharg. Batter. (2013)
36. Monroe, C., Newman, J.: The impact of elastic deformation on deposition kinetics at lithium/polymer interfaces. J. Electrochem. Soc. **152**(2), A396–A404 (2005). https://doi.org/10.1149/1.1850850
37. Wang, C., Lin, Y., Chiu, K., Leu, H., Ko, T.: Advanced carbon cloth as current collector for enhanced electrochemical performance of lithium-rich layered oxide cathodes. ChemistrySelect **2**(15), 4419–4426 (2017). https://doi.org/10.1002/slct.201700457
38. Lain, M.J., Brandon, J., Kendrick, E.: Design strategies for high power vs. high energy lithium-ion cells. Batteries **5**(3), 64 (2019). https://doi.org/10.3390/batteries5030064
39. Lingappan, N., Kong, L., Pecht, M.: The significance of aqueous binders in lithium-ion batteries. Renew. Sustain. Energy Rev. **147**, 111227 (2021). https://doi.org/10.1016/j.rser.2021.111227
40. Lestriez, B.: Functions of polymers in composite electrodes of lithium-ion batteries. C. R. Chim. **13**(11), 1341–1350 (2010). https://doi.org/10.1016/j.crci.2010.05.005
41. Placke, T., Siozios, V., Schmitz, R., Lux, S.F., Bieker, P., Colle, C., Meyer, H.-W., Passerini, S., Winter, M.: Influence of graphite surface modifications on the ratio of basal plane to "Non-Basal Plane" surface area and on the anode performance in lithium-ion batteries. J. Power Sources **200**, 83–90 (2012). https://doi.org/10.1016/j.jpowsour.2011.10.097
42. Cheng, Q., Yuge, R., Nakahara, K., Tamura, N., Miyamoto, S.: KOH etched graphite for fast chargeable lithium-ion batteries. J. Power Sources **284**, 258–264 (2015). https://doi.org/10.1016/j.jpowsour.2015.03.011
43. Huang, X.: Separator technologies for lithium-ion batteries. J. Solid State Electrochem. **15**(4), 649–662 (2011). https://doi.org/10.1007/s10008-010-1100-x
44. Lee, H., Yanilmaz, M., Toprakci, O., Fu, K., Zhang, X.: A review of recent developments in membrane separators for rechargeable lithium-ion batteries. Energy Environ. Sci. **7**(12), 3857–3886 (2014). https://doi.org/10.1039/C4EE01432D
45. Arora, P., Zhang, Z.: Battery separators. Chem. Rev. **104**(10), 4419–4462 (2004). https://doi.org/10.1021/cr020738u

46. Winter, M., Barnett, B., Xu, K.: Before Li ion batteries. Chem. Rev. **118**(23), 11433–11456 (2018). https://doi.org/10.1021/acs.chemrev.8b00422
47. Wold, J., Marcicki, J., Masias, A.: Derived quantities uncertainty propagation in high precision battery testing. J. Electrochem. Soc. **164**(13), A2131–A2137 (2017). https://doi.org/10.1149/2.1071713jes
48. Duffner, F., Wentker, M., Greenwood, M., Leker, J.: Battery cost modeling: a review and directions for future research. Renew. Sustain. Energy Rev. **127**, 109872 (2020). https://doi.org/10.1016/j.rser.2020.109872
49. Ji, X., Lee, K.T., Nazar, L.F.: A highly ordered nanostructured carbon-sulphur cathode for lithium-sulphur batteries. Nat. Mater. **8**(6), 500–506 (2009). https://doi.org/10.1038/nmat2460
50. Bruce, P.G., Freunberger, S.A., Hardwick, L.J., Tarascon, J.-M.: Li–O_2 and Li–S batteries with high energy storage. Nat. Mater. **11**(1), 19–29 (2012). https://doi.org/10.1038/nmat3191
51. Wang, W., Luo, Q., Li, B., Wei, X., Li, L., Yang, Z.: Recent progress in redox flow battery research and development. Adv. Funct. Mater. **23**(8), 970–986 (2013). https://doi.org/10.1002/adfm.201200694
52. Simon, P., Gogotsi, Y.: Materials for electrochemical capacitors. Nat. Mater. **7**(11), 845–854 (2008). https://doi.org/10.1038/nmat2297
53. Palomares, V., Serras, P., Villaluenga, I., Hueso, K.B., Carretero-González, J., Rojo, T.: Na-ion batteries, recent advances and present challenges to become low cost energy storage systems. Energy Environ. Sci. **5**(3), 5884–5901 (2012). https://doi.org/10.1039/c1ee02717j

Recent Investigation of Na- and K-Ion Batteries

Narender Budhiraja and Santhosh Kumar Mahadevan

Abstract The escalating demand for renewable energy storage has driven profound investigation into the regime of battery technologies. On this perspective, sodium (Na)- and potassium (K)-ion batteries (NIB and KIB) have emerged as feasible and viable contenders to lithium-ion batteries (LIB), owing to their huge abundance, minimal expense, and analogous electrochemical properties. In the present study, an extensive scrutiny of the current status of KIB and NIB technology, encompassing electrode materials, electrolytes, and hybrid device architectures, has been elucidated. The electrochemical mechanisms underlying Na- and K-ion intercalation, conversion, and associated reactions are outlined, placing more attention on ion transport and electrode–electrolyte interfaces. With an emphasis on improving performance, safety, and scalability, specific energy and power density, cycle life, and stability aspects of NIB and KIB technologies have been discussed, along with their possible use in the defense sector, electric vehicles, medical equipment, and renewable energy storage. Lastly, future research and development prospects are described, underlining the necessity of ongoing innovation in device engineering, electrochemistry, and material science for fully exploiting the promise of Na- and K-ion batteries.

Keywords Renewable · Electrode · Sodium-ion batteries · Potassium-ion batteries · Defense sector

---

N. Budhiraja (✉)
Department of Physics, Satish Chander Dhawan Government College, Ludhiana, Punjab, India
e-mail: narenderarora1988@gmail.com

S. K. Mahadevan
Facultad de Ingeniería y Ciencias, Universidad Adolfo Ibanez, Vina del Mar, Valparaiso, Chile

R. K. Sonker et al. (eds.), *Low-Dimensional Multifunctional Hybrid Nanostructured Materials*, Engineering Materials, https://doi.org/10.1007/978-981-96-7952-2_5

1 Introduction

The rapid surge in energy demand increases the usage of non-renewable energy through burning fossil fuels and exploiting rare earth metals. Energy requirements in everyday life have always been increasing at an exponential rate since ancient times. This not only depletes finite resources but also contributes direct impact on environmental degradation and climate changes. To mitigate the impacts of global climate change, developed nations have proposed and committed to achieve carbon detachment by the end of twenty-first century. Therefore, thorough research into sustainable energy adaptation and storage technologies is necessary to congregate this goal. Currently, the energy market is dominated by lithium-ion batteries (LIBs) because of their extended life cycle and high energy density [1]. However, contemporary energy storage technologies such as LIBs and lead-acid batteries are insignificant enough to meet imminent demands. Due to the global energy crisis, it is imperative to foster and promote electrochemical energy storage devices that have long cycle spans and stability. LIBs have played a major mainstay in the sectors of electric vehicles and portable devices over the last few decades. For the past 35 years, LIBs have dominated the market for portable electronics since the Sony and Toshiba Company first commercialized these batteries in 1991, utilizing graphite as the inorganic negative electrode [2]. Even though lithium-ion battery technology is very advanced today, there are still a lot of issues with these batteries, including low-temperature performance, limited life, inadequate safety, and expensive cost [3–5]. Furthermore, the growing demand for lithium resources has resulted in price hikes because of the deprived relative abundance and heterogeneous distribution of lithium-ion nature. Their poor safety and expensive cost constrain their use in substantial energy storage systems. Presently available sustainably grown electrode materials for lithium-ion batteries are challenging to use in colossal electrical energy storage applications due to their high cost and resource constraints. Effective and affordable energy storage technologies have been developed continually to build a sustainable "energy web" due to the increasing demand for energy storage in transportation, renewable energy, and the numerous electronic gadgets [6]. With the rise of digital technologies and electronic gadgets, our contemporary society is currently witnessing the so-called fourth industrial revolution. Thus, it is imperative to create new energy storage systems that go well beyond LIBs. Figure 1 represents a diagrammatic representation of the "energy web." As a result, finding an LIB substitute has become a daunting endeavor. Potentially inexpensive alkali metal-ion devices, such as K-ion batteries (KIBs) and Na-ion batteries (NIBs) [7, 8], have received attention recently. Sodium-ion batteries can satisfy extensive grid energy storage requirements because of the profusion of sodium and the comparable electrochemical characteristics of lithium. Sodium-ion battery electrode materials are more widely available and more reasonably priced than lithium-ion battery electrode materials. However, sodium has a relatively high standard potential (−2.72 V vs. standard hydrogen electrode) compared to lithium (−3.05 V vs. standard hydrogen electrode), NIBs have a lower energy density than LIB. Because of their natural plenty and appropriate

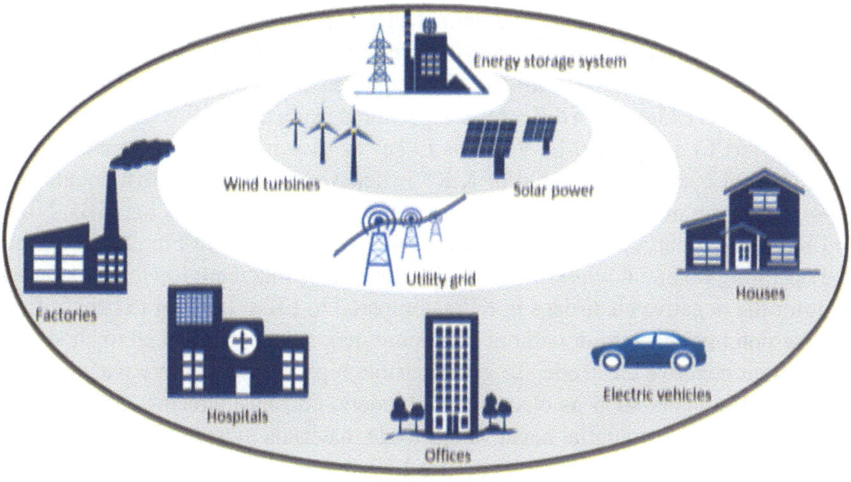

Fig. 1 A diagrammatic representation of the "energy web" [6]. Reproduced with permission @ copyright (2018) Springer

potential (−2.92 V vs. SHE) in comparison to lithium-ion battery, potassium-ion battery (KIBs) is another capable option for low-cost applications. The earth's crust contains 2.09 weight percent (wt%) potassium [9]. It is approximately 900 times more abundant than lithium. The comparatively low voltage of the K^+:K redox pair is an additional benefit [9, 10]. In contrast to Li^+:Li and Na^+:Na, the standard voltage of K^+:K in propylene carbonate is lower. The sodium-ion battery industry has shown a variety of suitable electrode materials in recent years.

This chapter depicted the state of research on KIBs and NIBs, covering cathode and anode materials, electrolytes, binders, prospects, and upcoming difficulties. First, a summary of the cathode and anode materials is provided based on the various types of reactions. Second, there is a discussion of the use and impact of binders and electrolytes for KIBs and NIBs, respectively. Furthermore, the primary future challenges and a few solutions are also suggested to support the advancement of both KIBs and NIBs.

2 Potassium-Ion Batteries (KIBs)

KIBs have the potential to combine the beneficial qualities of LIB's and NIB's technology without having their respective disadvantages. Gigantic concern in potassium-ion batteries employing non-aqueous electrolytes is now rapidly increasing. The primary cathode and anode resources for usual KIB with non-aqueous electrolytes are covered in this section, along with a research outlook for the future. Propylene carbonate along with ethylene carbonate is an example of carbonate ester solvent in which potassium has a smaller standard electrode potential (SEP) than $Li:Li^+$. The cyclic voltammetry of a Ni-lattice-ascribed material in KPF_6:EC solution in the standard potential window of −0.5 to +0.5 V is displayed in Fig. 2a, which refers to $Li:Li^+$ electrode that is isolated in a different compartment using a glass filter. K-anodization currents start to flow at −0.15 V versus $Li:Li^+$ as shown in aforementioned figure. The standard operating potential of $K:K^+$ is determined by the Nernst equation to be −0.12 V versus $Li:Li^+$. The operating potential stability ranges for LIBs, NIBs, and KIBs using carbonate ester-based electrolytes are shown in Fig. 2b. The negative operating potential window of $E^0(K:K^+)$ indicates that there is a lower potential limit for the negative electrodes in KIBs compared to LIBs. In order to benefit from the low-potential constraint without potassium deposition, it is crucial to choose an appropriate negative electrode. To get a feasible high-voltage battery performance, battery components such as electrolytes, binders, current collectors, and positive significant and low-potential negative electrode materials must be developed.

A comparison of the possible zones for LIB versus NIB versus KIB is shown in Fig. 2b. In disparity to the Li- and Na-ascribed systems 4.50 and 4.21 V, respectively, KIBs exhibit the broadest voltage window (4.60 V). The schematic representation of the potassium-ion battery [3] is depicted in Fig. 3a. KIB uses a "rocking chair" approach to charge and discharge by allowing K ions to migrate across electrodes. Figure 3 shows the components of KIBs that control charge/discharge and storage. KIB's redox reaction involves K-ion intercalation/de-intercalation occurring at the anode (negative electrode) and cathode (positive electrode). The separator, a porous membrane submerged in the electrolyte, connects both electrodes through ions. This separator ensures utter ionic transport between the anode and cathode while also averting short-circuiting between electrodes. During charging, the cathode's K atoms release electrons to create K ions, which flow to the anode through the separator and electrolyte [11]. When K ions arrive at the anode side, they join with electrons from the external circuit. The charging process converts electrical energy from a peripheral power source into chemical energy while during discharge, the aforementioned occurrences are reversed. When an outer load drains a battery array, K ions flow through the electrolyte from the anode to the cathode. At the same time, electrons relocate from the anode to the external load [12].

The KIB technology must fulfill a few fundamental parameters in order to store energy through the aforementioned processes. For example, to store and liberate K^+ ions reversibly, the cathode material has to contain appropriate lattice sites or gaps. In order to offer strong cycle stability and reversible capacity, the cathode electrode should also have a sturdy crystal structure and a large number of active spots. Another highly advantageous quality for a cathode material is a high and broad operating potential [14]. However, it is advantageous for the anode to have a lower working potential to generate a giant voltage and better energy. To have a vast specific capacity, the anode material must also have a large number of open sites to accord enough K ions. Figure 3b illustrates theoretical capacity of anode materials of various alloy-ascribed batteries. Anode material must also have a robust structure to withstand volume fluctuations during the reformable K-ion intercalation/de-intercalation mechanism to exhibit high cycling performance. Therefore, before

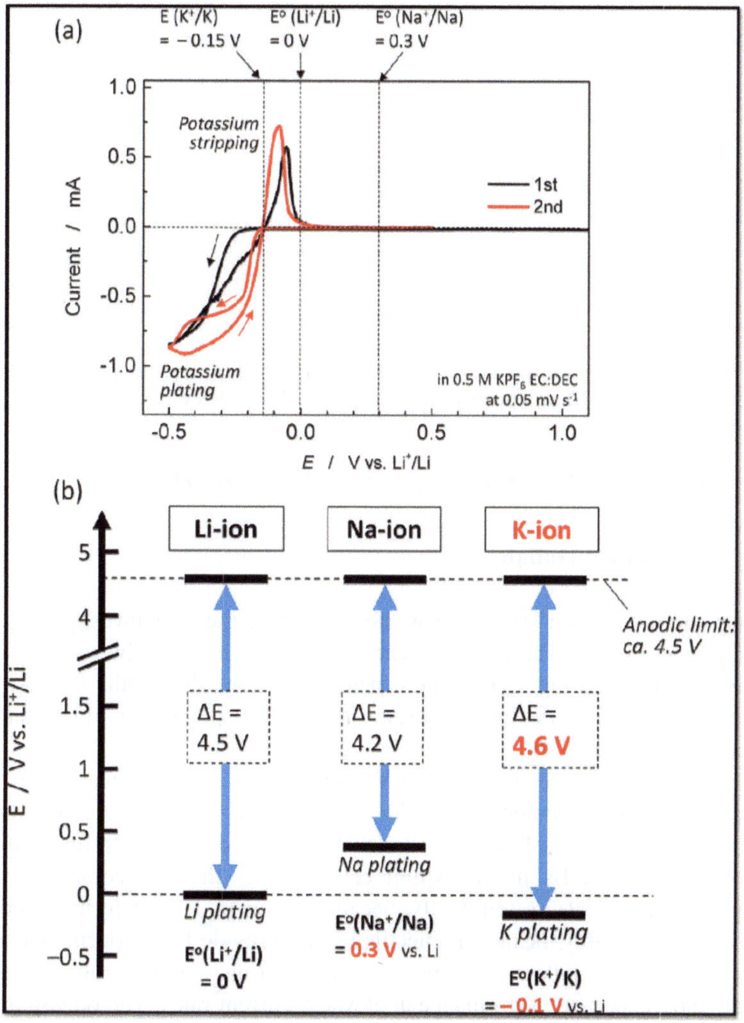

Fig. 2 **a** Ni electrode cyclic voltammetry [9], reproduced with permission @ copyright 2015 Elsevier. **b** Li-, Na-, and K-based battery potential windows, implying immiscible electrolyte solutions of carbonate esters as solvents [10]. Reproduced with permission @ copyright (2019) ACS

battery construction, it is crucial to comprehend the characteristics and challenges of each anode and cathode material [15].

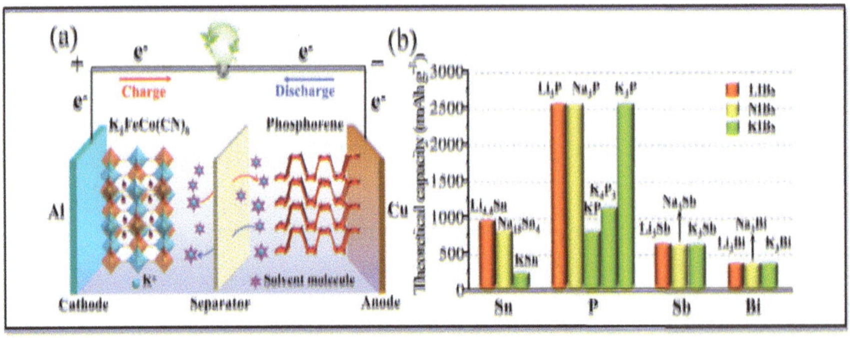

Fig. 3 **a** Schematic representation of K-ion batteries, **b** theoretical capacity of anode material of batteries [13]. Reproduced with permission @ copyright (2019) Wiley

2.1 Cathode Material for Potassium-Ion Batteries

2.1.1 Hexacyanometallates

Prussian blue, $KFe^{3+}Fe^{2+}(CN)_6$, was the subject of the first investigation into a cathode material for KIBs in 2004. The massive K^+ ion may be reabsorbed and released without causing significant structural tensions due to the metal–organic framework's wide spacing. In actuality, even if all of the potassium is eliminated, the structure is still stable [16].

Furthermore, compared to LIBs (3.1 V) and NIBs (3.2 V), hexacyanoferrate cathodes have a functional potential of 3.70 V against K:K^+ in KIBs. According to the conventional potentials of K/K^+, Prussian blue cathodes are therefore intriguing examples of how KIBs might exhibit cell voltages that are comparable to or even greater than LIBs. Additionally, because a pair of K^+ per formula unit may be traversed reversibly, hexacyanometallates can provide large specific capacities. Lastly, hexacyanometallates also provide great thermal stability, robust long-term cycling, and peeped-up performance at elevated current rates. The release of interstitial water in hexacyanoferrates during electrochemical cycling is a disadvantage that can result in some adverse reactions, safety issues, and detrimental impacts on the electrochemical efficiency of non-aqueous cells. Furthermore, hexacyanometallates also typically have a poor density of material (1.85 g cm^{-3} at 20 °C), which results in lower volumetric energy [17, 18]. The latter may result in greater costs as thicker hybrid electrodes are needed for a particular battery's power retention capacity, which raises the price of the electrolyte. Throughout the charge/discharge operation, the interstitial water content may be released. It would result in a number of adverse reactions, a decline in electrochemical performance, and potential safety issues with KIBs based on organic electrolytes [19, 20]. The quantity of K ions that may be transferred between the electrodes was also found to be reduced by high water content in the interstitial region. Hexacyanometallates are a promising contender for

KIB technology in the future, but numerous issues still need to be resolved before this material can provide the best and most practical full-cell performance.

2.1.2 Layered Oxides

Potential cathode materials for KIBs include layered metal oxides. Layered oxides are extensively studied for KIBs and contain the general formula Z_xMO_2 (Z = K, Li, Na, etc.); M represents metals such as Fe, Ni, and Co. The potassium layered oxides are superior to their sodium and lithium equivalents in several ways. Compared to their lithium-based counterparts, potassium-based layered oxides have several advantages, such as a lower propensity for spinel production [15, 21]. For example, during the delithiation process, the transition metal ions in many Li-ascribed layered oxide electrodes migrate to the Li layer, changing the crystal structure permanently and thus causing the energy density and capacity to fade. The rapid displacement of Mn in the Li sites in $LiMnO_2$ causes the layered structure to nearly completely change into a spinel structure. The larger ionic size of Na makes its electrodes more energy demanding than Li-ascribed electrodes. However, K-based electrodes are even more challenging due to their large radius compared to Na. Furthermore, in comparison to their Na and Li counterparts, the wider K slab spacing creates incredibly deformed voids in the alkali metal layer, which results in far less advantageous transition metal migration. There are still several issues that must be resolved to satisfy the particular specifications when using layered oxide compounds as cathode materials in real-world KIBs. Meticulously K^+–K^+ repulsion inside K^+ layer is caused by the greater K ion size, which is the main obstacle. Due to the capricious ionic radius of the Li^+ (0.75 Å), Na^+ (1.03 Å), and K^+ (1.37 Å) ions [22], discrepancies are found for comparable oxides. Furthermore, during charge, irreversible structural amendment at low potassium concentrations is more liable to happen, which results in decreased cycle stability. Actually, it has been found that only a small number of potassium layered oxides provide reversible cycling over 4.02 V versus $K{:}K^+$. Furthermore, stacked potassium-based oxides' potential profiles had more features (steps) and larger voltage hysteresis, suggesting a greater number of transitional stages. In conclusion, stacked oxides face a number of obstacles, including limited specific capacity, and low discharge potential. Furthermore, the K-based layered compounds are more vulnerable to water insertion or oxidation due to the wider inter-slab gap, necessitating the use of an inert environment for handling and storage. Yang Xu reported a graphene oxide/layered potassium titanium niobate nanohybrid material as a K-battery anode [23]. The potassium storage capability of KTNO/rGO was evaluated; it demonstrated a foremost charge capacity of 127.0 mAh/g and maintained 76.2% over 500 cycles at 20 mAh/g [24]. Examining the intercalation procedure revealed a material that might have less strain with both niobium and titanium redox activities assisting in the charge storage. Higher capillary energy is promised by the relatively high density of the material (approximately 4.0 g cm^{-3}), which explains the interest in this class of materials that needs more investigation.

2.1.3 Polyanionic Compound

Initially, polyanionic substances like phosphates (such as amorphous $FePO_4$ or sulfates ($KFeSO_4F$) were analyzed for their higher performance in KIB technology [25]. Due to the intense bonds and inducible action of the anionic groups, this material category is of keen interest because of its great thermal resiliency and trivial oxygen loss at elevated temperatures. Specifically, compared to the equivalent lithium (3.30 V against $Li:Li^+$) and sodium (3.40 V against $Na:Na^+$) compounds, $KFeSO_4F$ demonstrated a greater working potential value (3.60 V against $K:K^+$). However, there was a decrease in the number of alkali correspondents and a rise in potential hysteresis (approximately 0.60 V for K, 0.30 V for Na, and 0.40 V against Li metal). With a standard working potential of 3.5 V against $K:K^+$, $K_3V_2(PO_4)_3$:C nanohybrid material demonstrated an astonishing capacity sustainability of 96% subsequently over 100 cycles and provided capacities of about 50 mAh/g [26].

In conclusion, polyanionic compounds have been found to have favorable electrochemical performance along with elevated average operational potentials over 4.0 V. However, because of the hefty anion (XO_4) structure, these materials usually exhibit more porous material densities (2.6–3.6 g/cm^3) and frequently relatively poor specific capacities. Figure 4 provides a basic comparison of the benefits and drawbacks of the studied cathodes for KIBs as well as suggestions for resolving some of the current issues.

2.1.4 Organic Electrodes

Organic compounds have recently drawn attention as possible cathode materials for high-energy battery development because of their promising qualities, which include affordability, environmental friendliness, recyclability, etc. The growing emergence of redox-active organic substances as viable electrode materials for rechargeable batteries is ecstatic. Organic electrodes can achieve high gravimetric energy density and stability during long cycles in rechargeable batteries in addition to being structurally manipulable, plentiful on Earth, and environmentally benign. Choosing high energy density electrodes should start with organic materials that have redox centers of the appropriate potential. In the quest for high energy density organic electrodes, choosing light organic molecules with the capacity to use multiple electrons per unit formula is equally crucial. Moreover, the intrinsically lower rate capability of organic-based electrodes is a result of the low electronic conductivity of organic materials, which generally have relatively wide band gaps [6]. The majority of organic electrodes are heterocyclic semiconductors, which are classified as either n-type or p-type based on the various transport carriers (holes and electrons) found in organic solids. Their drawbacks and possible solutions are elucidated in Fig. 4. The specific energy, high efficiency, and cycle stability for the electrochemical performance are the main elements that should be taken into account as electrodes when developing a battery system with organic-based materials.

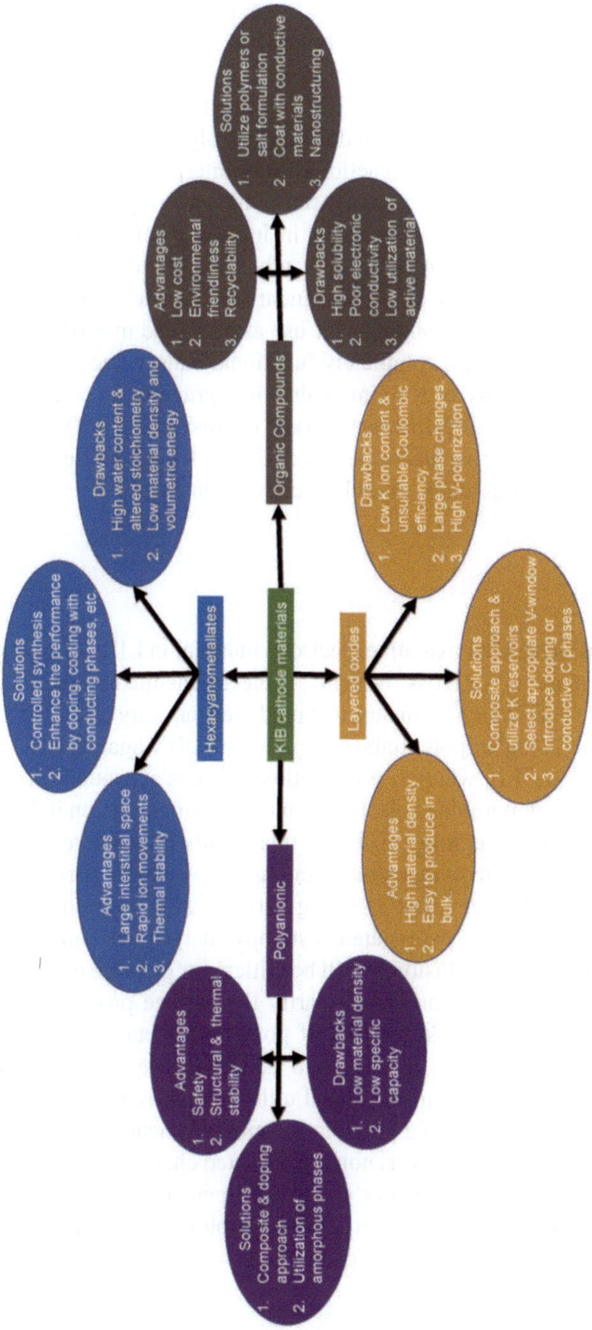

Fig. 4 Benefits and downsides of various cathode systems for KIBs, along with some potential solutions [3]. Reproduced with permission @ copyright (2020) Wiley

2.2 *Anode Materials for K-Ion Batteries*

The K-metal-ascribed electrode was initially selected as anode material for potent-energy KIB applications because of its profound and vast significant theoretical potential. However, due to the asymmetric electron density distribution caused by the K-metal anode, dendritic development and solid electrolyte interphase layer rupture may result in additional secondary reactions. New interphase layers are continuously formed on the electrode surface as a result of the electrolyte interphase layer's ongoing rupture. This procedure may result in increased interfacial polarization, excessive electrolyte consumption, and, ultimately, capacity loss. At this stage, the potassium metal's high reactivity (and consequent safety issues) and the aforementioned disadvantages render it unsuitable for use as an anode in KIB technologies. Therefore, in order to generate a completely functional battery, the appropriate host material is needed. The ability of K to intercalate into graphite, in contrast to Na, is a recent anode-side breakthrough that has sparked interest in KIBs. This section will cover research on KIB negative electrode materials, such as graphite, metal alloys, organics, intercalation compounds, as well as conversion anodes.

2.2.1 Graphite

Graphite is the most advanced oxidative electrode material in LIBs as well as viable negative electrode material for KIBs. When K^+ is electrochemically intercalated into graphite, KC_8 is created, which resonates to a predicted capacity of 280 mAh/g. Depotassiation happens at higher potentials (pertaining to 0.30 V against K:K^+ for K vs. 0.1 V against Li:Li^+ for Li), whereas potassiation happens at around 0.2 V against K:K^+ or at potentials comparable to electrochemical lithium insertion in LIBs (0.1 V against Li:Li^+) [27, 28]. In contrast to LIBs, this results in an increase in voltage hysteresis. However, graphite demonstrated an impressive power performance with a little capacity drop when exposed to 4 A/g current, which is comparable to its characteristic of Li-ion cells, if potassiated at stumpy charge rates and depotassiated at elevated charge rates. In conclusion, it will be critical for graphite to improve long-term charge–discharge rate performance, clarify the storage process, and, finally, investigate and eliminate possible hazards of potassium metal arranging at small potentials and higher charge rates, as this could lead to intense safety vulnerabilities and risk factors that would prevent the actual utilization of KIBs. Moreover, when the steady state of the plateau during potassiation moves beneath the lower cut-off, a comparatively poor rate capability was noted at elevated charge rates. In comparison to graphite, hard carbons exhibit better charge rate performance and much better prolonged cycling performance. As compared to graphite, the capacities measured at 28 mA/g were inferior, i.e., 273 mAh/g [29, 30].

2.2.2 Alloying and Conversion Materials

The incorporation of potassium compound into antimony–potassium (Sb:K) composite in a weight ratio of 70:30 was demonstrated by McCulloch et al. in 2015 [31]. About 435 mAh g^{-1} of capacity was supplied by the substance. A phosphorus-graphite synergistic material was reported to structure the alloy having specific capacity of 842 mAh/g. In addition to alloying materials, conversion materials, such as Co_3O_4–Fe_2O_3:C, were described. Conversion reactions, or the synthesis of K_2O and the corresponding metals, resulted in capacities of around 351 mAh/g at a standard depotassiation potential of 1.3 V against K:K^+. Additionally, it was reported that sulfide–carbon composites exhibited reformulation (CoS:C) or convergence alloying reactions (SnS_2:C and Sb_2S_3:C). CoS:C exhibited depotassiation capacities of 554 mAh/g at a functional potential of around 1.60 V against K:K^+, SnS_2 displayed nearly 311 mAh/g at 1.1 V, and Sb_2S_3:C had almost 555 mAh/g at about 1.2 V [32]. In Fig. 5, various pros and cons of respective anode materials are illustrated along with innovative solutions.

In Fig. 6, outline of materials for specific K-ion batteries' (a) positive and (b) negative electrode materials with their capacity against operating potential is demonstrated. Several material classes, including layered transition metal oxide substances, polyanionic molecules, and organic compounds, can be used to classify the cited positive electrode materials (Fig. 6a). The most exciting potential positive electrode materials for LIBs and NIBs are layered transition metal oxides, which is why potassium encrusted oxides have drawn interest from both the scientific community and the practical community for KIB applications.

The proposed electrode materials for the negative electrodes for KIBs (Fig. 6b) fall into the following categories: oxides, chalcogens, carbonaceous substances, K-alloying materials, multi/polyanionic substances, and organic materials. Due to their limited operating potential and adequately big capacity, graphite materials are extensively used in LIBs and have been technically refined to improve their performance which are viable choices as negative electrode materials for KIBs.

3 Sodium-Ion Batteries

Na-ion battery's electrode materials differ enough from their Li-ion rivals in regard to dynamics and electrochemistry. The French research organization CNRS CEA reports an 18,650-size cell as the first commercial Na-ion battery. The digit 18,650 refers to the size of a cylindrical battery encapsulated in the form of a metal container having dimensions 18 mm in diameter and 65 mm in height. This Na-ion battery having specific energy of 90 Wh/kg has been cycled over 1000 times [31, 33].

The sodium-ion battery schematic is shown in Fig. 7. O_3 and P_2 crystalline forms of layered TMO such as $NaYO_2$ are available. Na is found in octahedral positions in the O_3-$NaYO_2$ phase, while inside the P_2-phase prismatic sites include Na. Advanced NIBs with greater energy and power densities, improved operational life span and

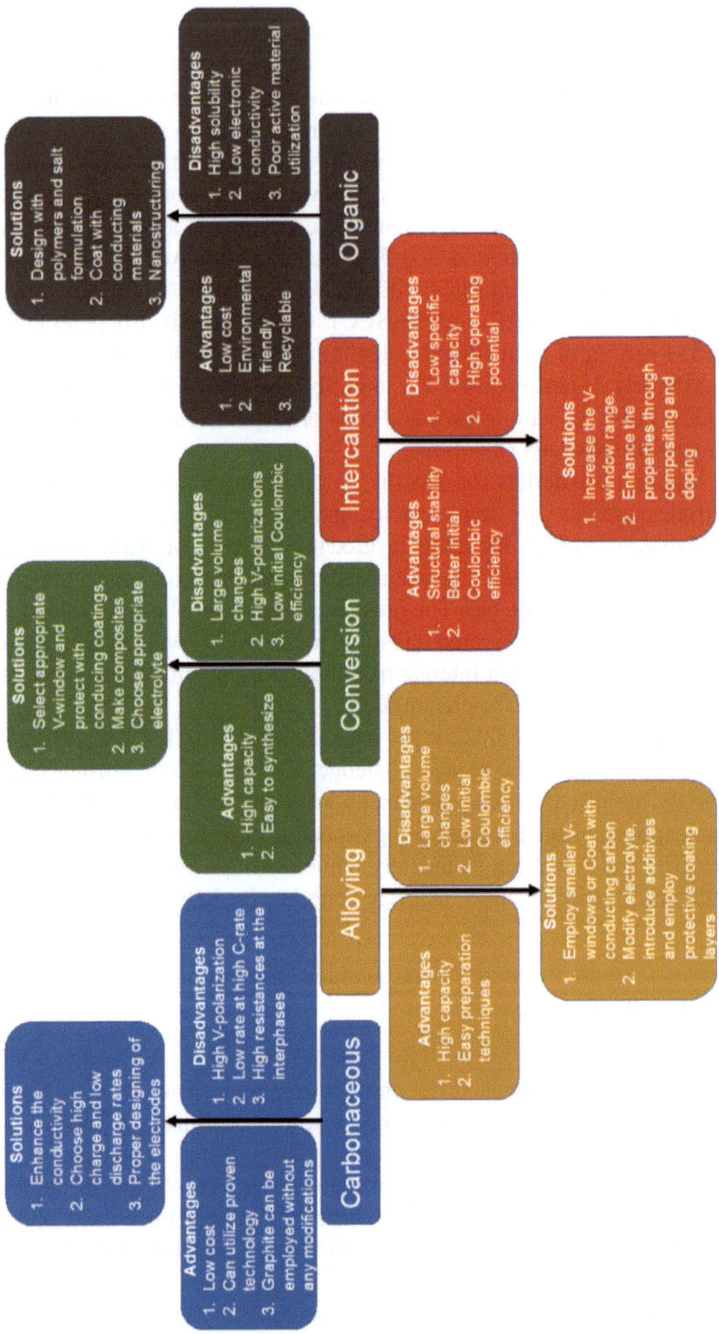

Fig. 5 The pros and cons of various anodes for KIBs, as well as some suggested methods to enhance the electrochemical characteristics [3]. Reproduced with permission @ copyright (2020) Wiley

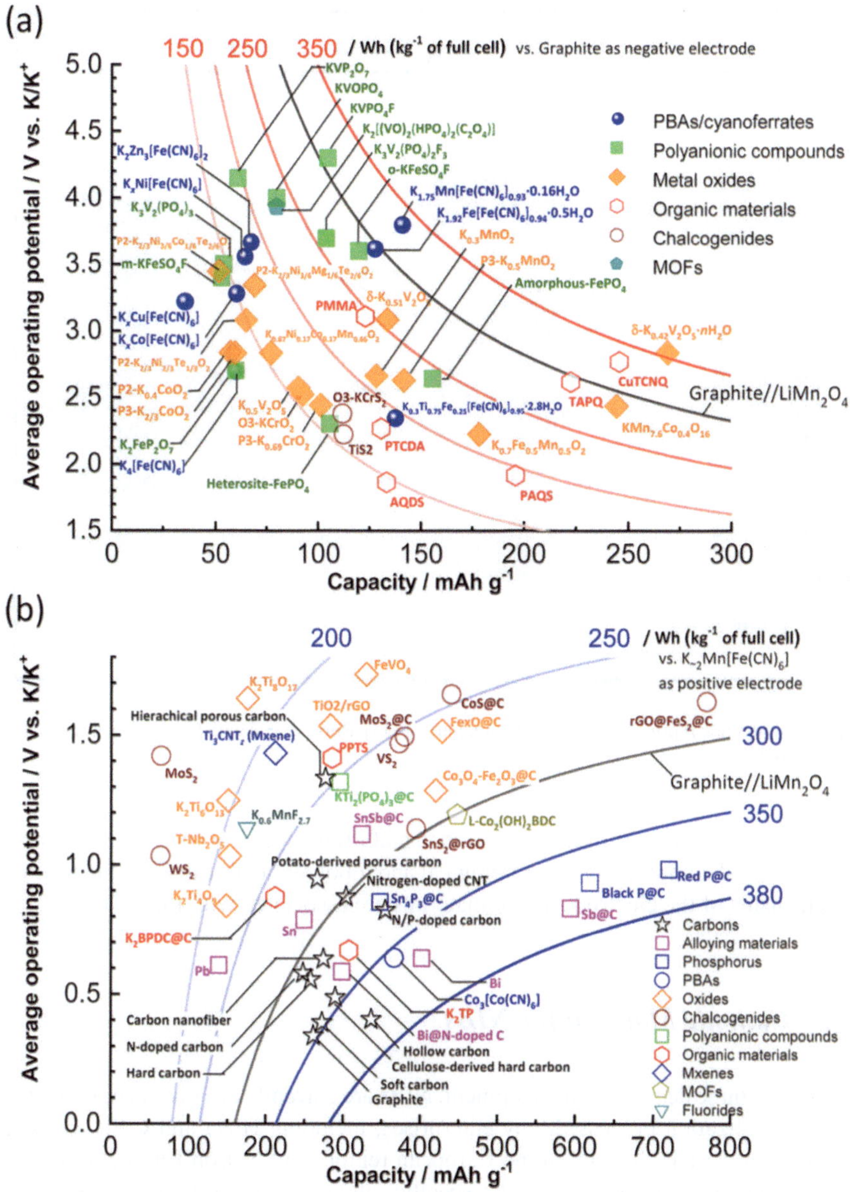

Fig. 6 Overview of materials for specific K-ion batteries' **a** positive and **b** negative electrode materials with their capacity versus operating potential [9, 10]. Reproduced with permission @ copyright (2020) ACS

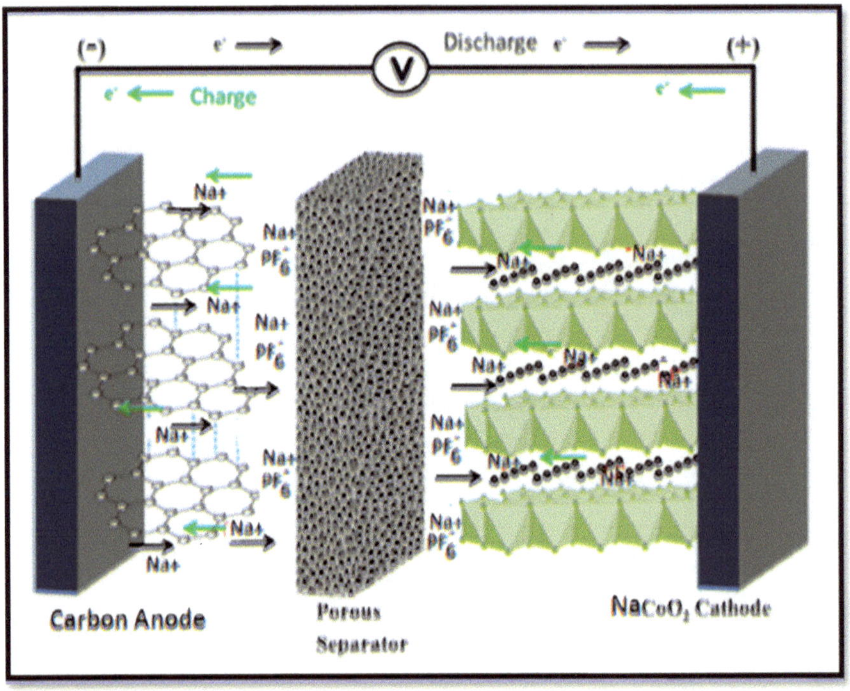

Fig. 7 Schematic representation of NIB's technology [4]. Reproduced with permission @ copyright (2020) ACS

reliability, and reduced costs are required to make their practical application possible. The chemical composition, morphological and topographical structure, and reaction mechanisms of the electrode active materials are associated with these factors.

3.1 *Cathode Material for NIBs*

Following three decades of development, NIBs are currently at a pivotal point in their commercialization process. Many enterprises, including HiNa and CATL in Asia, Faradion in Britain, Tiamat in the European region, and Natron Energy in the US, are commercializing NIBs with the specific intent of using Prussian white, vanadium phosphate, or sodium layered transition metal oxides (Na_xTMO_2) as cathode materials [34, 35]. In addition to Na_xTMO_2, Prussian blue/white, phosphates, and organic compounds are other interesting sodium-ion cathode materials. However, low specific capacity, low capacity retention, and inherent water/potential toxicity are the drawbacks of these sodium-ion cathode types. Na_xTMO_2-type materials experience irreversible phase changes, electrolyte breakdown, and high reactivity toward moisture; hence, the "optimal choice" for the positive active material is still contentious

[15, 36]. To achieve the best cathodes for commercial NIBs, a thorough assessment of Na_xTMO_2 chemical composition, electrochemical performance, and synthesis cost is essential. Regarding the battery electrolyte, NIBs must meet the same specifications as LIBs: high ion conductivity, inertness to chemicals, interfacial stability, redox stability, thermal stability, a wide potential window, ease of synthesis, minimal toxicity, and affordability. In most cases, an electrolyte consisting of dissociated sodium salts facilitates Na-ion transport between the two electrodes. The sodium salts $NaPF_6$ and $NaClO_4$ are frequently utilized for this purpose. In the electrolyte system of NIBs, $NaClO_4$ or $NaPF_6$ disaggregate in a concoction of binary or ternary carbonates was frequently utilized [37]. High redox potential, giant theoretical explicit capacity, enough ion diffusion channels to guarantee quick Na-ion incorporation and excavation, high electrochemical response activity and strong structural stability, ease of manufacture, plenty of resources, and environmental friendliness are all requirements for cathode materials, which are often sodium-intercalated compounds. Presently, layered TMOs, organic compounds, high charge density anionic compounds, and hexacyanoferrates have been identified as NIB cathodes and have made significant strides in enhancing electrochemical characteristics and clarifying the process of Na-ion storage [9, 38].

The cathode responses in a Na-particle battery using hard carbon (C_6) anode and metal oxide, $NaYO_2$ cathode are portrayed in Eq. (1):

$$Na_xC_6 + Na_{1-x}YO_2 \leftrightarrow NaYO_2 + C_6 \tag{1}$$

The valence states of the various metals are diminished and oxidized concurrently with cell discharge and charge that are evidenced in the additional voltage plateaus seen in mixed metal oxide cathodes of Na-ion cells. Metal phosphates and phosphate fluorides have been investigated as Na-ascribed cathodes in Na-ion batteries. Furthermore, hexacyanoferrates and its analogs, such as $KFe_2(CN)_6$ and $MnFe(CN)_6$, and $Fe_2(CN)_6$, have been explored as Na-ion battery cathodes. They exhibit reversible capacities of 82–121 mAh/g in sodium half-cells with voltages ranging from 3 to 3.5 V [39]. One disadvantage of Prussian blue and its equivalents is its low density (1.8 g/cm^3).

3.2 *Anode Material for Na-Ion Batteries*

3.2.1 Materials for Intercalation-Type Anodes

Ti-based oxides and carbonaceous compounds are the two categories of intercalation-type materials. Numerous materials have been thoroughly investigated as NIB anodes, including graphite, carbon nanotubes (CNTs), graphene, hard and soft carbon, and Ti-based compounds. Intercalation-type anodes' superior cycling capability, which is made possible by the host material's stable lattice structure and allows for reversible intercalation and de-intercalation of Na ions, is its most alluring property. Concurrent intercalation, incorporation of heteroatom, lattice defect disruption, morphological modification, nanostructure engineering, thermo-phase state modification, and hybrid composite fabrication have all been used in recent years to improve Na-ion storage capabilities.

Most experiments have used hard carbon anodes, although alternative Na intercalating hybrid materials, specifically $Na_2Ti_3O_7$, $Na_3Ti_2(PO_4)_3$, and Na-ascribed polyalloys, have also been studied. A hard carbon anode has a capacity of 250 mAh/g at a standard voltage of 0.25 V [40, 41]. Previously referred to as non-graphitizable carbon, hard carbon is made via the thermal decomposition of materials and has flaws in graphene sheets as well as randomly arranged graphitic nanodomains. Hard carbon's disordered structure results in reduced conductivity. Because of the unique Na-ion hybrid storage procedure of hard carbon, glucose was thermally decomposed to create sodium-embedded carbon material having a huge reversible specific capacity (300 mAh g^{-1}) that is significantly much larger as compared to graphite and soft carbon [42, 43]. Researchers have focused more on hard carbon due to its low redox potential and large specific capacity for storing Na. Carbon obtained from oligomers and monomers, carbon derived from single polymer precursors, and carbon collected from biotic and natural organic resources are all included in the precursor selection of hard carbon. Particularly, performance of hard carbon can be effectively enhanced by heteroatom doping. When heteroatoms (N, B, S, or P) are added to the carbon lattice structure, imperfection sites are created that can ingest Na^+ and change the carbon morphology. It was discovered that nitrogen doping could control volume expansion linked to Na-ion insertion or extraction in addition to increasing porosity and adjusting flaws. The huge surface area of nanoparticles can also enhance rate performance. Therefore, nitrogen insertion and nano-crystallization are efficient methods for enhancing carbon-based materials' Na-ion storage capabilities.

Choosing the right electrolytes is crucial for making Na-ion batteries work well. To maintain the equilibrium between positive and negative electrodes during cycling, alkyl/aryl carbonate-based electrolytes including sodium salts (e.g., $NaPF_6$ and $NaClO_4$) are combined with modest quantities of additives [44].

Carbon-based materials are consistently the most popular option among potential anodes due to their high conductivity, low platform voltage for discharge and charging, affordability, and environmental friendliness [45]. Furthermore, carbon-based materials have a variety of customizable structures, which helps to design matrix structures and, ultimately, create stable composites in addition to allowing for the tweaking of their electrochemical performance. Graphite and non-graphite are currently the two primary categories of carbon-based materials. The results of density functional theory (DFT) verified that the full charge transport from alkali compound to carbon metal characterize the chemical connection between both. The twofold Na-intercalated graphite compound is transient because the energy liberated due to transfer of electrons from Na to C atoms is not enough, consequential in a low planar binding force of Na-ion. This makes it intricate for Na-ion to nest into a graphite matrix. In 2014, it was demonstrated that the graphite having a lattice distance of 4.21 Å may be used as an anode for NIBs [46, 47]. If ternary-graphite hybrid compounds are stable against oxidation, alkali ions can undergo intense solvation and intercalate into graphite, hence providing capacity. The pathway procedure of the hybridization reaction was thoroughly examined, and it was confirmed that, in contrast to a standard graphite intercalation reaction, interacted ions rather than charged species were injected into graphite and thus peeled off crystalline lamella instead of delamination. The reaction is extremely reversible, meaning that the charge capacity and potential distribution don't change for many cycles even with the significant volume growth. Graphite was a promising anode for NIBs due to the co-intercalation process, but it still had a high redox potential (0.60–0.80 V against Na^+:Na utilizing different ether-ascribed electrolytes), which will drastically reduce the whole cell's energy density [2].

3.2.2 Organic Material as Anode for NIB

Tiny organic molecules and organic polymers are the general categories for the organic materials investigated as NIB anodes. The domain spots of reversible reaction with Na ion, such as C–O (benzoquinones, carboxylates, and imides variants), C–N, and N–N (azo derivatives) bond compounds, can be used to identify small organic molecules. Significant sources, a flexible structure, a relatively high theoretical specific capacity, and predictable fabrication techniques are the benefits of organic electrode materials [32, 48]. Its low conductivity and high solubility in the electrolyte were drawbacks. Furthermore, rapid capacity deterioration may arise from the disintegration of particles brought on by a significant volume shift during the intercalation/de-intercalation of Na ions. Understanding the free radicals produced by the Na-ion insertion/extraction mechanism is essential to address these issues properly. The primary redox processes in sodium storage take place at unsaturated C–O, C–N, and N–N bonds [49–51]. These events result in the formation of an odd electron on the carbon atom and the reactive intermediate radical with an anionic charge on the N and O atoms. This type of intermediate is typically unstable and quickly transforms into inactive compounds through reactions with free radicals or other molecules in the electrolyte, which ultimately results in poor cycle performance and irreversibility of the redox process.

4 Conclusion

The technologies of KIBs and NIBs have shown giant promise as LIB substitutes, providing a potential solution for acute energy storage and sustainability problems. NIBs and KIBs have been thoroughly investigated as charge carriers in battery systems due to their abundance, affordability, and comparable electrochemical characteristics to lithium. Even though there are a few issues, such as the requirement for additional advancements in scalability, safety, and energy density. It will take ongoing innovation in electrochemistry and related device engineering to meet these concerns.

Notwithstanding these obstacles, KIBs and NIBs have a lot of potential for broad use in a range of applications, such as household appliances, tech gadgets, consumer electronics, mobile devices, electric cars, digital displays, medical equipments, and renewable energy storage. These batteries have the potential to be extremely important in the shift to a more sustainable and energy-efficient future as long as research and development continue in the aforementioned sector.

References

1. Kim, H., Kim, J.C., Bianchini, M., Seo, D.H., Rodriguez-Garcia, J., Ceder, G.: Recent progress and perspective in electrode materials for K-ion batteries. Adv. Energy Mater. **8**, 1–19 (2018). https://doi.org/10.1002/aenm.201702384
2. Hu, J., Hong, Y., Guo, M., Hu, Y., Tang, W., Xu, S., Jia, S., Wei, B., Liu, S., Fan, C., Zhang, Q.: Emerging organic electrodes for Na-ion and K-ion batteries. Energy Storage Mater. **56**, 267–299 (2023). https://doi.org/10.1016/j.ensm.2023.01.021
3. Rajagopalan, R., Tang, Y., Ji, X., Jia, C., Wang, H.: Advancements and challenges in potassium ion batteries: a comprehensive review. Adv. Funct. Mater. **30**, 1–35 (2020). https://doi.org/10.1002/adfm.201909486
4. Abraham, K.M.: How comparable are sodium-ion batteries to lithium-ion counterparts? ACS Energy Lett. **5**, 3544–3547 (2020). https://doi.org/10.1021/acsenergylett.0c02181
5. Vaalma, C., Buchholz, D., Passerini, S.: Non-aqueous potassium-ion batteries: a review. Curr. Opin. Electrochem. **9**, 41–48 (2018). https://doi.org/10.1016/j.coelec.2018.03.031
6. Lei, Y., Huang, W., Huang, Q., Zhang, A.: A novel polysiloxane elastomer based on reversible aluminum-carboxylate coordination. New J. Chem. **43**, 261–268 (2019). https://doi.org/10.1039/c8nj04761h
7. Li, Y., Yang, C., Zheng, F., Pan, Q., Liu, Y., Wang, G., Liu, T., Hu, J., Liu, M.: Design of TiO_2eC hierarchical tubular heterostructures for high performance potassium ion batteries. Nano Energy **59**, 582–590 (2019). https://doi.org/10.1016/j.nanoen.2019.03.002
8. Zuo, W., Innocenti, A., Zarrabeitia, M., Bresser, D., Yang, Y., Passerini, S.: Layered oxide cathodes for sodium-ion batteries: storage mechanism, electrochemistry, and techno-economics. Acc. Chem. Res. **56**, 284–296 (2023). https://doi.org/10.1021/acs.accounts.2c00690
9. Hosaka, T., Kubota, K., Hameed, A.S., Komaba, S.: Research development on K-ion batteries. Chem. Rev. **120**, 6358–6466 (2020). https://doi.org/10.1021/acs.chemrev.9b00463
10. Qiao, S., Zhou, Q., Ma, M., Liu, H.K., Dou, S.X., Chong, S.: Advanced anode materials for rechargeable sodium-ion batteries. ACS Nano **17**, 11220–11252 (2023). https://doi.org/10.1021/acsnano.3c02892
11. Liu, S., Kang, L., Zhang, J., Jun, S.C., Yamauchi, Y.: Carbonaceous anode materials for non-aqueous sodium- and potassium-ion hybrid capacitors. ACS Energy Lett. **6**, 4127–4154 (2021). https://doi.org/10.1021/acsenergylett.1c01855
12. Ghahari, A., Raissi, H.: Architectural design of anode materials for superior alkali-ion (Li/Na/K) batteries storage. Sci. Rep. **14**, 1–16 (2024). https://doi.org/10.1038/s41598-024-54214-6
13. Dhir, S., Wheeler, S., Capone, I., Pasta, M.: Outlook on K-Ion batteries. Chem. **6**, 2442–2460 (2020). https://doi.org/10.1016/j.chempr.2020.08.012
14. Jian, Z., Liang, Y., Pérez, I.A.R., Yao, Y., Ji, X.: Poly(anthraquinonyl sulfide) cathode for potassium-ion batteries. Electrochem. Commun. **71**, 5–8 (2016). https://doi.org/10.1016/j.elecom.2016.07.011
15. Bie, X., Kubota, K., Hosaka, T., Chihara, K., Komaba, S.: A novel K-ion battery: hexacyanoferrate(II)/graphite cell. J. Mater. Chem. A **5**, 4325–4330 (2017). https://doi.org/10.1039/c7ta00220c

16. Hwang, J.Y., Myung, S.T., Sun, Y.K.: Recent progress in rechargeable potassium batteries. Adv. Funct. Mater. **28**, 1–45 (2018). https://doi.org/10.1002/adfm.201802938
17. Bin, D.S., Lin, X.J., Sun, Y.G., Xu, Y.S., Zhang, K., Cao, A.M., Wan, L.J.: Engineering hollow carbon architecture for high-performance K-ion battery anode. J. Am. Chem. Soc. **140**, 7127–7134 (2018). https://doi.org/10.1021/jacs.8b02178
18. Hong, S.Y., Kim, Y., Park, Y., Choi, A., Choi, N.S., Lee, K.T.: Charge carriers in rechargeable batteries: Na ions vs. Li ions. Energy Environ. Sci. **6**, 2067–2081 (2013). https://doi.org/10.1039/c3ee40811f
19. Adams, R.A., Varma, A., Pol, V.G.: Temperature Dependent Electrochemical Performance of Graphite Anodes for K-Ion and Li-Ion Batteries, pp. 1–31. Davidson School of Chemical Engineering, Purdue University, West Lafayette, IN, USA
20. Nason, C.A.F., Vijaya Kumar Saroja, A.P., Lu, Y., Wei, R., Han, Y., Xu, Y.: Layered potassium titanium niobate/reduced graphene oxide nanocomposite as a potassium-ion battery anode. Nano-Micro Lett. **16**, Article 59 (2024). https://doi.org/10.1007/s40820-023-01222-2.
21. Olsson, E., Yu, J., Zhang, H., Cheng, H.M., Cai, Q.: Atomic-scale design of anode materials for alkali metal (Li/Na/K)-ion batteries: progress and perspectives. Adv. Energy Mater. **12**, 2200662 (2022). https://doi.org/10.1002/aenm.202200662
22. Xu, T., Sun, W., Kong, T., Zhou, J., Qian, Y.: Stable graphite interface for potassium ion battery achieving ultralong cycling performance. Acta Phys.-Chim. Sin. **40**, 1–8 (2024). https://doi.org/10.3866/PKU.WHXB202303021.
23. Shin, K., Zheng, Y., Zhang, F., Wu, S., Tang, Y.: Facile ion-exchange strategy for Na^+/K^+ hybrid-ion batteries with superior rate capability and cycling performance. ACS Appl. Energy Mater. **3**, 7030–7038 (2020). https://doi.org/10.1021/acsaem.0c01060
24. Zhang, W., Sun, P., Sun, S.: A theoretical method to predict novel organic electrode materials for Na-ion batteries. Comput. Mater. Sci. **134**, 42–47 (2017). https://doi.org/10.1016/j.commatsci.2017.03.031
25. Zhou, Y., Yin, L., Xiang, S., Yu, S., Johnson, H.M., Wang, S., Yin, J., Zhao, J., Luo, Y., Chu, P.K.: Unleashing the potential of MXene-based flexible materials for high-performance energy storage devices. Adv. Sci. **11**(3), 2304874 (2024)
26. Min, X., Xiao, J., Fang, M., Wang, W., Zhao, Y., Liu, Y., Abdelkader, A.M., Xi, K., Kumar, R.V., Huang, Z.: Potassium-ion batteries: outlook on present and future technologies. Energy Environ. Sci. **14**, 2186–2243 (2021). https://doi.org/10.1039/d0ee02917c
27. Lee, W., Kim, J., Yun, S., Choi, W., Kim, H., Yoon, W.S.: Multiscale factors in designing alkali-ion (Li, Na, and K) transition metal inorganic compounds for next-generation rechargeable batteries. Energy Environ. Sci. **13**, 4406–4449 (2020). https://doi.org/10.1039/d0ee01277g
28. Luo, W., Wan, J., Ozdemir, B., Bao, W., Chen, Y., Dai, J., Lin, H., Xu, Y., Gu, F., Barone, V., Hu, L.: Potassium ion batteries with graphitic materials. Nano Lett. **15**, 7671–7677 (2015). https://doi.org/10.1021/acs.nanolett.5b03667
29. Kim, H., Seo, D.H., Bianchini, M., Clément, R.J., Kim, H., Kim, J.C., Tian, Y., Shi, T., Yoon, W.S., Ceder, G.: A new strategy for high-voltage cathodes for K-ion batteries, stoichiometric $KVPO_4F$. Adv. Energy Mater. **8**, 1–33 (2018). https://doi.org/10.1002/aenm.201801591
30. McCulloch, W.D., Ren, X., Yu, M., Huang, Z., Wu, Y.: Potassium-ion oxygen battery based on a high capacity antimony anode. ACS Appl. Mater. Interfaces **7**(45), 26158–26166 (2015). https://doi.org/10.1021/acsami.5b08037
31. Chayambuka, K., Mulder, G., Danilov, D.L., Notten, P.H.L.: From Li-ion batteries toward Na-ion chemistries: challenges and opportunities. Adv. Energy Mater. **10**(32), 2001310 (2020). https://doi.org/10.1002/aenm.202001310
32. Dahbi, M., Yabuuchi, N., Kubota, K., Tokiwa, K., Komaba, S.: Negative electrodes for Na-ion batteries. Phys. Chem. Chem. Phys. **16**(29), 15007–15028 (2014). https://doi.org/10.1039/c4cp00826j
33. Goikolea, E., Palomares, V., Wang, S., de Larramendi, I.R., Guo, X., Wang, G., Rojo, T.: Na-ion batteries—approaching old and new challenges. Adv. Energy Mater. **10**(34), 2002055 (2020). https://doi.org/10.1002/aenm.202002055

34. Palomares, V., Serras, P., Villaluenga, I., Hueso, K.B., Carretero-González, J., Rojo, T.: Na-ion batteries, recent advances and present challenges to become low cost energy storage systems. Energy Environ. Sci. **5**(3), 5884–5901 (2012). https://doi.org/10.1039/c2ee02781j
35. Wu, Y., Shuang, W., Wang, Y., Chen, F., Tang, S., Wu, X.L., Bai, Z., Yang, L., Zhang, J.: Recent Progress in Sodium-Ion Batteries: Advanced Materials, Reaction Mechanisms and Energy Applications. Springer Nature Singapore, Singapore (2024)
36. Ponrouch, A., Marchante, E., Courty, M., Tarascon, J.M., Palacín, M.R.: In search of an optimized electrolyte for Na-ion batteries. Energy Environ. Sci. **5**(9), 8572–8583 (2012). https://doi.org/10.1039/c2ee22258b
37. Zhao, L.F., Hu, Z., Lai, W.H., Tao, Y., Peng, J., Miao, Z.C., Wang, Y.X., Chou, S.L., Liu, H.K., Dou, S.X.: Hard carbon anodes: fundamental understanding and commercial perspectives for Na-ion batteries beyond Li-ion and K-ion counterparts. Adv. Energy Mater. **11**(5), 2002704 (2021). https://doi.org/10.1002/aenm.202002704
38. Liang, H.J., Hou, B.H., Li, W.H., Ning, Q.L., Yang, X., Gu, Z.Y., Nie, X.J., Wang, G., Wu, X.L.: Staging Na/K-ion de-/intercalation of graphite retrieved from spent Li-ion batteries: in operando X-ray diffraction studies and an advanced anode material for Na/K-ion batteries. Energy Environ. Sci. **12**(11), 3575–3584 (2019). https://doi.org/10.1039/c9ee02759a
39. Loaiza, L.C., Monconduit, L., Seznec, V.: Si- and Ge-based anode materials for Li-, Na-, and K-ion batteries: a perspective from structure to electrochemical mechanism. Small **16**(23), 1905260 (2020). https://doi.org/10.1002/smll.201905260
40. Kulova, T.L., Gavrilin, I.M., Skundin, A.M., Kovtushenko, E.V., Kudryashova, Y.O.: New electrochemical systems for sodium-ion batteries. Russ. J. Phys. Chem. A **98**(4), 771–776 (2024). https://doi.org/10.1134/S0036024424040150
41. Luo, W., Shen, F., Bommier, C., Zhu, H., Ji, X., Hu, L.: Na-ion battery anodes: materials and electrochemistry. Acc. Chem. Res. **49**(2), 231–240 (2016). https://doi.org/10.1021/acs.accounts.5b00482
42. Massé, R.C., Uchaker, E., Cao, G.: Beyond Li-ion: electrode materials for sodium- and magnesium-ion batteries. Sci. China Mater. **58**(9), 715–766 (2015). https://doi.org/10.1007/s40843-015-0084-8
43. Kubota, K., Komaba, S.: Review—practical issues and future perspective for Na-ion batteries. J. Electrochem. Soc. **162**(14), A2538–A2550 (2015). https://doi.org/10.1149/2.0151514jes
44. Kubota, K., Dahbi, M., Hosaka, T., Kumakura, S., Komaba, S.: Towards K-ion and Na-ion batteries as "Beyond Li-Ion." Chem. Rec. **18**(4), 459–479 (2018). https://doi.org/10.1002/tcr.201700057
45. Tarascon, J.M.: Na-ion versus Li-ion batteries: complementarity rather than competitiveness. Joule **4**(8), 1616–1620 (2020). https://doi.org/10.1016/j.joule.2020.06.003
46. Li, S., Xie, L., He, M., Hu, X., Luo, G., Chen, C., Zhu, Z.: Metal-organic frameworks-derived bamboo-like CuO/In_2O_3 heterostructure for high-performance H_2S gas sensor with low operating temperature. Sens. Actuators, B **310**, 127828 (2020). https://doi.org/10.1016/j.snb.2020.127828
47. Wang, M., Wang, H., Zhang, H., Li, X.: Aqueous K-ion battery incorporating environment-friendly organic compound and Berlin green. J. Energy Chem. **48**, 14–20 (2020). https://doi.org/10.1016/j.jechem.2019.12.019
48. Liu, X., Sun, Y., Tong, Y., Wang, X., Zheng, J., Wu, Y., Li, H., Niu, L., Hou, Y.: Exploration in materials, electrolytes and performance towards metal ion (Li, Na, K, Zn and Mg)-based hybrid capacitors: a review. Nano Energy **86**, 106070 (2021). https://doi.org/10.1016/j.nanoen.2021.106070
49. Liu, Y., Liu, Y., Wu, X., Cho, Y.R.: High performance aqueous zinc battery enabled by potassium ion stabilization. J. Colloid Interface Sci. **628**, 33–40 (2022). https://doi.org/10.1016/j.jcis.2022.08.046
50. Razavi, R.: Potential of nanocages as anode electrodes in Na- and K-ion batteries. Ionics **27**, 1159–1163 (2021). https://doi.org/10.1007/s11581-021-03915-3
51. Saadh, M.J., Mustafa, M.A., Batoo, K.M., Chandra, S., Kaur, M., Hussain, S., Alsaadi, S.B., Al-Tameemi, A.R., Kadhum, E.H., Falih, K.T., Alzubaidi, L.H., Liu, Y., Su, G.: Performances

of nanotubes and nanocages as anodes in Na-ion battery, K-ion battery, and Mg-ion battery. Ionics **30**, 2657–2664 (2024). https://doi.org/10.1007/s11581-024-05440-5

Flexible Nanogenerators Based on Low Dimensional Nanomaterials

Shakti Singh and Abhisikta Bhaduri

Abstract Nanogenerators are the newest edition in the field of Renewable energy devices. These devices' applications range from energy harvesting from mechanical force to driving advanced sensors and electronic devices. There are many challenges in the fabrication process like the selection of electrodes, materials, type of contact wire, etc., which significantly affects the output power. Flexible nanogenerators based on various polymeric materials are highly efficient and robust. Self-powered sensors based on these nanogenerators are highly efficient, sensitive, and cost-effective. Different smart materials based on metal oxides, MXENEs, carbon-based materials are used in nanogenerators for high output power and self-powered sensing properties. Smart materials possess greatly enhanced electromechanical features, including exceedingly high piezoelectric and triboelectric coefficients thereby broadening the scope of their application to sustainable energy. Moreover, these materials possess sufficient structural flexibility and rigidity to enable the seamless incorporation of nanogenerators into wearable electronics, medical devices, and IoT systems. In addition, the capability of smart materials to respond to external forces such as mechanical stress and temperature makes these nanogenerators even more reliable and functional in practical applications.

1 Introduction to Nanogenerators and Self-Powered Sensors

Nanogenerators are a new type of energy-harvesting devices that have the potential to harness otherwise wasted mechanical energy and convert it to electrical energy. A nanogenerator is a device that changes tiny amounts of physical energy, heat, or

S. Singh (✉)
University of Seoul, Seoul, South Korea
e-mail: shakti665@gmail.com

A. Bhaduri
Department of Physics, North Bengal St. Xavier's College, Rajganj, West Bengal, India

R. K. Sonker et al. (eds.), *Low-Dimensional Multifunctional Hybrid Nanostructured Materials*, Engineering Materials, https://doi.org/10.1007/978-981-96-7952-2_6

other types of energy into electrical energy using nano materials using the piezoelectric effect, triboelectric effect, or pyroelectric effect. It is extensively used for energy harvesting in self-sustaining systems where energy can be supplied continuously without the help of external batteries. A self-powered sensor is a sensor which sense external stimuli like humidity, temperature, gas etc. without an external power supply by capturing energy from its environment like mechanical motion, vibrations, or temperature changes using nanogenerators. Human movement, water waves, wind, and vehicle movement are all sources of this mechanical energy [1]. Energy is vital for daily life and sustainable progress, powering sectors from transportation and communication to electronics like sensors and smartphones. It is no doubt that rapid industrialization and population explosion has led to an invisible energy crisis and its consumption is being exhausted at a rate that will soon diminish fossil fuel reserves. This scenario suggests that timelines and approaches are going to be critical in preserving these non-renewable resources. In addition, mounting concerns with the greenhouse emission, prompt extensive pursuit of the renewable energy alternatives. Wang et al. in 2006 developed [2] a pioneering piezoelectric generator or nanogenerator which changed the landscape of this sector, nanogenerator is a device made from ZnO nanowires and works on the piezoelectric effect. A significant leap forward in the development of nanogenerator was witnessed in 2012 when Wang and colleagues demonstrated a 3D nanostructured piezoelectric nanogenerator that can convert mechanical energy to electric energy [3]. Piezoelectric nanogenerators (PENG) and triboelectric nanogenerators (TENG) are meant to turn mechanical energy to electrical energy, whereas pyroelectric nanogenerators (PyENG) do the opposite by turning thermal energy to electricity [4]. The hybrid nanogenerator (HNG) wears the crown as a cutting-edge technology since it takes different types of nanogenerators along with different energy sources into account for better energy efficiency during harvesting [5].

In brief NGs provides following advancements:

- Energy Solution: With a reduction in fossil fuel utilization, nanogenerators can simultaneously assist global energy requirements.
- Technological Advancement: Further development of the device is likely to improve upon self-powering and renewable energy technologies.
- Portable Applications: Optimally suited for compact, portable, and wearable devices which promises a greener future.
- Energy Harvesting: Collect energy from microwaves, mechanical vibrations, moving waves, moving air, and varying temperatures.
- Efficiency: Devise can economically harness nanotechnology and material science to produce high energy outputs at minimal energy inputs.

Before discussion of self-powered sensor, it is important to discuss various types of nanogenerators as shown in Fig. 1.

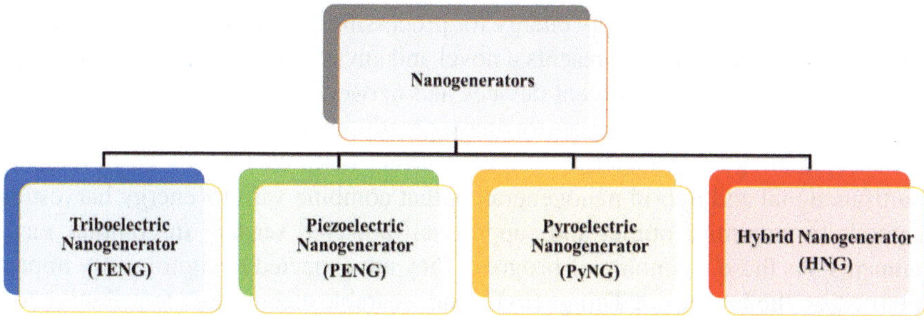

Fig. 1 Different types of nanogenerators based on recent advancement

1.1 *Types of Nanogenerators*

As mentioned earlier, the nanogenerators are of 4 types:

(a) Piezoelectric Nanogenerator (PENG)
(b) Triboelectric Nanogenerator (TENG)
(c) Pyroelectric Nanogenerator (PyNG)
(d) Hybrid Nanogenerator (HNG).

Self-powered sensors based on nanogenerators represent a transformative approach in the field of energy harvesting and sensor technology. These devices harness ambient energy from their surroundings to power themselves, eliminating the need for external power sources such as batteries.

By utilizing nanogenerators, which convert various forms of mechanical, thermal, or vibrational energy into electrical energy, self-powered sensors can operate continuously and autonomously. This capability is especially valuable for applications in remote, hard-to-access locations, or in systems where frequent maintenance and battery replacement are impractical, such as environmental monitoring, structural health diagnostics, and wearable biomedical devices.

Self-powered sensors operate by using nanogenerators to detect various external stimuli like pressure, temperature, chemicals without using external power sources [6–8]. Different kinds of nanogenerators, including piezoelectric, triboelectric, and pyroelectric nanogenerators, are capable of converting stress, contact, and heat into power. This power can be utilized to operate sensors directly or stored for use at a later time. For example, Piezoelectric Nanogenerators (PENG) cause electric charge generation in a material that is mechanically stressed. Triboelectric Nanogenerators (TENG) work by generating electricity through the physical contact and separation of certain materials, and Pyroelectric Nanogenerators (PyENG) utilize temperature differences along with the pyroelectric effect to produce energy [7, 9]. When integrated with other devices such as sensors, the nanogenerators do not require external energy to work, increasing their performance. These systems are able to detect many parameters in the environment like pressure, movement, temperature and chemical composition, and able to supply energy for processing signals and data communication [10]. This combination presents a novel and environmentally friendly approach to supplying energy to intelligent devices and networks. The efficiency, sensitivity and durability of self-powered sensors have greatly improved with the development of nanotechnology and material science. Current research focuses on developing multifunctional and hybrid nanogenerators that combine various energy harvesting methods to maximize output and support self-powered sensing in complex environments. As these technologies progress, they are expected to significantly impact smart cities, the Internet of Things (IoT), and wearable electronics, contributing to a more autonomous and energy-efficient future [11].

TENG and PENG are most common types of Nanogenerators, which are used for self-powered sensing applications because of their efficiency and output power. Output power in case of TENG and PENG is highest among all nanogenerators.

2 Overview of Triboelectric Nanogenerator

Triboelectric nanogenerators (TENGs) convert mechanical energy into electrical energy using the triboelectric effect and electrostatic induction. This effect occurs when two materials come into contact and separate, leading to an electron transfer that generates opposite charges on the materials. Upon separation, the resulting electric potential difference can produce current through an external circuit. TENGs effectively capture energy from sources such as human motion, vibrations, water waves, and airflow, offering a versatile solution for self-powered systems [12, 13]. TENGs are designed by combining materials with different triboelectric properties, such as polymers, metals, or composites, to maximize charge generation during contact and separation. They offer unique advantages, including lightweight, flexibility, operation across various frequencies, and the ability to harvest energy from low-frequency and irregular mechanical movements, which are challenging for traditional energy harvesting technologies [14]. TENGs are ideal for powering small, portable devices like wearable sensors, health monitors, and environmental systems. They also have potential for larger applications, such as harvesting energy from ocean waves, recovering energy from industrial machinery, and supporting smart city infrastructure. Material selection is crucial for optimizing performance, with materials from opposite ends of the triboelectric series generating more charge [15]. Surface modifications, such as nano-structuring or coatings, can further enhance contact area and energy conversion efficiency. Advanced materials and hybrids enable TENGs to adapt to diverse conditions and deliver higher outputs [16]. Current research focuses on improving durability, energy storage, and integration into existing systems. Hybrid nanogenerators combining triboelectric, piezoelectric, or electromagnetic effects are also being developed to enhance efficiency and versatility. Triboelectric nanogenerators (TENGs) are emerging as key players in energy harvesting, enabling self-powered systems and sustainable energy solutions. By converting everyday mechanical energy into electricity, they support advancements in IoT, smart cities, and wearable devices, promoting autonomous operation and reducing reliance on traditional energy sources. TENGs work by combining contact electrification, which creates static charges, with electrostatic induction, which converts mechanical energy into electrical power through changes in electric potential. As the technology evolves, TENGs are expected to significantly impact renewable energy and energy-efficient systems globally. Four different modes for triboelectric nanogenerators (TENGs) have been proposed to optimize energy harvesting efficiency [3], which are as follows:

- Vertical contact-separation mode,
- Lateral-sliding mode,
- Freestanding triboelectric-layer mode,
- Single-electrode mode.

The vertical contact-separation mode is widely used for its ability to generate high output and adapt to various applications. It works by utilizing perpendicular motion between two material surfaces, with the potential difference between electrodes depending on their separation distance [17]. The working mechanism of vertical contact separation mode can be understood by Fig. 2, which is as follows:

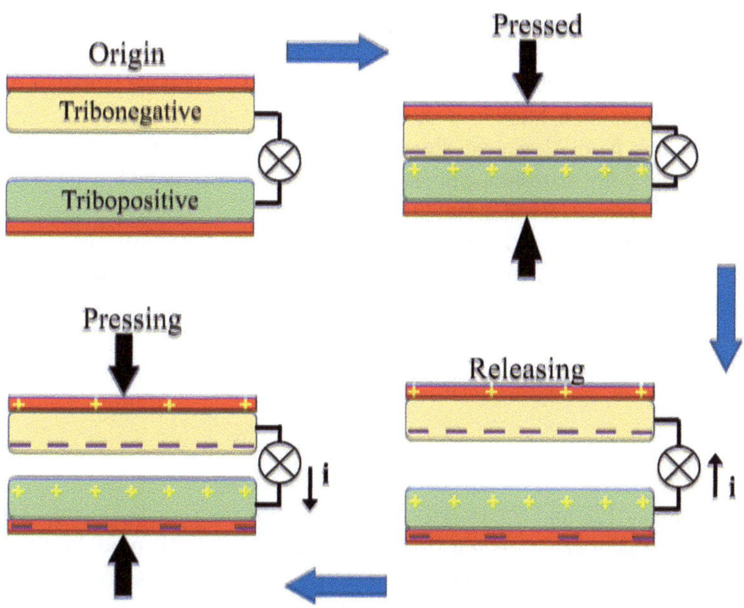

Fig. 2 Working mechanism and energy generation process of TENG based on vertical contact separation mode during pressing and releasing

The theoretical framework for describing the operation of TENGs is based on Maxwell's equations, with Maxwell's displacement current specifically defined as follows:

$$J_{\mathrm{D}} = \frac{\partial D}{\partial t} = \varepsilon \frac{\partial E}{\partial t} + \frac{\partial P_{\mathrm{s}}}{\partial t} = \frac{\partial \sigma_{\mathrm{Tribo}}(z, t)}{\partial t} \tag{1}$$

Here, ε, D, E, P_{s} represent the permittivity of the medium, displacement field, electric field, and the polarization arising from the polarization charges generated on the surface due to the triboelectric or piezoelectric effect, respectively. In triboelectric nanogenerators (TENGs), the surface charge density, denoted as σ_{Tribo}, plays a critical role. The first term in Eq. (1) refers to a time-dependent electric field, which generates electromagnetic waves. The final term highlights the effect of surface polarization, an essential mechanism for the functionality of nanogenerators. TENGs operate by utilizing electrostatic charges generated through contact electrification. When two different materials repeatedly come into contact and then separate, a dynamic surface polarization occurs. This process is central to the mechanism through which TENGs convert mechanical energy into electrical energy [14].

3 Overview of Piezoelectric Nanogenerator

The piezoelectric effect involves a direct relationship between mechanical stress or strain and electrical responses. When certain dielectric materials experience deformation from an external force, internal polarization generates opposite charges on their surfaces, a phenomenon known as the direct piezoelectric effect [18]. Conversely, when an electric field is applied along the polarization direction, these materials deform, and the deformation disappears once the field is removed. This is referred to as the inverse piezoelectric effect [19].

Piezoelectric nanogenerators (PENGs) operate by converting mechanical energy into electrical power via the piezoelectric effect. This process is governed by constitutive equations

$$\sigma_{\mathrm{p}} = c_{\mathrm{pq}} \varepsilon_{\mathrm{q}} - e_{\mathrm{kp}} E_{\mathrm{k}} \tag{2}$$

$$D_{\mathrm{i}} = e_{\mathrm{iq}} \varepsilon_{\mathrm{q}} + k_{\mathrm{ik}} E_{\mathrm{k}} \tag{3}$$

The above Eqs. (2), and (3) describe the relationship between the electric field E_{k}, electric displacement D_{i}, stress tensor σ_{p}, strain tensor ε_{p}, elastic modulus tensor c_{pq}, piezoelectric tensor e_{kp}, and dielectric tensor k_{ik}. The equations establish a strong interdependence between the electric field and mechanical deformation, making them inherently complex. Solving them analytically is difficult and often necessitates the use of advanced approximation techniques. Gao et al. [20] utilized a perturbation expansion approach to address this coupling, yielding an analytical expression for the distribution of the piezoelectric potential in PENGs under specific conditions. This study involved bending a nanowire using an atomic force microscope (AFM) tip and comparing the analytical approximations with finite element method (FEM) simulations. The results showed strong consistency, with a deviation of under 6%, confirming the accuracy of the perturbation method.

The mechanism of PENGs can be summarized as follows: upon application of an external force, body and surface charges are generated in the piezoelectric material, creating a potential difference across the PENG's electrodes. When the electrodes are connected to an external load, these potential drives electron flow, thereby generating electrical energy.

$$D_{\mathrm{Ri}} = e_{\mathrm{ip}}\varepsilon_{\mathrm{p}} \tag{4}$$

$$\rho^{\mathrm{R}} = \frac{\partial D_{\mathrm{Ri}}}{\partial x_{\mathrm{i}}} \tag{5}$$

$$\Sigma^{\mathrm{R}} = n_{\mathrm{i}} D_{\mathrm{Ri}} \tag{6}$$

Electric displacement D_{Ri} as shown in Eq. (4), is determined by both the strain and the piezoelectric tensor. According to Eqs. (5), and (6), under uniform strain distribution within the piezoelectric material, only remnant surface charges remain, allowing the PENG to behave similarly to a charged capacitor. The equivalent circuit for such a PENG resembles a voltage source. The voltage source value can be defined by Eq. (7):

$$V_{\mathrm{o}} = \frac{e_{\mathrm{ip}}\varepsilon_{\mathrm{p}}}{C_{\mathrm{p}}} \tag{7}$$

where C_{PENG} is the capacitance of the PENG. When connected to an external load, a portion of the voltage is dropped across C_{p}, reducing the output voltage below the open-circuit voltage. The combination of C_{p} and the external load forms a high-pass RC filter.

Subsequent research by Wang et al. [21] identified the role of displacement current in the mechanical-to-electrical energy conversion process in PENGs. The displacement current density, represented by Eq. (8), depends on both the strain rate and piezoelectric coefficient.

$$J = \frac{\partial D_{\mathrm{R}}}{\partial t} = e_{\mathrm{ip}} \frac{\partial \varepsilon_{\mathrm{p}}}{\partial t} \tag{8}$$

In this model, the equivalent circuit for a PENG is represented as a current source i_{p} in parallel with a capacitance C_{P}. Through circuit analysis, either a current source or voltage source can be used, and according to Norton's theorem, these are interchangeable representations. Equation (8) correlates the strain rate and piezoelectric coefficient with the PENG output, indicating that higher strain rates and piezoelectric coefficients result in increased output.

In this equivalent circuit model, a fraction of the generated charge is stored within C_P, and is unavailable for external loads, limiting the usable output. Thus, enhancing PENG performance can be achieved by optimizing the piezoelectric coefficient, increasing the device area, and designing effective circuitry for efficient energy extraction, given that strain rates and device areas are typically limited by the working environment. These improvements offer a pathway to higher output efficiencies for PENG technology.

4 Fabrication Methods and Characterization Techniques of Nanogenerators

The fabrication of flexible and wearable electronics significantly impacts their performance, cost, and durability. NGs, consisting of electrode, triboelectric, piezoelectric layers, and substrates, utilize diverse manufacturing techniques. This overview addresses the common methods for producing conductive materials and other components, as well as strategies for integrating distinct functional layers.

4.1 Spinning Techniques

A wide range of methods exists for the preparation of conductive materials, among which spinning techniques are particularly notable for their ability to continuously produce fiber materials. The primary spinning methods include wet spinning, dry Spinning, Melt spinning, Gel spinning, electrospinning, etc.

4.2 Coating Techniques

Coating techniques are vital in electronic fabrication, with methods such as dip coating, spin coating, blade coating, and seam coating offering distinct advantages. Dip coating is cost-effective but wasteful, spin coating ensures uniformity for small devices, blade coating minimizes waste but limits uniformity, and seam coating provides high speed and precision at higher costs [22, 23].

4.3 *Plating Techniques*

Metallic coatings improve textile conductivity, achieved through electroplating, which provides uniform and cost-effective coverage, or electroless plating, which operates without an electric field but may result in uneven thickness [24].

4.4 *Printing Techniques*

Advances in printing technologies have enabled efficient fabrication of conductive materials. Inkjet printing is eco-friendly but unsuitable for large-scale production. Offset and flexographic printing are cost-effective for high volumes. Screen printing supports mass production but has environmental drawbacks, while gravure printing offers high speed with similar pollution concerns [25].

The advancement of these spinning techniques such as wet spinning, dry spinning, melt spinning, gel spinning, and electrospinning enable the continuous production of conductive fibers which are cheaper and easier to scale. Dip coating is inexpensive but wasteful, and spin coating is material efficient but only works on small devices, while blade coating is cost effective but not very precise, seam coating allows for fast and precise painting but comes at a high cost. All plating techniques improve conductivity, although electroplating is the most economical and provides the best coverage. Electroless does not require an electric field, but the resulting thickness is less reliable. Inkjet printing is eco-friendly but not efficient for mass production and flexography and offset printing are cost effective but not for the environment, while screen printing can produce in bulk but is bad for the environment and gravure printing is fast but does the same pollution. Some techniques are better suited than others, and if chosen wisely, optimal efficiency, sustained energy consumption, and scalability are achievable.

4.5 *Fabrication of Fiber-Based Materials*

Beyond spinning, fiber-based materials can be created through knitting or weaving techniques:

- **Knitted Fabrics**: Characterized by their elastic and flexible nature, knitted fabrics consist of interconnected loops that provide excellent breathability. However, their complex structures complicate manufacturing processes.
- **Woven Fabrics**: Constructed from two interlaced sets of yarns (warp and weft), woven fabrics are notable for their dimensional stability, wash resistance, and suitability for mass production due to their simpler architecture.

4.6 Integration of Functional Layers

The integration of various functional layers in nanogenerators is a critical aspect of their design. Common integration methods include:

- **Binder Addition**: The incorporation of binders facilitates adhesion between different layers.
- **Intermolecular Interactions**: Utilizing specific molecular interactions can enhance layer bonding.
- **External Forces**: Techniques such as thermal bonding and ultrasonic bonding leverage heat or ultrasonic energy to achieve effective integration.

In such fabrication, according to studies by Park and his counterparts, the addition of a TiOx intermediate layer in TENG impedes interfacial charge recombination, improving charge retention. These increases yield a remarkable rise in the values of V_{OC} (272 V), I_{SC} (9.1 μA), and charge density (30 μC) [26]. Kim et al. showcased that adding a PDMS interlayer in PVDF or nylon based TENGs enhances their charge trapping and storage capabilities. Initial measurements of surface potentials indicated that PVDF by itself quickly lost ability to store charge, while PDMS and PVDF/PDMS retained half of their potential after 100 s. KPFM showed that PDMS, which has a greater work function than aluminum, helps in electron trapping in deep energy levels and ensures that charge recombination does not take place, resulting in improved charge storage capacity [27]. In summary, the diverse fabrication methods available for both conductive and non-conductive materials are pivotal in the development of advanced flexible and wearable electronics. Continued exploration and refinement of these techniques will enhance the performance and applicability of nanogenerators in various applications.

5 Smart Material Selection for Enhanced Output of Nanogenerators

Material selection is a critical factor in the design and performance of triboelectric nanogenerators (TENGs) and piezoelectric nanogenerators (PENGs). The performance of these devices in energy conversion depends on the material properties, with TENGs particularly reliant on the selection of frictional contact materials. Materials that occupy different positions on the triboelectric series generate greater charge separation when they come into contact and then separate. Selecting materials with high differences in their triboelectric properties maximize charge generation. Materials with higher surface roughness can enhance the effective contact area, leading to increased charge density. Optimizing the surface morphology through material selection and processing can significantly improve TENG efficiency. Since TENGs rely on repeated contact and separation, selecting materials that exhibit high durability and resistance to wear is vital for long-term stability and performance.

In TENGs, the material's physical properties impact the efficiency of charge transfer and, consequently, the overall output performance. Factors such as triboelectric polarity should be considered along with material selection; electron-accepting materials, such as PTFE and FEP, and electron-donating materials such as nylon and PDMS effectively separate charges. Also, materials with a high dielectric constant can help in charge retention and electrostatic induction. Surface roughness, as well as micro and nano structuring enhances effective contact area which results in better charge generation and transfer efficiency. Moreover, mechanical flexibility and toughness are vital for enduring a cyclic mechanical load without structural degradation.

For PENGs, the ability of the material to convert mechanical stress into energy is primarily determined by the intrinsic piezoelectric coefficient (d_{33}, d_{31}). Performance ferroelectric materials based on lead-free ceramics with barium titanate ($BaTiO_3$) and piezoelectric polymers (PVDF, PVDF-TrFE) are selected for their outstanding features of dipole orientation, elasticity, and superior toughness.

5.1 Materials for TENG

Triboelectrification occurs when electrons are transferred between two materials based on their electron affinities, with the material having higher affinity accepting electrons and the other donating them, as illustrated in Fig. 3. These roles can be identified using the triboelectric series. The triboelectric series guides the selection of materials to optimize TENG performance, advancing energy harvesting technologies and wearable devices. Ongoing research focuses on improving efficiency and adaptability for diverse applications and environments.

5.1.1 Electron Acceptors

A total of 14 electron acceptor materials were identified, including polyvinylidene fluoride (PVDF), polyolefins, polystyrene (PS), polyester (PET), and silicone. Other materials include polytetrafluoroethylene (PTFE) at 34%, polydimethylsiloxane (PDMS) at 22%, and fluorinated ethylene propylene (FEP) at 18%. Additional materials are bacterial nanocellulose (BNC), alginate sodium, nitrocellulose, graphene, acrylics, rubber, and polyvinyl alcohol (PVA).

5.1.2 Electron Donors

A total of 20 electron donor materials were identified, with key contributors including gold, PET, acrylics (PMMA), and polyurethane. Other notable materials are aluminum (26%), copper (20%), skin (10%), and nylon (8%). Additional materials used include Kapton, cellulose (wood, paper, nanofibrils), indium-tin oxide

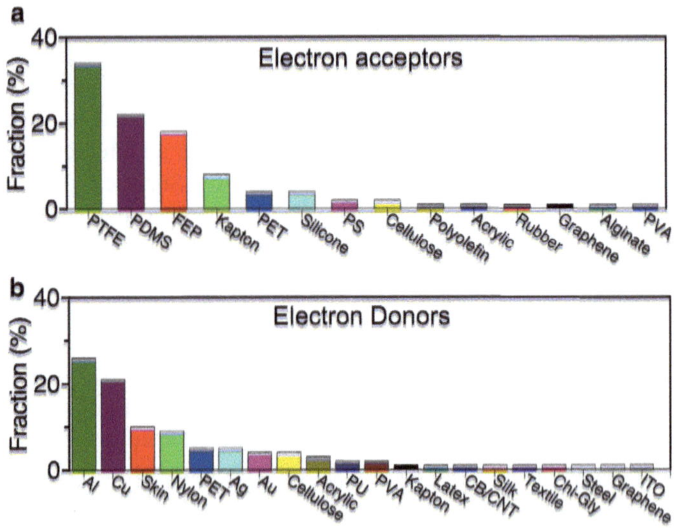

Fig. 3 Different types of material used in TENG representing their electron affinity. **a** Series of electron acceptors and, **b** electrons donors [28]. Reproduced with permission @ copyright (2020) Wiley

(ITO), latex, carbon black, carbon nanotubes, textiles, graphene, chitosan-glycerol, silk, polyvinyl alcohol (PVA), and steel. Interestingly, metals such as Al, Cu, Ag, and Au are frequently utilized as both triboelectric layers and electrodes, which contrasts with the conventional approach of selecting materials with significantly different charge affinities. The rationale behind this choice will be further discussed.

For PENGs, the efficiency of PENGs is directly related to the piezoelectric coefficient of the materials used. Materials with higher piezoelectric coefficients, such as polyvinylidene fluoride (PVDF) or certain ceramics (e.g., PZT), generate more electrical output under mechanical stress. For wearable applications, flexible and lightweight materials are essential. The mechanical properties, such as tensile strength and elasticity, must be compatible with the intended application to ensure effective energy harvesting from dynamic motions. The operational environment can influence piezoelectric performance. Materials that maintain their piezoelectric properties across a range of temperatures and humidity levels are crucial for reliable operation.

5.2 *Materials for PENG*

Enhancing the piezoelectric coefficient has been identified as a significant method to improve the output performance of PENGs. Current advancements in PENG

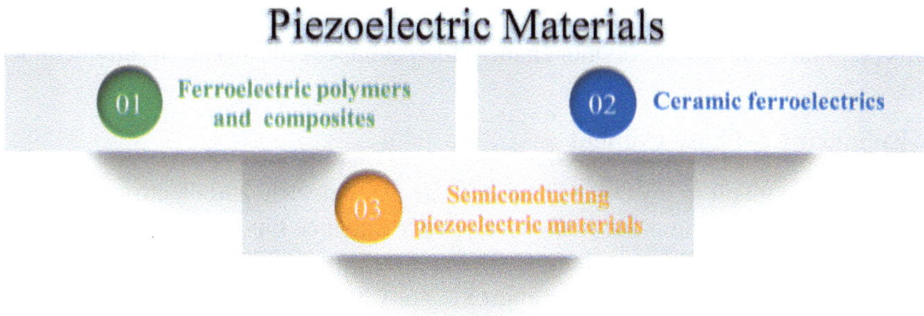

Fig. 4 Different types of piezoelectric materials based on recent advancement used in NGs

technology have focused on three main categories as shown in Fig. 4 of piezoelectric materials that have garnered substantial research interest. The first category encompasses ceramic ferroelectrics such as barium titanate (BTO), sodium potassium niobate (KNN), sodium bismuth titanate (NBT), lead zirconate titanate (PZT), and lead magnesium niobate titanate (PMN-PT). These materials exhibit notably high piezoelectric coefficients, particularly in lead-based ceramics such as PZT and PMN-PT, which contribute to their efficacy in energy harvesting applications.

The second category comprises of ferroelectric polymers and ferroelectric composites, which, despite having relatively lower piezoelectric coefficients than ceramic ferroelectrics, offer greater compatibility with current manufacturing processes. This compatibility, along with their flexibility and biocompatibility, make them ideal for integration into biomedical devices and wearable electronics, where non-toxicity and mechanical adaptability are crucial.

Finally, semiconducting piezoelectric materials, including zinc oxide (ZnO), gallium nitride (GaN), and cadmium sulfide (CdS), combine piezoelectric and semiconducting properties. These materials are valuable in applications that extend beyond traditional energy harvesting, such as in the emerging field of piezotronics, which leverages their dual functionalities for innovative electronic applications.

5.2.1 Ferroelectric Polymers and Composites

Ferroelectric polymers and composites represent a versatile alternative to ceramic piezoelectric materials, particularly in applications requiring lightweight and flexible materials. This is due to their inherent properties, which stem from distinct material structures and behaviors. Polyvinylidene fluoride (PVDF) is a prime example, distinguished by its unique negative longitudinal piezoelectric coefficient. Unlike traditional piezoelectric materials, PVDF contracts under an electric field along the c-axis, limiting its suitability as a matrix for ferroelectric composites with ceramic fillers. PVDF exists in several crystalline phases—α, β, γ, δ, and ε—with the β-phase displaying superior piezoelectricity due to its molecular alignment of –CF_2–CH_2– units perpendicular to the polymer chain, which enhances the dipole alignment under stress.

- **Enhancing β-Phase Content**

The β-phase content in PVDF is critical for its piezoelectric and electromechanical performance, with various methods developed to enhance this phase. Sencadas et al. found that mechanically stretching PVDF at a ratio of 500% and a temperature of 80 °C could increase the β-phase to approximately 80%, albeit with a 10% reduction in crystallinity. Another technique, the press-and-fold (P&F) method developed by Meng et al., involved folding the polymer, pressing it under 300 kN at 165 °C, and then quenching in cold water, achieving up to 98% β-phase content. This method enhances β-phase formation through effective stress transfer, compared to simple uniaxial stretching. Other methods, like electrospinning, also promote β-phase development, as the extensional forces and high voltage applied between the syringe tip and collector pole the fiber and orient the molecular dipoles.

Boosting the amount of β-phase present PVDF is essential for improving its piezoelectric property, which is very important for the effective functioning of the piezoelectric nanogenerators. Poling, where an electric field is applied during the process to dipole and energize the material, is one possible way to improve the β-phase content. The dipole alignment produced during energization increases the output and energy harvesting efficiency. Nanocomposites formed by the addition of some nanoparticles of ZnO and $BaTiO_3$ help in increasing the β-phase formation which in turn increases the piezoelectric charge generation and improves the mechanical properties. Utilizing dimethylformamide, a solvent vapor treatment, aids in changing the α-phase to the β-phase and improves the piezoelectric coefficients. Furthermore, mechanical tension applied to the material increases electromechanical response due to change of phase to the β-phase. All these methods improve the energy conversion efficiency, mechanical strength, and aid in the better functioning of the piezoelectric nanogenerators.

- **Interfacial Effects on Composites**

In nanogenerators, interfacial polarization plays a vital role in enhancing the charge separation and generation process. This kind of polarization at the boundary of the piezoelectric matrix, like PVDF, and fillers embedded inside it, for example, nanoparticles or nanofillers, would result in the accumulation of charge carriers, hence forming dipoles at the interfaces, which enhance the piezoelectric response when mechanical stress or deformation is applied. This local charge accumulation may increase the total output of the nanogenerator, reflecting that interfacial polarization bears a great role in energy conversion efficiency [29].

This interfacial polarization is really influenced by several factors because the nature of the fillers then decides the extension in performance enhancement. For instance, some examples are nanoparticles: zinc oxide or barium titanate, being able to obviously enhance the piezoelectric response. These fillers are characterized by a high dielectric constant and high polarizability, which may enhance the interfacial charge distribution when they are incorporated in the PVDF matrix. Therefore, these fillers facilitate better piezoelectric energy harvesting by increasing the surface area available for charge accumulation. Besides this, nanoparticles of high surface energy can align dipoles in the PVDF matrix, hence increasing the β-phase content to further optimize piezoelectric performance [5, 30].

The enhancement of interfacial polarization by fillers, however, is dependent on filler size, shape, and dispersion within the matrix. Well-dispersed fillers provide a more homogeneous interface, thus making the polarization even within the material. In cases of aggregation or poor dispersion, regions with weak interfacial polarization could be formed within the material, which may cause a decrease in the overall piezoelectric efficiency of the material. Another very important factor is the interaction between filler and matrix, since strong interactions between the filler and matrix will lead to better stress transfer and charge separation, enhancing the piezoelectric response.

In balancing the mechanical properties, the piezoelectric performance takes into consideration other critical factors, including the volume fraction of fillers. The increased content of fillers might be beneficial regarding interfacial polarization, while such high values can reduce the flexibility and mechanical strength of materials, if not optimized. Consequently, filler selection and further optimization of filler dispersion and its interaction with the matrix will be an indispensable strategy in achieving high-performance piezoelectric nanogenerators.

- **Copolymerization to Induce Ferroelectric and Relaxor Behavior**

Copolymerizing PVDF with trifluoroethylene (TrFE) forms P(VDF-TrFE), a relaxor ferroelectric polymer that introduces a unique relaxor behavior. This effect arises from conformational disorder rather than atomic disorder, as seen in inorganic relaxors. The asymmetric carbon in TrFE introduces chain chirality, resulting in a 3/1-helical structure in the ground state, and the distortion in gauche torsional angles triggers disorders in this conformation, which leads to relaxor characteristics. Epitaxial growth of P(VDF-TrFE) on substrates such as PTFE, by exploiting lattice matching and subsequent heat treatment, improves crystallinity and piezoelectric properties. Methods that use an applied shear force during heat treatment also enhance crystallinity and electrical polarization without the need for a specific interlayer. Pressing with a removable PTFE template further improves d_{33} values, owing to the confined lamellar growth.

- **Impact of Fillers and Composites on Piezoelectric Performance**

Incorporating fillers such as carbon nanotubes, graphene, or metal nanowires into PVDF enhances β-phase nucleation by surface interactions. Functionalized carbon nanotubes, for instance, interact with CF_2 groups, inducing nucleation and promoting β-phase growth. The optimal filler concentration (e.g., 2% for ester-functionalized multiwalled carbon nanotubes) can yield nearly 100% β-phase in solvent-cast PVDF, achieving substantial piezoelectric enhancement.

Ferroelectric composites composed of nano-ceramic fillers like barium titanate (BTO) within polymer matrices harness both the flexibility of polymers and the piezoelectric properties of ceramics. For effective performance, even dispersion and filler connectivity are essential. Unevenly dispersed fillers can reduce the piezoelectric potential due to aggregation, as illustrated in BTO-PDMS composites where aggregated BTO nanoparticles exhibit lower piezoelectric output. To prevent aggregation, scaffolds (e.g., graphene or metal nanowires) facilitate even dispersion by providing attachment sites that reduce filler interactions.

- **Connectivity and Stress Transfer in Composites**

Stress transfer efficiency from the polymer matrix to the fillers is crucial for effective composite piezoelectricity. Filler length plays a significant role, with longer fillers improving stress transfer across the composite. Xie et al. demonstrated that freezing cast methods could create dendritic PZT channels in PDMS matrices, forming a connected filler network that enhances stress transfer. A gradient structure in freeze-cast PZT fillers maximizes piezoelectric output, particularly in regions with high filler concentration, due to more efficient stress transfer. Additionally, electric field alignment of PZT nanowires using dielectrophoresis in a PDMS matrix further improves filler alignment and dispersion, yielding an even distribution beneficial for uniform stress application.

- **Advanced Fabrication Techniques and Interfacial Effects**

Recent advancements such as 3D printing allow for highly customized piezoelectric architectures. Chen et al. used de-binding and sintering processes to achieve near-bulk density in printed $BaTiO_3$, while Li et al. integrated an electric-field-assisted printing technique that simultaneously poles the piezoelectric material, preventing poling failure in complex structures. This technique has enabled the fabrication of structures like artificial arteries with tubular piezoelectric architectures.

Interfacial polarization, often characterized by the Maxwell–Wagner effect, plays a role in dielectric behavior by promoting charge accumulation at the filler-matrix interface due to permittivity and conductivity differences. Kim et al. utilized this in a triboelectric nanogenerator (TENG) by combining poly(2,2,2-trifluoroethyl methacrylate) with PDMS to achieve high charge density. For PENGs, however, high dielectric constants are detrimental as they increase internal capacitance, reducing power transfer to external circuits due to higher displacement currents within the device. In composites with metal fillers, power dissipation occurs due to the conductive nature of the fillers, reducing the efficiency of the PENG.

Overall, the interplay between polymer crystallinity, filler distribution, and interface effects is central to optimizing the piezoelectric response in ferroelectric polymers and composites. Each aspect—from phase content to filler connectivity—affects the electromechanical properties and applicability of these materials in energy harvesting, medical devices, and flexible electronics.

5.2.2 Ceramic Ferroelectrics

Ceramic Ferroelectric materials are dielectrics with reversible spontaneous polarization, induced by phase transitions. Their high dielectric constant and hysteresis make them ideal for energy storage, enabling high-density capacitors for electronic devices [31]. Perovskite-type ceramics, like PZT and PMN-PT, are especially recognized for their high piezoelectric constants.

- **Theory of Ceramic Ferroelectrics**

As discussed earlier, ferroelectric materials are dielectrics with spontaneous, reversible polarization caused by phase transitions. Their high dielectric constant and energy storage capacity make them ideal for capacitors in electronic devices [31]. High piezoelectric performance in ceramic ferroelectrics is driven by polarization rotation, where the polarization vector shifts within the material, boosting its response. This process depends on the material's energy landscape; for instance, in barium titanate, flatter energy paths enable easier polarization rotation under an external electric field, enhancing its efficiency [31]. The longitudinal piezoelectric constant (d_{33}) can be expressed as:

$$d_{33}^{*}(\theta) = \cos\theta\left(d_{15}\sin^2\theta + d_{31}\sin^2\theta + d_{33}\cos^2\theta\right) \tag{9}$$

where θ is the angle between the [001] crystallographic axis and the direction in which the piezoelectric response is being evaluated, d_{15} is the shear piezoelectric constant, d_{31} and d_{33} are the longitudinal piezoelectric constants. Enhancing piezoelectric performance in ceramic ferroelectrics involves strategies like tuning compositions to create multiphase regions, such as morphotropic phase boundaries, and introducing structural heterogeneity, like polar nanoregions, to promote polarization rotation. These approaches optimize materials for energy harvesting and sensing applications.

Each component type offers advantages and limitations on its own, therefore, material selection is crucial for optimal performance of TENG and PENG devices. Materials with high differences in triboelectric properties and surface roughness are best for TENG as they maximize charge generation and density. PVDF, PTFE, and PDMS as electron acceptors provide strong negative triboelectric effects while aluminum, copper, and nylon as electron donning materials contribute positively. Metal based materials also act as electrodes, boosting charge affinity and conductivity. PZT and PMN-PT materials have a high piezoelectric coefficient and can easily convert energy, however their brittleness makes them undesirable. Polymer based materials such as PVDF are more flexible and biocompatible, but their piezoelectric shock output isn't as high. Ferroelectric polymers such as P(VDF-TrFE) enhance engineering adaptability while composite materials help to overcome property limitations of ferroelectric materials. Fillers like graphene and BTO also improve mechanical properties and charge alignment. ZnO and GaN, inorganic semiconducting materials, combine the two great functions of piezoelectric and electronic materials, widening their usage from energy harvesting to advanced devices called piezotronics. While ceramics exhibit high dielectric constants and polarization rotation benefits, their rigid structure limits use in flexible applications. Polymer-based alternatives, though mechanically adaptable, require strategies like β-phase enhancement and nanofiller dispersion to optimize performance. Overall, selecting the right material involves balancing efficiency, durability, mechanical flexibility, and environmental adaptability, ensuring optimal performance for diverse TENG and PENG applications.

5.3 Conductive Materials for Electrodes

For TENGs and PENGs, highly conductive materials like silver, copper, or polymers enhance energy output by minimizing resistive losses. These materials must also ensure strong adhesion and compatibility with active layers to maintain device integrity and performance.

5.4 Cost and Availability

- **Economic Feasibility**: The availability and cost of materials can influence the scalability and commercial viability of TENGs and PENGs. Selecting materials that are abundant and cost-effective can facilitate widespread adoption of these technologies.

6 Basics of Self-Powered Sensors and Design

Sensors are devices engineered to detect and transduce physical, chemical, or biological stimuli into quantifiable electrical signals, enabling the precise monitoring of diverse environmental parameters [32–38]. They serve as critical components in the advancement of stretchable and wearable smart electronics, driving innovations in multiple technological domains. Among the emerging sensor technologies, self-healing materials have garnered significant attention in the field of electronics due to their potential to enhance device durability and functionality. For example, Huynh et al. reported the development of a self-healing sensor comprising a disulfide-cross-linked polyurethane matrix and a composite structure integrating polyurethane with silver microparticles (μAg). A self-powered sensor based on a nanogenerator utilizes mechanical energy from its environment to generate electricity, eliminating the need for external power sources. By integrating triboelectric or piezoelectric mechanisms, these sensors enable real-time monitoring and energy-efficient operation, making them ideal for applications in healthcare, environmental monitoring, and wearable electronics [39, 40].

6.1 Types of Sensors

Sensors can be categorized into three primary types:

- **Physical Sensors**: Physical sensors convert physical stimuli like temperature, pressure, and light into signals with high precision and stability. Their flexibility and reliability make them ideal for wearable electronics and electronic skin applications [41].
- **Chemical Sensors**: Chemical sensors detect specific substances by converting chemical interactions into electrical signals. They are widely used for gas detection, humidity measurement, and identifying harmful ions in water, addressing key environmental and public health challenges.
- **Biosensors**: Biosensors integrate biological elements with transducers to detect specific analytes, offering high selectivity and sensitivity. They are crucial in biomedicine, diagnostics, food safety, and environmental monitoring.

6.2 Nanogenerators and Self-Powered Sensors

Self-powered sensors represent an emerging class of advanced devices that utilize nanogenerator technology to harvest energy from their environment, eliminating the need for external power sources. To understand the significance of self-powered sensors, it is essential to examine the principles and classifications of nanogenerators [12, 13]. Nanogenerators convert ambient mechanical energy, such as motion or vibrations, into electricity, addressing the growing need for sustainable energy solutions amid environmental challenges. Introduced in 2006 by Wang et al., piezoelectric nanogenerators (PENGs) utilize ZnO nanowires, while triboelectric nanogenerators (TENGs), developed in 2012, harness the triboelectric effect. These technologies enable self-powered sensors, which operate without external power sources by harvesting energy from movements, water waves, or wind. TENG-based sensors have versatile applications, including motion detection, fluid monitoring, and gas sensing, paving the way for energy-efficient and autonomous systems [42, 43]. Based on their operational principles, TENG-based self-powered sensors are classified into two main types as shown in Fig. 5:

- **Self-powered Active Sensors**—In these nanogenerator-based sensors, external stimuli directly interact with the nanogenerator for detection [8].
- **Self-powered Passive Sensors**—In these sensors, external stimuli such as humidity, chemicals, or light do not directly interact with the nanogenerator. Instead, they interact with a chemiresistive film connected to the nanogenerator. This interaction alters the film's resistance, thereby changing the output across it [44].

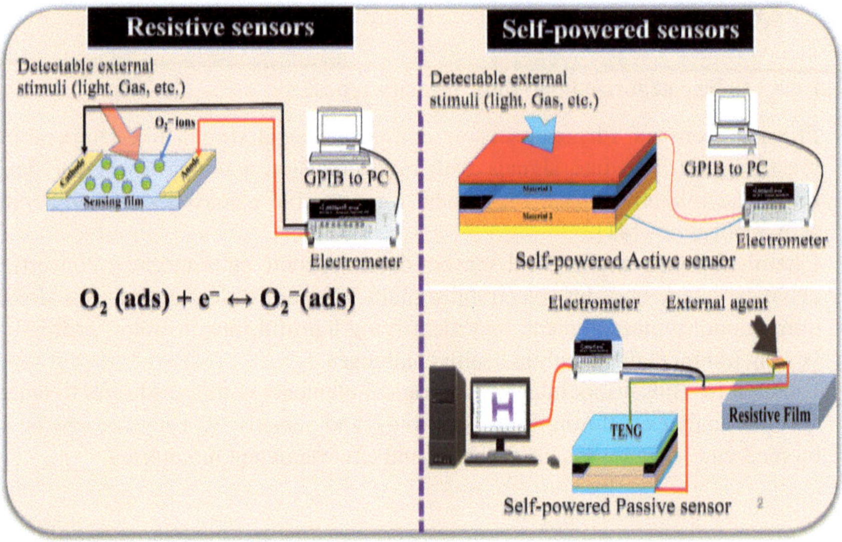

Fig. 5 Image showing difference between resistive and self-powered sensors with their commonly used experimental setup

This classification underscores the diverse design approaches and potential applications of self-powered sensors, paving the way for next-generation sensing technologies.

6.2.1 Self-Powered Active Sensors

Active self-powered sensors use nanogenerators to detect external stimuli, such as gases or pressure, by directly altering their output power. In TENG-based sensors, chemisorption on triboelectric layers triggers electron transfer, modifying voltage or current output. The sensors make for an efficient but inexpensive scanning device as they allow for detecting weak signals to a certain degree [8].

6.2.2 Self-Powered Passive Sensors

The sensors make for an efficient but inexpensive scanning device as they allow for detecting weak signals to a certain degree. In SSM, TENG is connected to a resistive film, and instead of applying pressure to the nanogenerator, it is applied to the film directly. Passive self-powered sensors depend on TENGs as power suppliers and the TENG with the series resistor performs the sensing depending on the type of the recruited film (p-type or n-type). It consists of a resistive film that is connected to the TENG, which changes the resistance of the film when light or gas is in use and the film measures the change in voltage [7, 8, 40, 44, 45].

It divides the circuits operating voltage, and the output voltage depends on the film's resistance in relation to the shown circuit system in Fig. 5. In this system:

- R_2: Resistance of the sensing element (resistive film)
- R_1: Internal resistance of the electrical circuit
- V_{IN}: Maximum output voltage supplied by the TENG to the resistive film.

The output voltage across R_2 is given by the voltage division equation:

$$V_{OUT} = \{R_2 / (R_1 + R_2)\} \times V_{IN}$$

This equation indicates that the output voltage V_{OUT} is directly proportional to the resistance of the sensing element R_2. Whenever the R_2 shifts as a result of external stimulus, V_{OUT} shifts as well. This is further supported by $V = IR$, which shows that current and voltage are interdependent variables and thus represent a dynamic sensing model.

7 Applications of Smart Materials Based Flexible Nanogenerators

Smart materials-based flexible nanogenerators (FNGs) based on PENG and TENG have emerged as cutting-edge energy harvesting devices capable of converting diverse forms of mechanical energy, such as vibrations, motions, and environmental forces, into electrical energy through piezoelectric, and triboelectric effects as shown in Figs. 6 and 7. Their lightweight, flexible, and adaptable design allows integration into textiles, implants, and structural components. As a result, FNGs are poised to revolutionize multiple domains by providing self-sustained power solutions. This chapter delves into the scientific principles, mechanisms, and technological applications of flexible nanogenerators across key fields.

FNGs have been effectively employed as biomechanical sensors due to their ability to convert mechanical stimuli into electrical signals. These devices can monitor physiological parameters, including heart rate, respiratory patterns, and body movements. For example, integrating PENGs into shoe insoles enables real-time monitoring of gait and locomotion dynamics. As the foot strikes the ground, the mechanical pressure applied to the insole generates corresponding electrical signals, reflecting the intensity and frequency of motion. PENGs allow for motion analysis, as well as pressure monitoring due to their ability to generate electrical signals while running, walking or squatting. They can be put inside insoles in order to correct joint alignment as well as facilitate rehabilitation. They can also be fixed to joint for accurate biomechanical evaluations that provide insight into movement patterns. PENG sensors are

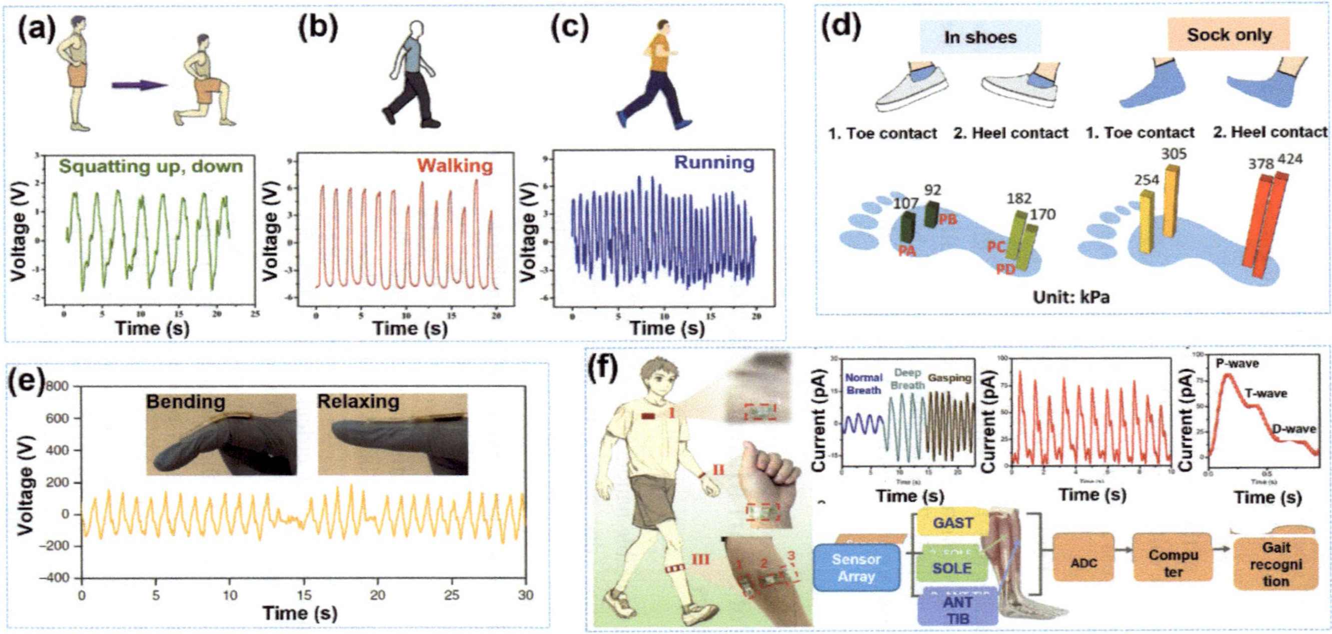

Fig. 6 Applications of flexible NGs: **a–c** attached in shoes for human motion monitoring [46]. Reprinted with permission @ copyright (2020) Elsevier. **d** Pressure mapping of shoes worn and sock only conditions depending on the PENG outputs in different foot positions [47]. Reprinted with permission @ copyright (2019) American Chemical Society. **e** Output performance of PENG device under finger bending and relaxing conditions. Reprinted from [48], under the licence CC BY 4.0, Springer Nature. **f** Application schematics and performance of a flexible PENG as piezoelectric sensors attached on different body parts on the body [49]. Reprinted with permission @ copyright (2020) Elsevier

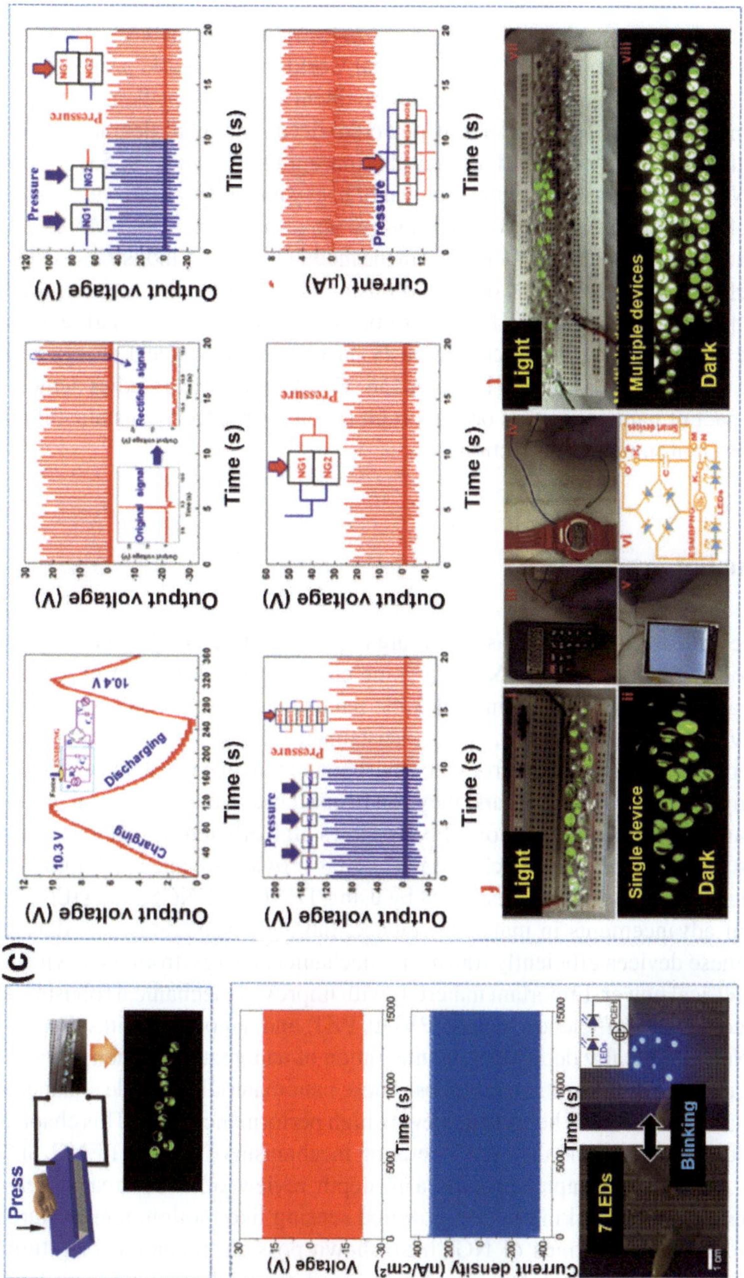

Fig. 7 Applications of flexible NGs: **a** lighting of 10 LED bulbs using NG [50]. Reprinted with permission @ copyright (2020) Elsevier; **b** reliability test [51]. Reprinted with permission @ copyright (2018) American Chemical Society. **c** Output performance and applications of a PENG [52]. Reprinted with permission @ copyright (2020) Elsevier

also useful in prosthetics and exoskeletons allowing for smooth interactions with users.

In the context of practical application of piezoelectric nanogenerators (PENGs) and triboelectric nanogenerators (TENGs) in real world, considerable progress has been achieved in the self-remediation of the environment as well as medicine. TENG-based systems have been applied for self-powered electrochemical degradation of organic pollutants, which include dyes and pharmaceutical residues [8, 53]. Additionally, PENGs have been used for the removal of heavy metal ions from wastewater using piezoelectric driven adsorption and ion-exchange processes [54]. Moreover, TENGs can serve as self-powered sterilizers in the medical and food industries as they can effectively disrupt membranes of bacteria using electric fields in antimicrobial applications [55]. TENG and PENG have also been incorporated into smart textiles for wearable health monitoring to continuously harvest biomechanical energy and track body systems without external batteries [56, 57]. All these case studies demonstrate the impact of nanogenerators on important problems related to sustainability, healthcare, or environmental protection.

8 Conclusion

Energy harvesting and autonomous systems are becoming more efficient because of flexible nanogenerators (FNGs). FNG is a game-changing breakthrough in renewable energy and self-powered sensing. Flexible nanogenerators (FNGs), which are covered previously in this chapter, offer a sustainable energy harvesting option, which converts mechanical energy from the environment into electrical power. These devices are lightweight, adaptable, and they can be used use in environmental monitoring, IoT, and healthcare applications. FNGs shows high efficiency, scalable fabrication, using advanced materials technologies. FNGs provide advance technologies like wearable electronics and smart sensing by using TENGs, PENGS, and HENGs. By the use of advancements in materials science, fabrication methods, and device integration, these devices efficiently transform mechanical energy from the environment into electrical power. Important materials with improved mechanical robustness and energy conversion efficiency include PVDF, PZT, and nanocomposites. Performance of FNGs is typically dependent on integration of nanomaterials such as metal nanowires, graphene, and carbon nanotubes. These nanomaterials provide enhanced surface charge density, which helps in achieving high performance NGs. This chapter helps the reader to understand the basic operation mechanism of TENG, PENG, and hybrid NG. Also, these chapter provide a in depth review of wide area of self-powered sensing. Different kind of self-powered sensing like motion, temperature, chemical etc. using these kinds of NGs have shown possible potential for future. This urgently calls for further development in smart material technology to help in overcoming certain setbacks in the NG technology like energy conversion efficiency, mechanical robustness, and adaptability. More opportunities were provided by self-healing polymer materials, shape memory alloy materials, and piezoelectric composite materials in providing higher performances with longer lives. These advancements create an advantage of higher capability to respond to the NG in a more dynamic state. Coupled with materials such as graphene and carbon nanotubes, these ensure flexibility in use and, hence, wearability on the human body for continuous energy harvesting from motions. Besides this, sustainable, biodegradable materials are used to reduce possible environmental impacts even further and enhance their potential in miscellaneous applications. It is these changes that are really paving the way ahead for NG toward more reliable and efficient solutions for environmental sustainability. Flexible nanogenerators (NGs) represent a novel approach towards self-powered systems, but gaps still exist for further research and development. Being a rapidly evolving field, incorporation of modern technologies such as AI, IoT, and smart autonomous energy systems can result in drastic improvements. Sophisticated biomedical implants, environmental sensors, and advanced wearables serve as a few of the appealing prospects. Some aspects such as durability of the materials, scalability of the fabrication methods, and reliable long term usage in aggressive conditions need to be solved. Moreover, improving medical device biocompatibility, facilitating low cost production methods, and environmentally sustainable fabrication techniques are necessary the making technologies widespread. Solving these issues will expand the spectrum of practical applications of flexible NGs, making them suitable for a wider range of industries.

References

1. Fu, Y., Zang, W., Wang, P., et al.: Portable room-temperature self-powered/active H_2 sensor driven by human motion through piezoelectric screening effect. Nano Energy **8**, 34–43 (2014). https://doi.org/10.1016/j.nanoen.2014.05.012
2. Wang, Z.L., Song, J.: Piezoelectric nanogenerators based on zinc oxide nanowire arrays. Science **312**, 242–246 (2006). https://doi.org/10.1126/science.1124005
3. Fan, F.R., Tian, Z.Q., Wang, Z.L.: Flexible triboelectric generator. Nano Energy **1**, 328–334 (2012). https://doi.org/10.1016/j.nanoen.2012.01.004
4. Yang, Y., Guo, W., Pradel, K.C., et al.: Pyroelectric nanogenerators for harvesting thermoelectric energy. Nano Lett. **12**, 2833–2838 (2012). https://doi.org/10.1021/nl3003039
5. Hansen, B.J., Liu, Y., Yang, R., Wang, Z.L.: Hybrid nanogenerator for concurrently harvesting biomechanical and biochemical energy. ACS Nano **4**, 3647–3652 (2010). https://doi.org/10.1021/nn100845b
6. Singh, S., Yadav, P., Gupta, M.K., et al.: Gigantic stimulation in response by solar irradiation in self-healable and self-powered LPG sensor based on triboelectric nanogenerator: experimental and DFT computational study. Sens. Actuators B Chem. **359**, 131573 (2022). https://doi.org/10.1016/j.snb.2022.131573
7. Singh, A., Singh, S., Yadav, B.C.: In_2O_3 nanocubes and $ZnWO_4$ nanorod-based triboelectric nanogenerator for self-powered humidity sensors. Sens. Actuators B Chem. **398**, 134721 (2024)
8. Singh, S., Bhan, C., Bhaduri, A. et al.: Self-powered fluoride detection and removal system using waste material utilization and self-healing capabilities. Mater. Today Commun. 111061 (2024)

9. Lin, Z., Chen, J., Yang, J.: Recent progress in triboelectric nanogenerators as a renewable and sustainable power source. J. Nanomater. **2016**, 5651613 (2016). https://doi.org/10.1155/2016/5651613
10. Parida, K., Xiong, J., Zhou, X., Lee, P.S.: Progress on triboelectric nanogenerator with stretchability, self-healability, and bio-compatibility. Nano Energy **59**, 237–257 (2019). https://doi.org/10.1016/j.nanoen.2019.01.077
11. Liu, S., Yuan, G., Zhang, Y., et al.: A self-powered gas sensor based on coupling triboelectric screening and impedance matching effects. Adv. Mater. Technol. **6**, 2100310 (2021). https://doi.org/10.1002/admt.202100310
12. Wang, Y., Zhang, J., Jia, X., et al.: TENG-based self-powered device—the heart of life. Nano Energy **119**, 109080 (2024)
13. Xu, Q., Wen, J., Qin, Y.: Development and outlook of high output piezoelectric nanogenerators. Nano Energy **86**, 106080 (2021)
14. Li, Z., Zheng, Q., Wang, Z.L., Li, Z.: Nanogenerator-based self-powered sensors for wearable and implantable electronics. Research **2020**, 8710686 (2020). https://doi.org/10.34133/2020/8710686
15. Wang, Z.L.: On the first principle theory of nanogenerators from Maxwell's equations. Nano Energy **68**, 104272 (2020). https://doi.org/10.1016/j.nanoen.2019.104272
16. Aaryashree, Sahoo, S., Walke, P. et al.: Recent developments in self-powered smart chemical sensors for wearable electronics. Nano Res. **14**, 3669–3689 (2021). https://doi.org/10.1007/s12274-021-3330-8
17. Cai, C., Mo, J., Lu, Y., et al.: Integration of a porous wood-based triboelectric nanogenerator and gas sensor for real-time wireless food-quality assessment. Nano Energy **83**, 105833 (2021). https://doi.org/10.1016/j.nanoen.2021.105833
18. Xue, X., Nie, Y., He, B., et al.: Surface free-carrier screening effect on the output of a ZnO nanowire nanogenerator and its potential as a self-powered active gas sensor. Nanotechnology **24**, 225501 (2013). https://doi.org/10.1088/0957-4484/24/22/225501
19. Maity, K., Pal, U., Mishra, H.K., et al.: Piezo-phototronic effect in highly stable $CsPbI_3$–PVDF composite for self-powered nanogenerator and photodetector. Nano Energy **92**, 106743 (2022). https://doi.org/10.1016/j.nanoen.2021.106743
20. Gao, Y., Wang, Z.L.: Electrostatic potential in a bent piezoelectric nanowire. The fundamental theory of nanogenerator and nanopiezotronics. Nano Lett. **7**, 2499–2505 (2007)
21. Wang, Z.L.: On Maxwell's displacement current for energy and sensors: the origin of nanogenerators. Mater. Today **20**, 74–82 (2017)
22. Liu, Y.-Z., Zhang, Q., Wang, X.-X., et al.: Review of electrospinning in the fabrication of nanogenerators. ACS Appl. Nano Mater. **7**, 4630–4652 (2024)
23. Barhoum, A., Pal, K., Rahier, H., et al.: Nanofibers as new-generation materials: from spinning and nano-spinning fabrication techniques to emerging applications. Appl. Mater. Today **17**, 1–35 (2019)
24. Ali, A., Baheti, V., Militky, J.: Energy harvesting performance of silver electroplated fabrics. Mater. Chem. Phys. **231**, 33–40 (2019)
25. Wajahat, M., Kouzani, A.Z., Khoo, S.Y., Mahmud, M.A.P.: A review on extrusion-based 3D-printed nanogenerators for energy harvesting. J. Mater. Sci. **57**, 140–169 (2022)
26. Park, H.-W., Huynh, N.D., Kim, W., et al.: Electron blocking layer-based interfacial design for highly-enhanced triboelectric nanogenerators. Nano Energy **50**, 9–15 (2018)
27. Kim, D.W., Lee, J.H., You, I., et al.: Adding a stretchable deep-trap interlayer for high-performance stretchable triboelectric nanogenerators. Nano Energy **50**, 192–200 (2018)
28. Zhang, R., Olin, H.: Material choices for triboelectric nanogenerators: a critical review. EcoMat **2**, e12062 (2020)
29. Zhao, J., Li, F., Wang, Z., et al.: Flexible PVDF nanogenerator-driven motion sensors for human body motion energy tracking and monitoring. J. Mater. Sci. Mater. Electron. **32**, 14715–14727 (2021). https://doi.org/10.1007/s10854-021-06027-w

30. Mi, H.Y., Jing, X., Zheng, Q., et al.: High-performance flexible triboelectric nanogenerator based on porous aerogels and electrospun nanofibers for energy harvesting and sensitive self-powered sensing. Nano Energy **48**, 327–336 (2018). https://doi.org/10.1016/j.nanoen.2018.03.050
31. Sun, J., Li, Y.: Research on improving energy storage density and efficiency of dielectric ceramic ferroelectric materials based on $BaTiO_3$ doping with multiple elements. J. Compos. Sci. **7**, 233 (2023)
32. Bhaduri, A., Singh, S., Thapa, K.B., Yadav, B.C.: Improved room temperature liquefied petroleum gas sensing performance of $Ni_{0.5}Zn_{0.5}Fe_2O_4$@Cl-doped polypyrrole nanoweb. Mater. Sci. Eng. B Solid-State Mater. Adv. Technol. **279**, 115660 (2022). https://doi.org/10.1016/j.mseb.2022.115660
33. Singh, S., Bhaduri, A., Tripathi, R.K. et al.: Improved sensing behaviour of self-healable solar light photodetector based on core-shell type $Ni_{0.2}Zn_{0.8}Fe_2O_4$@poly(urea-formaldehyde). Sol. Energy **188**, 278–290 (2019). https://doi.org/10.1016/j.solener.2019.06.003
34. Singh, S., Dzeranov, A., Bondarenko, L., et al.: Modified Fe_3O_4 magnetite core@shell type nanomaterials for highly-responsive LPG sensing: a comparative analysis. ECS Sens. Plus **2**, 13601 (2023)
35. Sonker, R.K., Singh, K., Sonkawade, R. (eds.): Advanced Functional Materials for Optical and Hazardous Sensing: Synthesis and Applications, vol. 27. Springer Nature (2023)
36. Bhaduri, A., Singh, S., Thapa, K.B., Yadav, B.C.: Visible light-induced, highly responsive, below lower explosive limit (LEL) LPG sensor based on hydrothermally synthesized barium hexaferrite nanorods. Sens. Actuators B Chem. **348**, 130714 (2021). https://doi.org/10.1016/j.snb.2021.130714
37. Kumar, U., Gautam, R., Sonker, R.K., Yadav, B.C., Chan, K.-L., Wu, C.-H., Huang, W.-M.: Micro and nanofibers-based sensing devices. In: Smart Nanostructure Materials and Sensor Technology, pp. 97–112. Springer Nature Singapore, Singapore (2022)
38. Sonker, R.K., Sikarwar, S., Sabhajeet, S.R., Yadav, B.C.: Spherical growth of nanostructures ZnO-based optical sensing and photovoltaic application. Opt. Mater. **83**, 342–347 (2018)
39. Singh, S., Bhan, C., Gupta, M.K., et al.: Waste material based self-healable and self-powered detection of hazardous fluoride ions and its removal with novel adsorbent: an unexplored and highly responsive detection method. SSRN Electron. J. (2022). https://doi.org/10.2139/ssrn.4125194
40. Bhaduri, A., Singh, S., Gupta, M.K.: Self-powered fluoride ion detectors based on piezoelectric nanogenerators with filler materials comprising Mn-doped $BaTiO_3$ nanostructures and carbon nanotubes. ACS Appl. Nano Mater. (2023)
41. Li, S., Zhang, Y., Wang, Y., et al.: Physical sensors for skin-inspired electronics. InfoMat **2**, 184–211 (2020). https://doi.org/10.1002/inf2.12060
42. Wu, C., Wang, A.C., Ding, W., et al.: Triboelectric nanogenerator: a foundation of the energy for the new era. Adv. Energy Mater. **9**, 1802906 (2019). https://doi.org/10.1002/aenm.201802906
43. Dhakar, L.: Triboelectric Devices for Power Generation and Self-Powered Sensing Applications. Springer (2017)
44. Singh, S., Tripathi, R.K., Gupta, M.K., et al.: 2-D self-healable polyaniline-polypyrrole nanoflakes-based triboelectric nanogenerator for self-powered solar light photodetector with DFT study. J. Colloid Interface Sci. **600**, 572–585 (2021). https://doi.org/10.1016/j.jcis.2021.05.052
45. Sonker, R.K., Sabhajeet, S.R., Yadav, B.C.: TiO_2–PANI nanocomposite thin film prepared by spin coating technique working as room temperature CO_2 gas sensing. J. Mater. Sci.: Mater. Electron. **27**, 11726–11732 (2016)
46. Guan, X., Xu, B., Gong, J.: Hierarchically architected polydopamine modified $BaTiO_3$@P(VDF-TrFE) nanocomposite fiber mats for flexible piezoelectric nanogenerators and self-powered sensors. Nano Energy **70**, 104516 (2020)
47. Zhu, M., Shi, Q., He, T., et al.: Self-powered and self-functional cotton sock using piezoelectric and triboelectric hybrid mechanism for healthcare and sports monitoring. ACS Nano **13**, 1940–1952 (2019)

48. Khan, H., Mahmood, N., Zavabeti, A., et al.: Liquid metal-based synthesis of high performance monolayer SnS piezoelectric nanogenerators. Nat. Commun. **11**, 3449 (2020)
49. Yang, T., Pan, H., Tian, G., et al.: Hierarchically structured PVDF/ZnO core-shell nanofibers for self-powered physiological monitoring electronics. Nano Energy **72**, 104706 (2020)
50. Mondal, S., Paul, T., Maiti, S., et al.: Human motion interactive mechanical energy harvester based on all inorganic perovskite–PVDF. Nano Energy **74**, 104870 (2020)
51. Zhang, Y., Sun, H., Jeong, C.K.: Biomimetic Porifera skeletal structure of lead-free piezocomposite energy harvesters. ACS Appl. Mater. Interfaces **10**, 35539–35546 (2018)
52. Karan, S.K., Maiti, S., Paria, S., et al.: A new insight towards eggshell membrane as high energy conversion efficient bio-piezoelectric energy harvester. Mater. Today Energy **9**, 114–125 (2018)
53. Gao, S., Chen, Y., Su, J., et al.: Triboelectric nanogenerator powered electrochemical degradation of organic pollutant using Pt-free carbon materials. ACS Nano **11**, 3965–3972 (2017)
54. Kumar, M., Choudhary, S., Sharma, S.K., Randhawa, J.K.: Piezoelectric nanogenerators with hybrid nanofibers: a dual approach for energy generation and wastewater treatment. Environ. Sci. Nano (2025), in press
55. Reda, A.T., Weldemhret, T.G., Park, J.Y., et al.: Synthesis and characterization of zinc basic salt-loaded PVA–PEI polymeric composite for antimicrobial activity and triboelectric nanogenerator applications. Sens. Actuators A Phys. **370**, 115197 (2024)
56. Priyadarshini, B.S., Mitra, R., Manju, U.: Titania nanoparticle-stimulated ultralow frequency detection and high-pass filter behavior of a flexible piezoelectric nanogenerator: a self-sustaining energy harvester for active motion tracking. ACS Appl. Mater. Interfaces **15**, 45812–45822 (2023)
57. Febriola, N.A., Singh, S., Jeong, E., Yoon, J.: Highly deformable triboelectric nanogenerators fabricated using high-dielectric elastomers and double-network ionic hydrogels for use in energy harvesting and motion sensing. Chem. Eng. J. 158641 (2024)

Metal–Organic Framework (MOF)-Derived Flexible Energy Storage

Suprimkumar D. Dhas, Pragati N. Thonge, Tushar T. Bhosale, Amar M. Patil, Maqsood R. Waikar, Rajendra G. Sonkawde, Muhammad Ramzan Khawar, Dongwhi Choi, and Annasaheb V. Moholkar

Abstract Due to their well-ordered and flexible composition and structure, metal–organic frameworks (MOFs) have been the focus of intense interest as multifunctional nanostructures for electrochemical energy storage systems (ESS). The chapter explores the basic engineering approaches that increase the potential of MOF-derived materials in flexible ESS, including metal-ion batteries such as lithium-ion, sodium-ion, and zinc-ion batteries and supercapacitors (SCs), by controlled constituent design and state-of-the-art nanostructuring techniques. Specific focus is laid on MOF-based nanostructured arrays, which, when grown on conductive and flexible substrates with or without binders, enable fast electron and ion transportation, accelerated reaction kinetics, and exposure of more active sites. These 3-D nanoarray electrodes preserve structural integrity under bending or twisting, this nanoarray provides more active sites during charge/discharge cycling and significantly improves the device's specific capacity/capacitance. The chapter compares common substrates of MOF-based nanoarrays and addresses the crucial correlation between material composition, structure, and electrochemical activity. A systematic summary of MOF-based and derived nanostructures (NSs) such as metal selenides, oxides, nitrides, phosphides, sulfides, and nitrogen-doped carbons summarizes their performance and application

S. D. Dhas (✉) · T. T. Bhosale · A. V. Moholkar
Thin Film Nanomaterials Laboratory, Department of Physics, Shivaji University, Kolhapur, Maharashtra, India
e-mail: suprimkumardhas@gmail.com

P. N. Thonge
Department of Botany, Punyashlok Ahilyadevi Holkar, Solapur University, Solapur, Maharashtra, India

A. M. Patil
Nano-Electro-Mechanical Device Laboratory School of Mechanical Engineering, Yonsei University Seoul, Seoul, Republic of Korea

M. R. Waikar · R. G. Sonkawde
Radiation and Materials Research Laboratory, Department of Physics, Shivaji University, Kolhapur, Maharashtra, India

M. R. Khawar · D. Choi
Department of Mechanical Engineering (Integrated Engineering Program), Kyung Hee University, Yongin, Gyeonggi, Republic of Korea

R. K. Sonker et al. (eds.), *Low-Dimensional Multifunctional Hybrid Nanostructured Materials*, Engineering Materials, https://doi.org/10.1007/978-981-96-7952-2_7

prospects in flexible ESS. The chapter also presents the current technical challenges and viewpoints on the future growth of MOF-derived NSs. The chapter concludes with a discussion on current limitations and gives prospective strategies for the further growth of MOF-derived NSs for flexible ESS, to bridge current gaps and accelerate advances in next-generation energy storage technologies.

Keywords Metal–organic frameworks (MOF) · Flexible energy storage systems (FESS) · Derivatives · Functional materials · Electrochemistry

1 Introduction

The exponential growth in worldwide population and greater reliance upon fossil fuels have created an acute energy crisis, deterioration in the environment, and climatic issues in the form of global warming and frequent occurrences of extreme climatic phenomena. Accordingly, developing eco-friendly and sustainable energy storage systems is an absolute priority. Sustainable electrochemical energy storage systems that offer environmentally beneficial energy storage solutions include lithium–sulfur (Li–S) batteries (BTs), metal–air BTs, metal-ion (Li, Na, and K) BTs, and supercapacitors (SCs). These devices are suitable for use in power grids, electric vehicles, and portable electronics because of their efficient ability to store and release electrical energy [1–3]. Despite their advancements, these technologies continue to face significant challenges, including the shuttle effect in lithium–sulfur (Li–S) BTs, slow oxygen redox kinetics during recharging in metal–air (MA) BTs, uncontrolled dendritic formation in metal-ion (MI) BTs, and large volume changes in electrode materials. Reduced energy density, decreased rate capabilities, and decreased cycle stability are all caused by these problems [4–6]. The development of highly effective electrode materials with ideal electrochemical properties is crucial to the advancement of environmentally friendly energy conversion and storage technologies. Important issues in this field can be successfully addressed by modifying the composition and nanostructure of these materials. Incorporating various compositions can introduce multifunctionality within a single system. Various high-performance ESSs with superior flexibility, such as Li-ion BTs, sodium-ion BTs, K-ion BTs, Li–S BTs, MA BTs, Zn-ion BTs, and supercapacitors (SCs), have been extensively developed [7, 8]. In particular, nanostructured MOFs and MOF-derived NSs have developed as promising innovative electrode NSs for flexible electrochemical energy storage (FEES) applications (Fig. 1).

Moreover, materials with porous structures and good surface areas are advantageous, as they provide abundant accessible active sites. Therefore, careful design of these materials is crucial to achieving high efficiency and strength in practical ESS. Among the materials of interest, MOFs have paid important consideration because of their potential as advanced materials for efficient ESS applications. MOFs bid novel compositional and structural benefits compared to traditional materials, including extremely well-ordered, tunable metal nodes and organic linkers. Their inherent high

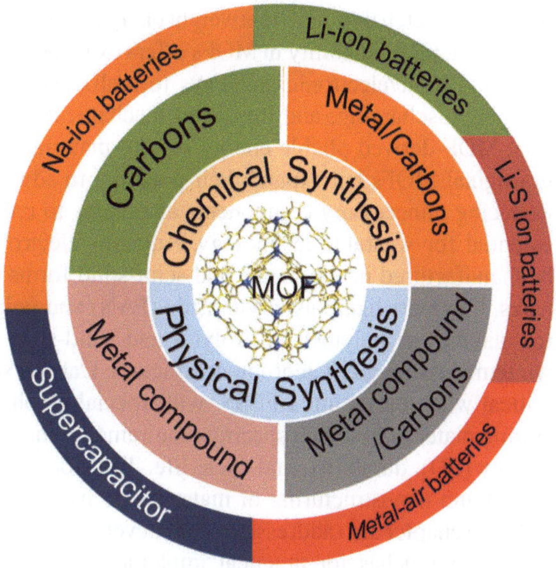

Fig. 1 An intangible overview of MOF-based NSs as advanced materials for FESS

porosity enhances accessibility to interior active species; facilitates reactant accumulation; and improves reaction kinetics through open channels, spatial confinement, and host–guest relations within the pores. By carefully selecting metal sites and ligands, it is possible to design pore size, shape, and surface environment to enhance these structural features [9].

Additionally, various MOF nanoarchitectures, containing 0-D, 1-D, 2-D, and 3-D NSs hierarchical NSs with highly porous, may be synthesized through precise control over MOF growth processes, such as recrystallization, surface modulation, and size confinement. Furthermore, the formation of superstructures by assembling MOF nanocrystals employing template-guided methods for MOF-based material development enhances the diversity of these structures. The resulting porous configurations and extensive surface areas of these innovative MOF-based NSs provide benefits for efficient ESS, including improved mass transport and greater exposure to active sites. In most aspects, MOFs possess vast potential in future energy storage systems because of tunable composition, high surface area, and potential to be hybridized. Their drawbacks in structural stability, electrical conductivity, and challenges in synthesis have to be tackled to achieve their potential in practical application. Structural reinforcement, proper current collector material selection, conductivity enhancement, and cost-effective modes of synthesis should be targeted in future work to reduce these limitations. Their performance can be improved further by incorporating advanced nanomaterials and novel electrode designs to achieve their vast potential in commercialized applications [10]. In an attempt to solve these problems, MOF-derived ESS systems have gained mechanical flexibility, high ED, and long-term durability. MOF-derived hierarchically porous matrices are composed of carbon, metal oxide/carbon composites, and conductive networks to support both bendable, stretchable, and lightweight energy-storing devices, in addition to. Furthermore, structural tunability in MOFs offers exact pore-size control, functional surface area, and ionic/electronic pathways to realize better ion diffusion, faster rates of charging/discharging, and improved mechanical strength upon repeated deformation. MOF-derived pliable energy-storing materials not only overcome limitations in original MOFs but expand their potential to advanced technological fields [11]. Reducing unnecessary components also helps to create lightweight ESS, which is essential for wearable technology. Aqueous electrolytes for these devices should be manufactured in solid-state and semi-solid forms to increase safety and prevent leaks. Polymers like polyacrylonitrile (PAN) and polyacrylamide (PAM) are useful for making gel polymer electrolytes and solid-state SCs without using separators while maintaining efficient ion conduction. Careful selection of each component is critical when designing a FESS, with special emphasis on optimizing the energy-storing material based on specific requirements such as energy density (ED), power density (PD), durability, and cycle life. This book chapter emphasizes the strategic design and nanostructuring of materials derived from MOFs for use in FESS [12, 13]. The chapter also addresses recent developments and existing challenges in these design approaches for practical implementation in FESS. It is intended that the content will offer valuable insights and inspire future advancements in the growth of high-performance NSs, contributing to effective and long-lasting FESS technologies.

2 MOFs in ESS

In this section, we will explore various synthetic methods aimed at optimizing the properties of MOFs for targeted ESS (as illustrated in Fig. 1). The discussion will center on the following key aspects: (i) modification of physicochemical characteristics, (ii) enhancement of porous nature and specific surface, (iii) improvement of charge transportation capabilities, and (iv) measurability and ease of processing.

2.1 Physical and Chemical Characteristics

MOFs are notorious for their versatility in modifying material properties when the linker and metal node are carefully chosen. Numerous papers have discussed a variety of synthetic methods that allow for the modification of MOF characteristics, including chemical composition, stability, particle size, and structural flexibility. To further alter the attributes of MOFs, components inside the framework can be added, changed, and removed after synthesis. Furthermore, crystal's size of MOF, shape, and crystallographic phases can all be altered to affect surface chemistry, which is crucial for applications like ESS [14, 15]. By adjusting these characteristics, MOFs can strategically regulate host–guest interactions, which makes them ideal for energy-related applications. For instance, selecting synthesis conditions correctly can increase electrochemical stability. MOFs with shorter, stiffer linkers, and redox-inactive nodes are more chemically and thermally stable. However, the high porosity and flexibility of MOFs lead to improved ion storage and transport capabilities. Additional methods to alter the mechanical characteristics of MOFs include the use of multi-metallic frameworks, altering host–guest interactions, incorporating flexible linkers, and regulating crystal size. Structural stiffness is beneficial for applications that are susceptible to dendritic development, which could result in short circuiting. However, in strategies where the active NSs undergo significant volume changes, a malleable MOF structure could help maintain the integrity of the material [16, 17]. Furthermore, flexible, stimuli-responsive MOFs could act as a safety mechanism, shutting down device operation when triggered by temperature, voltage, and mechanical changes, thus protecting both the device and the user. Distinguishing properties of MOFs are shown in Fig. 2.

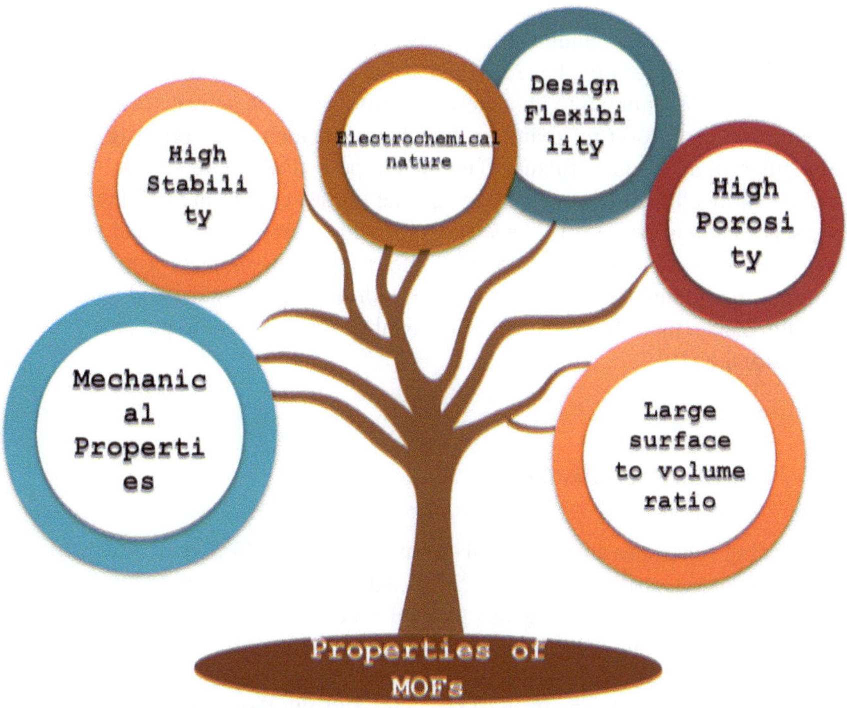

Fig. 2 Distinguishing properties of MOFs

2.2 *Porous with a Large Surface Area*

MOFs are well known for their remarkable porous nature and large surface area. By carefully selecting suitable metal nodes and linkers, it is likely to precisely regulate the pore size and structure of the framework. Isoreticular MOFs, which share a similar structural topology, can be synthesized by a variety of associated organic linkers with diverse lengths and functional groups. The high degree of resistor over the pore diameter and chemical atmosphere in reticular MOFs allows for the isolation and investigation of the structural and chemical components that influence electrochemical performance. Furthermore, the high porosity of MOFs, which enhance molecular mobility by combining micropores and mesopores, creates novel channels and pore spaces [18]. The increased surface area of these frameworks is particularly beneficial for catalytic reactions, like the oxygen reduction reactions (ORR) in lithium–oxygen BTs. This ability to modify porous nature and surface area is unparalleled in traditional inorganic materials and holds significant potential for a variety of FESS. Moreover, MOFs are recognized for having high porosity and larger surface areas than other nanomaterial. The general surface areas for MOFs are 1500 and 6000 m^2/g, while some have reached as far as 7000 m^2/g, and have as much as 90% porosities. Compared to this, there is a surface area in activated carbons in the range of 400 and 2,000 m^2/g and zeolites have as much as 1,500 m^2/g in surface and regular microporous structure. The regular pore structure in silicas like MCM-41 has a surface area of over 1,000 m^2/g. The porous crystalline substance classes known as covalent organic frameworks have good heat endurance and permanent pore space and have surface areas over zeolites and porous silicates. The unique pore space and functionality tunability coupled with good pore space and surface area in MOFs makes them highly suitable for application in gas storage, catalysis, and energy storage [19].

2.3 *Freestanding Electrodes*

Freestanding electrodes are a potential means of enhancing FESS performance due to their advantageous structural characteristics. MOF-based nanomaterials offer a range of metallic ions and organic ligands that enhance access to active sites and accelerate ion transport. They also have a highly porous nature and considerable specific surface. Consequently, it is anticipated that the synergistic effects of freestanding NSs and MOFs will enhance the performance of FESS. Freestanding MOF-derived electrodes may be broadly categorized into three classes depending on their structure and content: self-supporting MOF-based electrodes. The synthetic methods used have a major impact on the structural and electrochemical characteristics of these freestanding electrodes [20, 21]. The following are benefits of powder-based freestanding versions of conventional MOF-derived materials: They enhance electron transfer kinetics by eliminating inactive polymer binders, which makes it easier for active materials to make contact with conductive substrates and encourages effective electron transport. (i) They simplify the fabrication procedure and provide a uniform distribution of active elements by enabling the direct use of MOF-derived materials as binder-free electrodes.

The use of pure MOFs in ESS is typically constrained by small electrical conductivity. A promising approach to overcoming these limitations is the integration of freestanding architectures with MOF-based materials. This combination leverages the superior ion diffusion properties of MOFs, which are characterized by abundant porous structures. Additionally, freestanding MOF-based electrodes offer numerous reactive sites because of highest specific surface area and the absence of additives. This design improves the electrochemical performance by eliminating "dead mass," which in non-renewable powder-form MOF-derived materials can exceed 20%, thereby enhancing both volumetric and mass energy densities. In terms of cycle stability, energy storage devices must billet the volume extension of active materials throughout the redox process, which helps to prevent capacitance degradation. For energy transformation, open-structure arrays and strong substrate adhesion promote rapid diffusion of gases and mitigate material exfoliation during catalytic processes. Furthermore, freestanding MOF-based electrodes allow for a wider selection of conductive substrates, ranging from 1-D to 3-D forms, such as CNTF, carbon cloth, carbon paper, metal mesh, carbon-derived aerogels, and Ni and Cu foams. These substrates confer unique advantages to the electrodes, including low weight, reduced volume, flexibility, and cost-effectiveness [22, 23].

2.4 Synthesis of MOF NSs

Several porous MOFs have been manufactured in new years, as illustrated in Fig. 1. The synthesis methods chosen are based on the desired properties and final structures of the MOFs. Different techniques used for MOF synthesis include hydrothermal/solvothermal methods, electrochemical synthesis, mechanochemical processes, reaction–diffusion, microwave heating, and ultrasound irradiation. Each of these methods has been thoroughly reviewed in recent literature. Hydrothermal synthesis, typically performed at temperatures between 80 and 250 °C, is widely favored for its scalability, particularly in industrial production [15, 24]. However, it has certain drawbacks, such as the high costs of pressure-sealed heating systems and extended processing times. The reaction–diffusion process is effective for producing large single crystals but is limited by extremely slow reaction times, ranging from several days to months. The electrochemical approach is considered fast and environmentally friendly, with successful application in producing industrial-grade MOF HKUST-1. Nevertheless, it lacks versatility, as it cannot be used to synthesize a broad range of MOFs. Mechanochemical synthesis, which does not require solvents, offers an efficient method for MOF production but poses challenges for obtaining single crystals. Microwave-assisted synthesis has the advantage of high yields and improved control

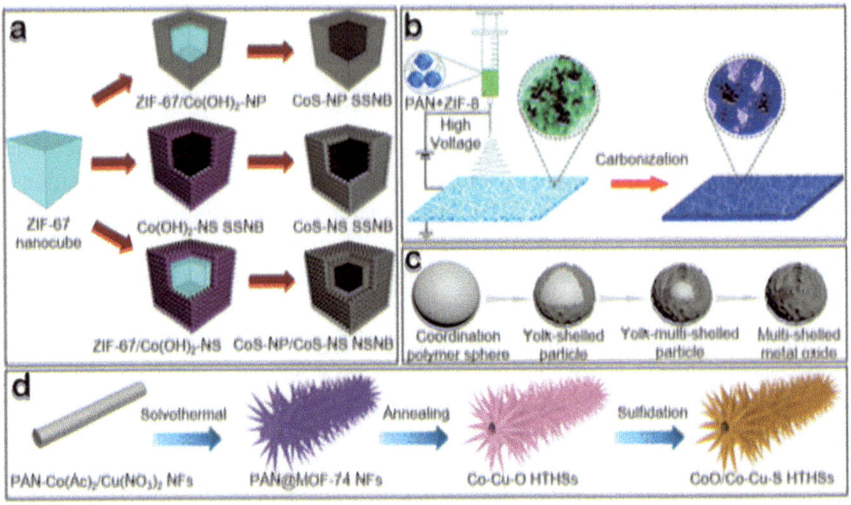

Fig. 3 Several common synthesis techniques for functional materials for SCs developed from MOFs. **a** To create complicated CoS NBs, sequentially alter ZIF-67's chemical interactions with water and Na_2S. **b** Carbonization after electrospinning for N-doped CNF based on hollow Nanoparticles (NPs). **c** Further heat treatment is to produce particles of multishelled metal oxide. The CoO/Co–Cu–S HTHSs are approached in a multistep manner [25]. Reproduce with Open access @ (2021) American Chemical Society

over morphology, particle distribution, and phase selectivity. However, it is not suitable for the production of large single crystals, and its slow processing speed limits industrial applications. Ultrasound irradiation offers rapid synthesis of phase-pure materials with uniform particle size and morphology, though it also presents challenges in producing large single crystals due to the disruption of seeding crystallites during the process. We have herein briefly presented a few typical synthetic methods for high-performance nanomaterials with novel NSs (Fig. 3) [25].

2.5 *Physical and Chemical Properties*

MOFs are widely recognized for tunable functionality, which can be achieved by carefully selecting linkers and metal nodes. Numerous literatures have outlined various synthetic approaches to adjust their chemical properties, cyclic life, particle size, and flexibility. Additionally, MOFs can suffer "post-synthetic modification," which permits additional customization by the removal, alteration, or exchange of linker and metal node components inside the framework. Adjusting the size and form of crystallites and adjusting certain crystallographic phases can also have an impact on the surface chemistry of MOFs. The tunability of MOFs is a critical feature that offers precise control over host–guest interactions and is particularly significant for energy storage applications [26].

The cyclic life span of MOFs can be boosted by meticulously choosing the synthesis conditions. For example, MOFs with redox-inactive metal nodes and shorter, stiffer linkers tend to be more chemically and thermally stable. However, for efficient ion transport and storage, frameworks with greater porosity and flexibility are more suited. Flexible linkers, multi-metallic framework design, host–guest contact strength modulation, and crystal size control can all change the mechanical properties of MOFs. For electrode materials that are susceptible to dendritic growth, which can cause short circuiting, structural stiffness may be essential. Nonetheless, MOF flexibility can support the preservation of structural reliability in systems where the active material experiences significant volume variations [27]. With a protective "shut-off" feature that is activated by changes in temperature, potential, and mechanical outputs to safeguard both the user and the apparatus, flexible, stimuli-responsive MOFs can also be utilized as safety devices. MOFs are particularly well known for exceptional high porous nature and huge specific surface. The pore diameter and framework architecture may be carefully controlled by choosing the right linkers and metal nodes. To create isoreticular MOFs, which have the same structural topology, a group of connected organic linkers with varying lengths and functional groups is employed [28, 29]. Because of their protective "shut-off" feature, which is activated by changes in temperature, voltage, or mechanical signals, flexible, stimuli-responsive MOFs can also be utilized as safety measures to safeguard both the user and the apparatus. Because of their exceptional porous nature and huge surface area, MOFs are particularly well known. It is feasible to accurately adjust the pore size and framework architecture by selecting the right linkers and metal nodes. Isoreticular MOFs are created using a collection of connected organic linkers with varying extents and functional groups, but they share the same structural topology [30, 31].

2.6 Charge Conduction

Efficient ion and charge conveyance is crucial for the performance of electrochemical devices. High internal resistances, leading to large overpotentials, can reduce the PD and overall performance of BTs and SCs. Emerging synthetic approaches to improve ion and charge conduction of MOFs not only create novel nanomaterials suitable for charge conveyance systems but also offer valuable molecular insights for advancing materials beyond porous MOF. Despite the fast growth of novel MOFs over the past 20 years, many of these frameworks remain insulating, primarily because of the prevalence of M–O linkages. To address this issue, extensive studies are focused on modifying the electronic structure by optimizing linkers and metal nodes using a strategy known as "through-bond" conductivity. For example, leveraging the loosely bound d-electrons in Fe^{2+} along with soft sulfur-based linkers resulted in a six order of magnitude increase in electronic conductivity associated to analogous manganese–sulfur bonds in M_2(dsbdc) MOFs, and a tenfold development over M–O linkages in M_2(dobdc). Another strategy for enhancing electron conductivity involves "through-space" mechanisms, such as π–π exchanges or electron hopping to guest molecules.

Several methods are also used to enhance ionic and proton conductivity in MOFs [32]. The incorporation of ionic guest species aids transport. Polar functional groups can be introduced post-synthetically into the framework, offering the possibility of controlling the direction and distribution of ion transport. Additionally, ionic frameworks with mobile counterions can significantly improve ionic conductivity. While conductivity of hole in MOFs has been a major focus, especially because of interest in hydrogen fuel cells, ion-conductive MOFs are expected to gain significant attention as the claim for solid electrolyte materials in ESS continues to grow. The wide range of approaches for adjusting the chemical, electronic, structural, and mechanical properties of MOFs makes them highly suitable for ESS [33]. However, different ESSs prioritize different material characteristics. Therefore, understanding the specific requirements of each device component is essential for choosing suitable design strategies and synthetic techniques. For instance, the highly porous nature of MOFs is advantageous for storing active species, which is mainly useful in systems that involve the chemical transportation of small molecules, such as polysulfides in Li–S BTs or O_2 in lithium–oxygen BTs. In these cases, enhancing electronic conductivity would favor "through-bond" mechanisms over "through-space" approaches.

2.7 Scalability and Process Stability

The rapid rise in MOF research over the past two decades can make it easy to overlook that this field is still in its initial phases. To transition these innovative materials from laboratory research to commercial applications, significant focus is needed on the scalability and process stability of MOFs. In terms of scalability, companies such as BASF, MOFWORX, MOF technologies, and NuMat have already demonstrated large-scale MOF production, utilizing chemical synthesis techniques, which involves physically mixing components, such as linkers and metal nodes, through grinding or ball milling without the need for solvents or extended heating, which helps to minimize both price and chemical loss [34]. This method is also well suited for producing mixed-metal MOFs by adjusting the reagents used. Using cost-effective metal precursors, such as salts, oxides, hydroxides, and carbon-based materials, can help in additional reduction of MOF production costs. Another innovative approach to improving MOF scalability is microwave irradiation, which may swiftly synthesize MOFs from precursors. MI needs less energy than conventional solvothermal techniques due to its focus on confined heating. MOFs, which are formed from a few basic chemical building blocks, are highly appealing for integration into ESS. Solution-phase layer-by-layer deposition, which gives rise to a subclass known as SURMOFs, enables precise control over molecular structure and film thickness. Electrochemical synthesis techniques facilitate uniform and rapid MOF growth, allowing for direct deposition onto conductive substrates with controlled layer thickness. Even greater control over film thickness can be achieved using vapor deposition techniques, such as ALD and CVD, which have facilitated the growth of MOFs on both flat surfaces and fibers. The variety of promising current collectors and the ability to finely control layer structures expand the potential of MOFs for electrochemical applications [35].

3 Component Design of MOF Composites for ESS

Pristine MOFs frequently have drawbacks that impair their functionality in real-world ESS. Their efficacy in various applications is diminished by these constraints, which usually include low conductivity and limited functional capabilities. Recent developments have investigated combining MOFs with functional materials to produce MOF composites in order to overcome these difficulties. This method mitigates the drawbacks of MOFs while maintaining their intrinsic advantages. Metal NPs, CNT, rGO, porous supports, molecular complexes, and conductive current collector are just a few of the functional elements that have been added to MOF composites to improve their properties and make them appropriate for advanced ESS applications (Fig. 4) [7].

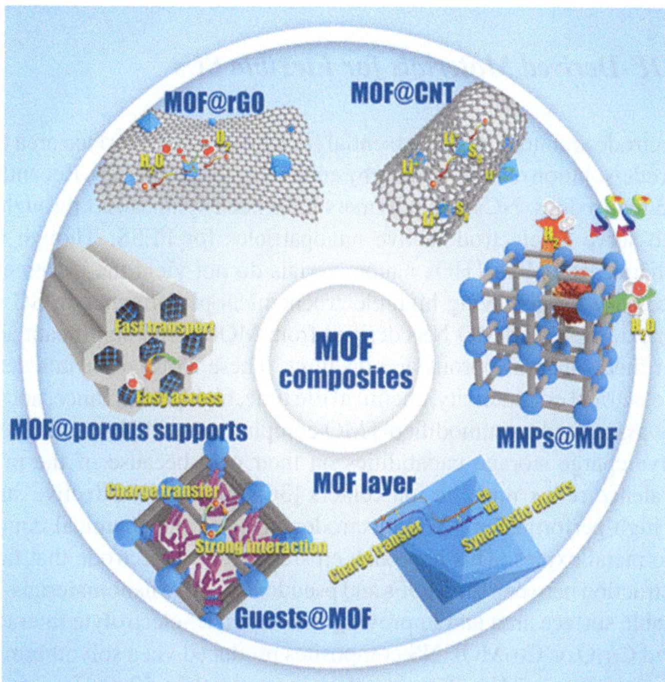

Fig. 4 Constituent design of MOF composites [7]. Reproduced with Open access @ (2021) American Chemical Society

4 Flexible Energy Storage Systems (FESS)

Every element of an FESS needs to be adaptable. This may be achieved via a flexible substrate, free-standing MOF material, and gel solution. Examples of the flexible substrates that are commonly used as mounting surfaces for electrode materials include carbon-based fabric, fibers, tapes, and metal foils and wires. The creation of free-standing electrode films is beneficial since it removes the requirement for polymer binders that don't add to the energy stored, increasing the device's overall ED [36]. Reducing superfluous parts is essential to developing energy storage solutions that are lightweight, which is important for wearable electronics. Solid-state or semi-solid aqueous-based electrolytes should be used in wearable electronics to address safety issues and prevent leakage. Solid-state SCs can be built without the need for separators by using gel polymer electrolytes composed of several polymers, such as PAN and PAM, which preserve efficient ion conduction pathways. This emphasizes how crucial it is to choose every part of an FESS wisely [8]. The primary goal must be to design the best energy storage material that meets specific requirements, including energy capacity, power output, shelf life, and cycle life.

4.1 *MOF-Derived Materials for Flexible SCs*

Porous electrode architectures are essential for a given a huge surface area that facilitates electrode/solution reactions, thereby enhancing the ED of SCs. Recently, various TMOs and hydroxides, NCs and polymers have been synthesized through different methods to serve as electrode-active nanoparticles for FESS. Though numerous of these chemicals TMOs/THOs nanomaterials do not yield the extremely porous nature necessary for achieving high electrochemical performance in SC materials. In this context, metal and MO NSs derived from MOFs offer significant advantages because of their inherent porous architectures. These TMO materials demonstrate improved electrical conductivity, chemical life time, high performance, and enhanced rate ability associated to unmodified TMO complements. MOFs have limited pseudocapacitive charge storage capabilities on their own because of the presence of variable valence redox metallic ion centers [36]. Additional effective strategy for emerging high-performance TMO electrodes involves the chemical conversion of MOFs into metal oxides. This approach creates a hybrid electrode that facilitates a close construction between the MOFs and pseudocapacitive nanomaterials, attracting the obtainable surface area and improving the electrode–electrolyte interactions [8].

A hybrid Co_3O_4@Co-MOF NS composites produced via a solvothermal process caused the formation of Co_3O_4 nanocubes approximately 50 nm in size, anchored onto leaf-like Co-MOF sheets. This structure exhibited a high surface area of 453.3 m^2/g, incorporating both microspores and mesopores. Such a hierarchical and porous architecture promotes enhanced electrolyte diffusion and improved redox reactions. In a 3 M KOH electrolyte, the electrodes displayed superior pseudocapacitive performance, achieving Cs of 1020 F/g at 0.5 A/g. An FESS built with Co_3O_4@Co-MOF and activated carbon electrodes achieved a potential of 1.5 V and an aerial capacitance of 192 mF/cm^2 at 0.5 mA/cm^2. Furthermore, the device maintained almost 99.7% of its capacitance after 400 cycles, indicating remarkable cycling stability and appropriateness for powering flexible electronic devices. In a related study, MOFs were chemically converted into metal oxides to create MHCF MOF nanocubes [37]. Furthermore, hollow and porous $NiCo_2O_4$ nanowall arrays were effectively produced using a straightforward two-step solution approach that comprised ion exchange, etching, and calcination. This process utilized 2-D Co-MOFs on a flexible CC substrate. The direct growth of the hollow, porous structures on the carbon fabric resulted in consistent coverage, improved surface area, higher mechanical stability, and enhanced ion transport, which made them appropriate for FESS. Flexible electrodes are better created without non-conductive binders since they can add too much weight and are not flexible, which can lead to surface cracks. The performance of three electrodes in a 2 M KOH electrolyte was tested; the Cs was 1055.3 F/g at 2.5 mA/cm^2. Nanostructured carbon and its hybrids, which are beneficial materials for energy storage, can be produced using MOFs [38]. Pseudocapacitive materials frequently have poor electrical conductivity, which leads to limited rate capacity and slow electron transport. A conductive matrix that promotes quicker electron transit can be made by including conductive carbons to overcome this difficulty. To optimize the use of the active material in the creation of carbon-based NSs from MOFs, pseudocapacitive elements should be combined into these carbon structures. As an anode material for asymmetric SCs, N_2-doped carbon NSs made from a Co-based MOF solution on CC, for instance, have demonstrated potential. Inert polymer binders, which might impair the electrode material's overall performance, are not required with this direct growing technique. Additionally, the carbon nanosheets' direct adherence to the CC improves structural integrity during flexible operation, enhancing the device's performance and longevity. Using a solution approach, 2-D sheet-shaped Co-based MOFs were created, and then they were thermally treated in a nitrogen atmosphere to become nitrogen-doped carbon NSs. Images of the resultant carbon NSs obtained using electron microscopy showed that they were around 70 nm thick and had a porous structure with a variety of mesopores and macropores. The Cs of the nanosheet was 321.9 F/g at 6.25 A/g, indicating typical non-Faradaic characteristics. As shown in Fig. 5a, the study team used a "one for two" approach to create positive and negative electrodes from the same MOF chemicals. The cathode material was formed by oxidizing the Co-based MOF in air to produce Co_3O_4 on CC. A flexible ASC was assembled using polyvinyl alcohol (PVA)-KOH electrolyte, which exhibited a quasi-rectangular shape in the CV plots in Fig. 5b. This configuration yielded a capacitance of approximately 116.8 F/g at 7.7 A/g [39]. The device was tested under numerous bending and winding conditions, yet it maintained structural integrity throughout the cycling process, demonstrating high stability with an impressive capacitance retention of 85.5% after 20,000 cycles, as displayed in Fig. 5c. The inset of Fig. 5c reveals how the gadget may be used practically to power LEDs by connecting three units. Other active materials, including MnO_2 and Bi_2O_3, were also deposited onto CC using MOF-derived CNS as substrates. The areal capacitances of the hydrothermally produced Bi_2O_3 and electrodeposited MnO_2 were 329.01 mF/cm^2 and 881 mF/cm^2, respectively. According to EIS investigations, active compounds showed greater resistance when directly put onto CC as opposed to when deposited on carbon NSs [40]. According to this investigation, using MOF-based carbon as the current collector for active material synthesis results in a notable improvement in the efficiency of charge transfer from the active material to the current collector. The completed whole cell demonstrated an aerial capacitance of 311 mF/cm^2 and a potential of 1.7 V, at 3 mA/cm^2. The cell also showed stable capacitance under different bending situations and an Ed of 124.8 μW h/cm^2 at a Pd of 2.55 mW/cm^2. Using a Cs of 381.2 F/g at 5 mV/s, electrochemical testing using a three-electrode setup revealed a virtually rectangular CV profile, which is indicative of EDLC. The electrode showed outstanding rate performance, retaining over 66% of its initial capacitance even at a high scan rate of 200 mV/s. The conductive network created by CNTs woven throughout the electrode layer, which improves electron transport, is responsible for this remarkable rate capability. Additionally, after 10,000 cycles, the porous film maintained 95% of its capacitance, demonstrating exceptional cycling stability. Using these films with a 6 M KOH electrolyte, a symmetric SC was built, exhibiting an ideal capacitive nature with a Cs of 157.9 F/g at 1 A/g. The gadget demonstrated its potential for energy storage applications by achieving an ED of 9.1 W h/kg at a PD of 3500 W/kg [41]. Hybrid electrodes, which leverage the complementary qualities of their respective components, can greatly increase the capacitance of SC devices The creation of ternary hybrid SC electrodes that combine metal oxides, MOFs, and nanocarbon materials has been the subject of recent research. MOFs are prized for their porous structure and vast surface area, whereas nanocarbon elements help to improve mechanical stability and electrical conductivity. Furthermore, by making an interconnected network that connects all of the active components in the electrode and promotes Faradaic reactions that aid in electrochemical charge storage, redox-active polymers can improve electrical conductivity. According to recent research, extremely compressible, three-dimensional porous hybrid electrodes including polyaniline, MOFs, and CNTs have been prepared for use in SC applications. A highly conductive and hydrophilic CNT sponge used as a substrate to create CNT sponge/MOF hybrids. To create a core/shell structure, MOF particles were successively coated onto the CNTs. The CNT sponge in this architecture serves as a current collector in addition to being a substrate for the growth of MOF particles. Ion diffusion is accelerated by the three-dimensional porous design's facilitation of hierarchical pores. Then, using an oxidative chemical polymerization technique, polyaniline was polymerized onto the CNT sponge/MOF hybrid NSs. In this procedure, ZIF-8 and ZIF-67 were both used; however, ZIF-67 gave the ZIF-8 crystals mechanical strength, preventing structural collapse after carbonization. By coating the CNT surfaces with MOF particles, which have an average size of about 200 nm, the accessible surface area for charge storage was greatly increased. At a scan rate of 5 mV/s, an FSSC with CNT sheets as the substrate revealed an aerial capacitance of 30 mF/cm^2. After 1000 cycles, the hybrid electrode

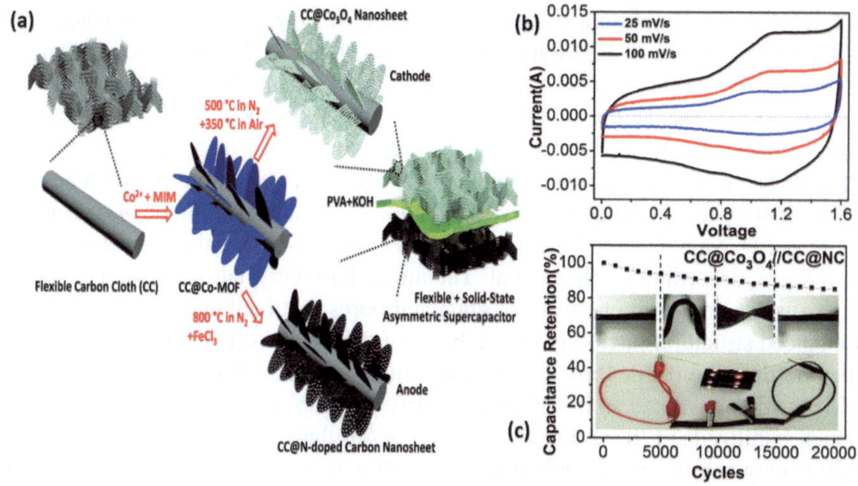

Fig. 5 **a** The "one for two" approach is seen schematically, with the N-doped carbon NS negative electrodes and 2-D Co_3O_4 NS positive electrodes combined into a flexible asymmetric SCs using the same Co-MOF precursors. **b** The constructed CV plots of ASSCs **c** display digital pictures of the cell's flat, bendy positions throughout the cycle test (top) and demonstrate how three complete cells, each measuring $1 \times 2.5 \times 0.08\ cm^3$, can illuminate eight orange LEDs [8]. Reproduced with permission @ Copyright (2017) Royal Society of Chemistry

retained 89% of its capacitance and demonstrated symmetrical and linear galvanostatic charge–discharge (GCD) behaviors [42]. The increased breakdown of the interlinked polymer components could be the cause of the observed stability reduction. The device's potential use in wearable electronics is indicated by its flexibility, which is displayed at diverse bending angles.

4.2 MOF-Derived Materials for Flexible Li-Ion BTs

Li-ion BTs are extensively utilized as the power source for manageable electronic storages. However, due to the mechanical requirements, conventional Li-ion BT designs are not appropriate for flexible electronics. The limits of traditional Li-ion BTs in these applications are demonstrated by the fact that, during vigorous daily use, wearable electronics can experience up to one million mechanical deformation cycles over a 100-day period. Because of this, a lot of research has gone into creating flexible Li-ion BTs, which need new electrode materials with different compositions and architectures in order to improve capacity and rate performance. Because MOFs have the capacity to design and synthesize novel electrode materials, they have demonstrated considerable promise in tackling these issues. Additionally, structural design, electrode and electrolyte materials, and enhanced assessment and characterization methods have been the main areas of recent advancements in flexible Li-ion

BT development. For instance, literature conducted by Ogale et al. exposed the excellent performance of an Fe-MOF anode with two ligands, demonstrating long-term stability over 1000 cycles at 2 A/g and reaching a Cs of 825 mA h/g at 250 mA/g after 100 cycles. This anode provided a 360 W h/kg ED at 500 mA/g for a full coin cell device, which could be charged quickly in 2 min. Furthermore, an electrothermal heater for drug delivery applications was effectively powered by a flexible Li-ion battery that included this Fe-MOF anode and a $LiCoO_2$ cathode [43]. Furthermore, a 2-D, wavy-layered MOF for direct use in electrochemical energy storage was developed, according to Zheng et al. The nickel-based crystalline structure offers a high surface area and a network of abundant pores that facilitate Li^+ intercalation. Enhanced structural flexibility and stability of the framework are achieved through the coordination of Me4bpz ligands with Ni(II) ions. Anodes fabricated from Ni–Me4bpz demonstrated an initial specific capacity (Csp) of 320 mA h/g at 50 mA/g, which stabilized at 120 mA h/g over 100 charge/discharge cycles. Furthermore, post-XRD analysis indicated the retention of the MOF's structural integrity, attributed to the wavy-layer design and ligand flexible capacity. Additionally, Ni–Me4bpz may be thermally converted into NiO NSs, which also exhibit potential for Li-ion storage with a alterable Csp of 200 mA h/g as revealed in Fig. 6a–f [44]. Peng and colleagues developed a method to incorporate active materials into CNT fibers, producing 1-D stretchable electrodes. Specifically, they fabricated a single coiled CNT fiber by twisting a bundle of CNT fibers that were spun from aligned CNT sheets. The coiled structure allows the CNT fiber to withstand significant tensile strain during elongation by untwisting, as demonstrated in their studies [37].

4.3 MOF-Derived Material for Flexible Li–S BTs

Li–S BTs are recognized for their high theoretical energy densities, positioning them as hopeful contenders for future energy storage technologies. However, their practical deployment faces significant challenges, including uncontrolled lithium dendrite formation, slow electrochemical reactions, and the polysulfide shuttle effect. To address these challenges, hierarchical matrices on CC were synthesized using MOFs by Jiang et al. For the anode, ZnO NS arrays were useful as hosts (CC-ZnO), promoting steady lithium deposition and removal. In symmetric cells, the CC-ZnO@Li anode demonstrated improved chemical strength, with a potential hysteresis of approximately 25 mV maintained for after 800 h at 1 mA/cm^2 and a capacity of 1 mA h/cm^2. To tackle issues on the cathode side, the authors advanced a multifunctional CC-NC-Co cathode host, offering physical detention, chemical interaction, and efficient electrocatalytic properties. Device equipped with CC-ZnO@Li anodes and CC-NC-Co@S positive electrode achieved impressive electrochemical performance, exhibiting a minimal capacity decay per cycle (0.02% over 900 cycles at 0.5 C and 0.03% over 600 cycles at 1 C) and superior rate capability (793 mA h/g at 4 C). Furthermore, a flexible pouch cell incorporating the CC-ZnO@Li anode and CC-NC-Co@S cathode maintained stability under various bending conditions

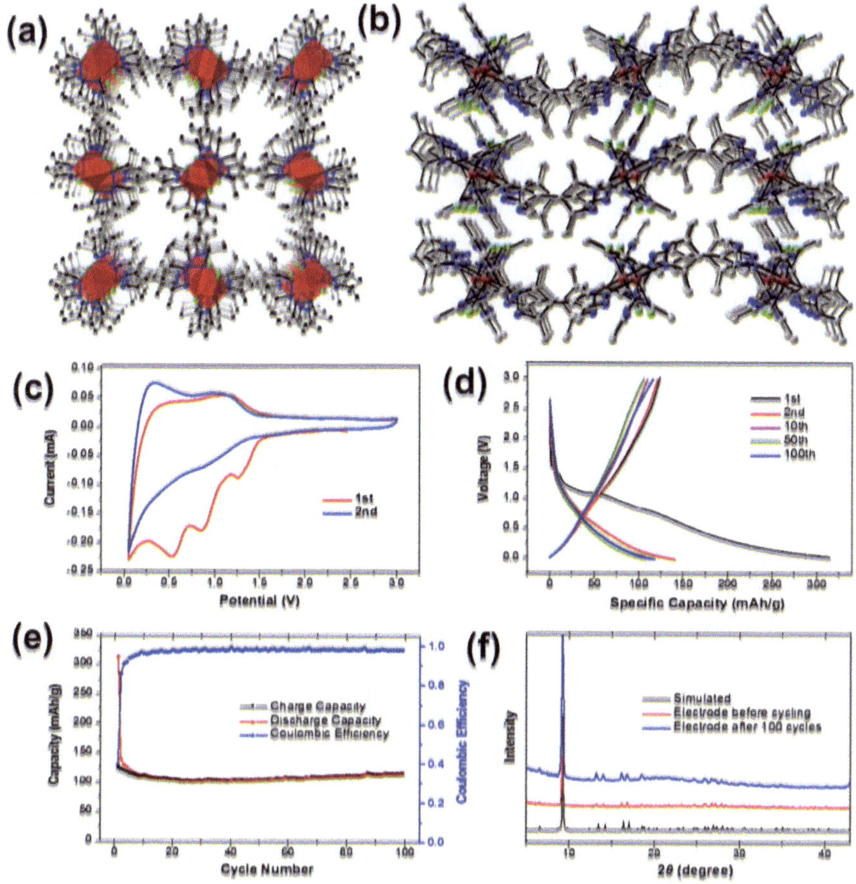

Fig. 6 Viewpoints of MOF (Ni–Me4bpz): **a** 1-D channel along the c-axis octahedron Cl, N, and C, in place of six-coordinated Ni ions; **b** a structure of layers from the a-axis with Ni, Cl, N, and C. Ni–Me4bpz electrochemical characteristics as LIB electrodes: GCD voltage plots at 50 mA/g; **c** CV plots for Ni–Me4bpz within the potential window of 0.01–3 V at 0.05 mV/s; **d** GCD plots for Ni–Me4bpz; **e** Csp and Coulombic efficiency with cycle number; and **f** Ni–Me4bpz PXRD patterns from simulation, electrode prior to cycling, and electrode following 100 cycles [44]. Reproduced with permission @ Copyright (2015) Elsevier

[45]. The scalable and controlled synthesis of graphene is essential for the practical utilization of graphene-derived materials. In 2004, Novoselov et al. successfully isolated graphene, a two-dimensional material with a layered structure, through the mechanical exfoliation of the original graphite. However, this method, being a low-efficiency physical process, is inadequate for large-scale manufacturing. Alternatively, chemical synthesis techniques, which are categorized into bottom-up and top-down approaches, offer more promising routes for mass production [46].

Approaches such as epitaxial growth of CVD, microwave irradiation, and silicon carbide which constitute the bottom-up method provide precise control over chemical structure, including edges, shapes, and the number of layers. Despite its precision, the high cost and hydrophobic nature of pristine graphene present challenges in constructing complex architectures for functional layered structures. On the other hand, a more cost-effective top-down method, such as Hummer's method, employs common oxidants like sulfuric acid H_2SO_4, HNO_3, and $KMnO_4$ to oxidatively exfoliate graphite, producing GO with oxygen-containing functional groups [47]. Zeng et al. developed a flexible, free-standing sulfur composite electrode (S@PCNFs-CNT) by incorporating sulfur into a microporous PCNFs-CNT composite. Schematic representation of the preparation technique for S@PCNFs-CNT NS is as presented in Fig. 7a, b. The resulting electrode demonstrated a superior cycle span and rate capability when utilized as a positive electrode material in Li–S BTs compared to sulfur electrodes without CNTs. Specifically, the S@PCNFs-CNT electrode maintained an alterable capacity of 637 mA h/g over 100 cycles at 50 mA/g and exhibited a rate capability of 437 mA h/g at 1 A/g (Fig. 7c, d). This enhanced capacity was accredited to the synergistic effects of the 3-D NSs and the incorporation of CNTs. This research underscores the potential of PCNFs-CNT as promising nanomaterials for high-performance, free-standing Li–S batteries (Fig. 7e, f) [48].

Jin et al. developed a self-supporting hybrid sulfur host through the in situ formation of a flexible CC structure, which incorporates vertically aligned G/CC and cobalt phosphide (CoP) resulting from an MOF anchored onto the graphene (CoP@G/CC). The sulfur cathodes produced from this material CoP@G/CC-S, with a sulfur loading of 2 mg/cm^2, demonstrate superior electrochemical performance, including high rate capability (930.1 mA h/g at 3.0 C) and exceptional cycle life, with a minimal capacity decay of 0.03% per cycle after 500 cycles. Additionally, at a higher sulfur loading of 10.83 mg/cm^2, an areal capacity of 8.81 mA h/cm^2 at 0.05 C was achieved. The flexible pouch cells based on CoP@G/CC-S also exhibit robust mechanical properties along with favorable rate and cycling performance [27]. Hu et al. developed an innovative heterogeneous composite, Ru/HfO_2-NC, consisting of Ru/HfO_2 NPs embedded in nitrogen-doped porous carbon. This material was synthesized through a co-precipitation and calcination method, utilizing ZIF as a template. The ZIF framework theaters a vital role in facilitating the in situ creation of the embedded structure and preventing nanoparticle agglomeration. The Ru/HfO_2-NC composite demonstrates exceptional ORR performance, including enhanced durability and methanol resistance, when tested in a 0.1 M KOH electrolyte, achieving a high half-wave probable ($E_{1/2}$) of 0.83 V and following a nearly four-electron reaction mechanism. Furthermore, a zinc–air battery utilizing Ru/HfO_2-NC as the cathode delivers a PD of 157.3 mW/cm^2, a Csp of 775 mA h/g, and a lifespan of 258 h at a discharge at 5 mA/cm^2 [49].

Li et al. explained how to make yolk-shelled micro-cages functionalized with Co-N_x-C (Co-N_x-YSC) on CC utilizing a self-assembly approach. This process involved creating cationic-modified MOFs with variables, which were subsequently deposited onto negatively charged carbon substrates. The ZIF-67/CC precursor was then thermally treated at 600 °C to create the Co-N_x-YSC-600/CC. Because of its remarkable functional oxygen activity, good cycling stability for 400 cycles at 10 mA/cm^2, and significant mechanical flexibility, this electrode is suitable for use as an air electrode

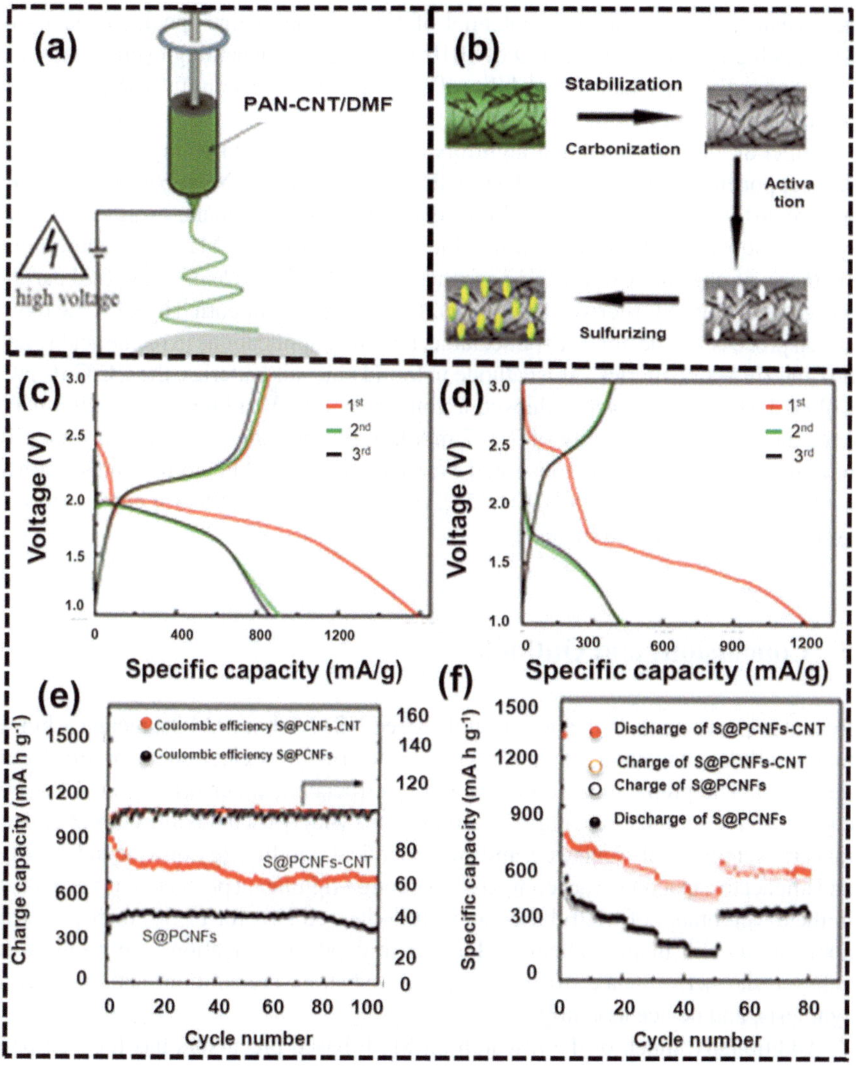

Fig. 7 **a**, **b** Diagrammatic representation of the S@PCNFs-CNT electrode synthesis procedure. S@PCNFs-CNT (**c**) and S@PCNFs (**d**) electrode voltage profiles at 50 mA/g. **e** Comparative charge capacity-cycle number at 50 mA/g. **f** Comparative capabilities of both the electrodes at of 0.05, 0.1, 0.25, 0.5, and 1 A h/g as a function of GCD cycles [48]. Reproduced with permission @ Copyright (2014) Royal Society of Chemistry

in flexible zinc–air BTs [50]. Zhong et al. introduced a novel structural design for air cathodes using an MOF-on-MOF approach, which enables the development of an efficient oxygen catalyst. The catalyst is composed of hierarchical Co_3O_4 NPs embedded in nitrogen-doped carbon nano-micro arrays on a flexible CC (Co_3O_4N-CNMAs/CC). When utilized as the cathode in zinc–air BTs, this material showed impressive performance data, including a potential of 1.461 V and a zinc-specific capacity of 815 mA h/g at 1 mA/cm^2, an ED of 1010 W h/kg Zn, outstanding cycling stability, and outstanding mechanical flexibility. These outcomes meaningfully surpassed the performance of cathodes based on Pt/C [51].

Liu et al. demonstrated a straightforward method for fabricating a flexible 3-D self-supporting bifunctional catalyst. In their study, ultrathin NiFe-based NiFe-LDH was robustly integrated with an MOF-derived carbon nanoflake array. This catalyst was subsequently utilized in flexible zinc–air BTs [52]. Ma et al. reported the synthesis of ultrathin CoFe-LDH nanosheets integrated onto leaf-like hybrid carbon nanoflakes (Co@NC) derived from bimetallic MOFs and supported by CC. The fabrication process was designed to be scalable, targeting applications in rechargeable and flexible zinc–air batteries. As a cathode in liquid zinc–air batteries, the CC/Co@NC/CoFe-LDH-700 composite achieved a voltage of 1.47 V, a high Pd of 129.3 mW/cm^2, and a specific capacity of 710.7 mA h/g. Furthermore, flexible zinc–air BTs utilizing this composite demonstrated excellent mechanical flexibility and sustained cycling stability for more than 40 h [53]. A few MOF-based NSs for energy storage systems are summarized in Table 1.

5 Conclusions and Outlook

Recent improvements in the use of MOF-based NSs (carbon, TMOs, metal carbides, metal sulfides, and their composites) as nanostructured electrodes for FESS are discussed. The preparation strategies and electrochemical performances of the NS electrodes are summarized. Furthermore, various widely used flexible current collectors are systematically reviewed, and their chemical stability, flexibility, conductivity, and impact to capacitance are compared. Although significant progress has been made in the development of MOF-based and MOF-derived NSs for FESS, their practical applications still meet challenges. Further in-depth investigations are required to enhance the performance of FESS, such as substrate preparation, active material synthesis, and device assembly.

Additionally, much of the research on MOF-based precursors has focused on a limited number of well-established MOFs, such as Prussian blue analogs (PBAs) and zeolitic imidazolate frameworks (ZIFs). Additionally, the resulting FESS products are often confined to metal oxides, sulfides, carbon, or their composites, with most structures being isotropic particles. There remains a need for further exploration of new MOFs, along with the synthesis of carbon-based materials. Constructing hollow nanostructures with higher complexity, hierarchical organization, and three-dimensional self-supported architectures is also highly desirable for the development of functional materials. While the self-templating properties of numerous MOF-derived techniques have simplified synthesis, there is still a demand for more versatile and scalable synthetic approaches with improved control, as much of the research to date has focused on isolated systems, limiting the potential for real-world applications.

Table 1 MOF-derived NSs as electrode materials for energy storage

MOF-precursors	Nanoarray electrode	Application	Specific capacitance/capacity	Stability/after cycles	Refs.
V-MOF	V-MOF-48	Zn-ion batteries	100.9 mA h/cm^3 at 0.1 A/cm^3	84.6%/400	[2]
Ni-MOF Ni-	Ni-MOF-74@CNTF		298.26 mA h/cm^3 at 0.5 A/cm^3	91.4%/1000	[3]
Mn-MOF	Od-Mn_3O_4@C		396.2 mA h/g at 0.2 A/g	95.7%/2000	[54]
V-MOF	Ni-V_2O_5NWs@CNT		90.3 mA h/cm^3 at 0.1 A/cm^3	89.8%/10000	[55]
Co-MOF	NC-Co_3O_4/CC		387.2 mA h/g at 25 mA/cm^3	Stably cycled for up to 60 cycles	[56]
	CC@Co_3O_4	Supercapacitor	900 F/g at 10 A/g	85.5%/20000	[40]
	CC@$NiCo_2O_4$		1055.3 F/g at 2.5 mA/cm^2	86.7%/20000	[57]
	$ZnCo_2O_4$@NCNWAs		2.244 F/cm^2 at 2 mA/cm^2	99.37%/10,000	[5]
	NiCo-DH	Ni–Zn batteries	303.6 mA h/g at 2 mA/cm^2	73%/850	[6]
	$Co(VO_3)_2Co(OH)_2$	Supercapacitor	522 mF/cm^2 at 0.5 m/A cm^2	90%/15,000	[29]
	CC/MOFC/MnO_2		651.7 mF/cm^2 at 1.0 mA/cm^2	50%/5000	[58]
	CC/CW/Fe_3O_4@C		463 F/g at 1 A/g	92.8%/10000	[9]
Zn-MOF	CF@NiCoZn-LDH/Co_9S_8-QD		350.6 mA h/g at 1 A/g	95.3%/8000	[10]
	CF@NiCoZn-S/NiCo2S4		194 mA h/g at 1 A/g	70.1%/10000	[11]
Co-MOF	P-CoS_2@P, S, N–C		689 C/g at 2 A/g	95.5%/5000	[15]
	P-doped$CoSe_2$/N–C	Zn–air batteries	None	Stably cycled for up to 80 cycles	[13]

(continued)

Table 1 (continued)

MOF-precursors	Nanoarray electrode	Application	Specific capacitance/capacity	Stability/after cycles	Refs.
	NC-Co/CoNx		41.5 mW/cm^3 at 96.6 mA/cm^3	Stably cycled for up to 1200 cycles	[13]
	CC/CW/p-VN@C	Supercapacitor	604.8 F/g at 1 A/g	90.5%/10000	[14]
Co/Fe-MOF	Fe-CoN_4@N–C	Zn–air batteries	806 mA h/g at 5 mA/cm^2	Stably cycled for up to 220 cycles	[15]
Co-MOF	NC-CoSAC		0.9 mW/cm^3 at 31.0 mA/cm^2	Stably cycled for up to 125 cycles	[16]
	CC@CoP/C-S	Li–S batteries	1535 mA h/g at 0.1 C	96.9%/500	[20]
	CoP/NC		1482 mA h/g at 0.1 C	81.6%/560	[15]
	CoP@G/CC-S		8.81 mA h/cm^2 at 0.05 C	85%/500	[26]

Furthermore, a deeper thought of the creation mechanisms of MOF-derived NSs, along with the relationship of composition, NSs, and output performance in FESS, remains insufficient. In this context, theoretical modeling and computational studies, in conjunction with experimental work, are predictable to show a vital role in future research. Overall, recent advancements in the use of MOF-derived NSs for FESS, including BTs and SCs, are reviewed. It is hoped that this chapter will inspire further exploration of the substantial potential that MOF-derived materials hold for FESS.

References

1. Takeuchi, E.S., Doeff, M.M.: An argument for basic battery science: each time an application demands a new battery chemistry to achieve previously unrealized functionality, a new fundamental understanding is required. Acc. Chem. Res. **51**(3), 573–574 (2018)
2. He, B., Zhang, Q., Man, P., Zhou, Z., Li, C., Li, Q., Xie, L., Wang, X., Pang, H., Yao, Y.: Self-sacrificed synthesis of conductive vanadium-based metal-organic framework nanowire-bundle arrays as binder-free cathodes for high-rate and high-energy-density wearable Zn-ion batteries. Nano Energy **64**, 103935 (2019). https://doi.org/10.1016/j.nanoen.2019.103935
3. Man, P., He, B., Zhang, Q., Zhou, Z., Li, C., Li, Q., Wei, L., Yao, Y.: A one-dimensional channel self-standing MOF cathode for ultrahigh-energy-density flexible Ni-Zn batteries. J. Mater. Chem. A **7**, 27217–27224 (2019). https://doi.org/10.1039/c9ta11759h

4. Gurav, S.R., Waikar, M.R., Rasal, A.S., Sonker, R.K., Sonkawade, R.G.: Current development and challenges in textile-based flexible supercapacitors. In: Smart and Flexible Energy Devices, pp. 539–550. CRC Press (2022)
5. Cai, D., Yang, Z., Tong, R., Huang, H., Zhang, C., Xia, Y.: Binder-free MOF-based and MOF-derived nanoarrays for flexible electrochemical energy storage: progress and perspectives. Small **20**, 1–28 (2024). https://doi.org/10.1002/smll.202305778
6. Chen, H., Shen, Z., Pan, Z., Kou, Z., Liu, X., Zhang, H., Gu, Q., Guan, C., Wang, J.: Hierarchical micro–nano sheet arrays of nickel–cobalt double hydroxides for high-rate Ni–Zn batteries. Adv. Sci. **6** (2019). https://doi.org/10.1002/advs.201802002
7. Qiu, T., Liang, Z., Guo, W., Tabassum, H., Gao, S., Zou, R.: Metal-organic framework-based materials for energy conversion and storage. ACS Energy Lett. **5**(2), 520–532 (2020)
8. Cherusseri, J., Pandey, D., Sampath Kumar, K., Thomas, J., Zhai, L.: Flexible supercapacitor electrodes using metal-organic frameworks. Nanoscale **12**(34), 17649–17662 (2020)
9. Peng, Z., Huang, J., Wang, Y., Yuan, K., Tan, L., Chen, Y.: Construction of a hierarchical carbon-coated Fe_3O_4 nanorod anode for 2.6 V aqueous asymmetric supercapacitors with ultrahigh energy density. J. Mater. Chem. A **7**, 27313–27322 (2019). https://doi.org/10.1039/c9ta10195k
10. Yang, Q., Wang, Q., Long, Y., Wang, F., Wu, L., Pan, J., Han, J., Lei, Y., Shi, W., Song, S.: In situ formation of Co_9S_8 quantum dots in MOF-derived ternary metal layered double hydroxide nanoarrays for high-performance hybrid supercapacitors. Adv. Energy Mater. **10**, 1–12 (2020). https://doi.org/10.1002/aenm.201903193
11. Yang, Q., Liu, Y., Deng, C., Yan, M., Shi, W.: MOF-derived 3D hierarchical nanoarrays consisting of NiCoZn–S nanosheets coupled with granular $NiCo_2S_4$ nanowires for high-performance hybrid supercapacitors. J. Mater. Chem. A **7**, 26131–26138 (2019). https://doi.org/10.1039/c9ta09692b
12. Ren, J., Huang, Y., Zhu, H., Zhang, B., Zhu, H., Shen, S., Tan, G., Wu, F., He, H., Lan, S., Xia, X., Liu, Q.: Recent progress on MOF-derived carbon materials for energy storage. Carbon Energy **2**(2), 176 (2020)
13. Zhang, H., Wang, T., Sumboja, A., Zang, W., Xie, J., Gao, D., Pennycook, S.J., Liu, Z., Guan, C., Wang, J.: Integrated hierarchical carbon flake arrays with hollow P-doped $CoSe_2$ nanoclusters as an advanced bifunctional catalyst for Zn–air batteries. Adv. Funct. Mater. **28**, 1–9 (2018). https://doi.org/10.1002/adfm.201804846
14. Guan, C., Sumboja, A., Zang, W., Qian, Y., Zhang, H., Liu, X., Liu, Z., Zhao, D., Pennycook, S.J., Wang, J.: Decorating Co/CoNx nanoparticles in nitrogen-doped carbon nanoarrays for flexible and rechargeable zinc-air batteries. Energy Storage Mater. **16**, 243–250 (2019). https://doi.org/10.1016/j.ensm.2018.06.001
15. Huang, J., Peng, Z., Xiao, Y., Xu, Y., Chen, L., Xiong, Y., Tan, L., Yuan, K., Chen, Y.: Hierarchical nanosheets/walls structured carbon-coated porous vanadium nitride anodes enable wide-voltage-window aqueous asymmetric supercapacitors with high energy density. Adv. Sci. **6** (2019). https://doi.org/10.1002/advs.201900550
16. Xu, Q., Jiang, H., Li, Y., Liang, D., Hu, Y., Li, C.: In-situ enriching active sites on Co-doped Fe–Co_4N@N–C nanosheet array as air cathode for flexible rechargeable Zn–air batteries. Appl. Catal. B Environ. **256**, 117893 (2019). https://doi.org/10.1016/j.apcatb.2019.117893
17. Zang, W., Sumboja, A., Ma, Y., Zhang, H., Wu, Y., Wu, S., Wu, H., Liu, Z., Guan, C., Wang, J., Pennycook, S.J.: Single Co atoms anchored in porous N-doped carbon for efficient zinc-air battery cathodes. ACS Catal. **8**, 8961–8969 (2018). https://doi.org/10.1021/acscatal.8b02556
18. Cao, Q., Gao, H., Gao, Y., Yang, J., Li, C., Pu, J., Du, J., Yang, J., Cai, D., Pan, Z., Guan, C., Huang, W.: Regulating dendrite-free zinc deposition by 3D zincophilic nitrogen-doped vertical graphene for high-performance flexible Zn-ion batteries. Adv. Funct. Mater. **31**(37), 1–11 (2021)
19. Reddy, M.S.B., Ponnamma, D., Sadasivuni, K.K., Kumar, B., Abdullah, A.M.: Carbon dioxide adsorption based on porous materials. RSC Adv. **11**, 12658–12681 (2021). https://doi.org/10.1039/d0ra10902a
20. Liu, S., Gao, D., Li, J., Hui, K.S., Yin, Y., Hui, K.N., Jun, S.C.: Phosphorus dual-site driven CoS_2@S, N Co-doped porous carbon nanosheets for flexible quasi-solid-state supercapacitors. J. Mater. Chem. A **7**, 26618–26630 (2019). https://doi.org/10.1039/c9ta09646a

21. Ji, D., Fan, L., Li, L., Peng, S., Yu, D., Song, J., Ramakrishna, S., Guo, S.: Atomically transition metals on self-supported porous carbon flake arrays as binder-free air cathode for wearable zinc-air batteries. Adv. Mater. **31**(16), 1–8 (2019)
22. Xiao, J.D., Jiang, H.L.: Metal-organic frameworks for photocatalysis and photothermal catalysis. Acc. Chem. Res. **52**(2), 356–366 (2019)
23. Li, J., Huang, W., Wang, M., Xi, S., Meng, J., Zhao, K., Jin, J., Xu, W., Wang, Z., Liu, X., Chen, Q., Xu, L., Liao, X., Jiang, Y., Owusu, K.A., Jiang, B., Chen, C., Fan, D., Zhou, L., Mai, L.: Low-crystalline bimetallic metal-organic framework electrocatalysts with rich active sites for oxygen evolution. ACS Energy Lett. **4**(1), 285–292 (2019)
24. Zhong, H., Ly, K.H., Wang, M., Krupskaya, Y., Han, X., Zhang, J., Zhang, J., Kataev, V., Büchner, B., Weidinger, I.M., Kaskel, S., Liu, P., Chen, M., Dong, R., Feng, X.: A phthalocyanine-based layered two-dimensional conjugated metal–organic framework as a highly efficient electrocatalyst for the oxygen reduction reaction. Angew. Chem., Int. Ed. **58**(31), 10677–10682 (2019)
25. Lu, X.F., Fang, Y., Luan, D., Lou, X.W.D.: Metal-organic frameworks derived functional materials for electrochemical energy storage and conversion: a mini-review. Nano Lett. **21**(4), 1555–1565 (2021)
26. Jin, J., Cai, W., Cai, J., Shao, Y., Song, Y., Xia, Z., Zhang, Q., Sun, J.: MOF-derived hierarchical CoP nanoflakes anchored on vertically erected graphene scaffolds as self-supported and flexible hosts for lithium-sulfur batteries. J. Mater. Chem. A **8**, 3027–3034 (2020). https://doi.org/10.1039/c9ta13046b
27. Guan, B.Y., Kushima, A., Yu, L., Li, S., Li, J., Lou, X.W.D.: Coordination polymers derived general synthesis of multishelled mixed metal-oxide particles for hybrid supercapacitors. Adv. Mater. **29**(17), 1–8 (2017)
28. Huang, Y., Fang, Y., Lu, X.F., Luan, D., Lou, X.W.D.: Co_3O_4 hollow nanoparticles embedded in mesoporous walls of carbon nanoboxes for efficient lithium storage. Angew. Chem. **132**(45), 20086–20090 (2020)
29. Zhang, Y., Chen, H., Guan, C., Wu, Y., Yang, C., Shen, Z., Zou, Q.: Energy-saving synthesis of MOF-derived hierarchical and hollow $Co(VO_3)_2$–$Co(OH)_2$ composite leaf arrays for supercapacitor electrode materials. ACS Appl. Mater. Interfaces **10**, 18440–18444 (2018). https://doi.org/10.1021/acsami.8b05501
30. Sonker, R.K., Shastri, R., Johari, R.: Superficial synthesis of CdS quantum dots for an efficient perovskite-sensitized solar cell. Energy Fuels **35**(9), 8430–8435 (2021)
31. Chen, Z., Hanna, S.L., Redfern, L.R., Alezi, D., Islamoglu, T., Farha, O.K.: Reticular chemistry in the rational synthesis of functional zirconium cluster-based MOFs. Coord. Chem. Rev. **386**, 32–49 (2019)
32. Zhou, H., Zhang, L., Zhang, D., Chen, S., Coxon, P.R., He, X., Coto, M., Kim, H.K., Xi, K., Ding, S.: A universal synthetic route to carbon nanotube/transition metal oxide nanocomposites for lithium ion batteries and electrochemical capacitors. Sci. Rep. **6**, 1–11 (2016)
33. Wang, T.C., Hod, I., Audu, C.O., Vermeulen, N.A., Nguyen, S.T., Farha, O.K., Hupp, J.T.: Rendering high surface area, mesoporous metal-organic frameworks electronically conductive. ACS Appl. Mater. Interfaces **9**(14), 12584–12591 (2017)
34. Ren, J., Dyosiba, X., Musyoka, N.M., Langmi, H.W., Mathe, M., Liao, S.: Review on the current practices and efforts towards pilot-scale production of metal-organic frameworks (MOFs). Coord. Chem. Rev. **352**, 187–219 (2017)
35. Rubio-Martinez, M., Avci-Camur, C., Thornton, A.W., Imaz, I., Maspoch, D., Hill, M.R.: New synthetic routes towards MOF production at scale. Chem. Soc. Rev. **46**(11), 3453–3480 (2017)
36. Isaeva, V.I., Kustov, L.M.: Microwave activation as an alternative production of metal-organic frameworks. Russ. Chem. Bull. **65**(9), 2103–2114 (2016)
37. Zhang, Y., Bai, W., Cheng, X., Ren, J., Weng, W., Chen, P., Fang, X., Zhang, Z., Peng, H.: Flexible and stretchable lithium-ion batteries and supercapacitors based on electrically conducting carbon nanotube fiber springs. Angew. Chem., Int. Ed. **53**(52), 14564–14568 (2014)
38. Zheng, S., Li, Q., Xue, H., Pang, H., Xu, Q.: A highly alkaline-stable metal oxide@metal–organic framework composite for high-performance electrochemical energy storage. Natl. Sci. Rev. 7(2), 305–314 (2020)

39. Guan, C., Zhao, W., Hu, Y., Lai, Z., Li, X., Sun, S., Zhang, H., Cheetham, A.K., Wang, J.: Cobalt oxide and N-doped carbon nanosheets derived from a single two-dimensional metal-organic framework precursor and their application in flexible asymmetric supercapacitors. Nanoscale Horiz. **2**(2), 99–105 (2017)
40. Liu, X., Guan, C., Hu, Y., Zhang, L., Elshahawy, A.M., Wang, J.: 2D metal-organic frameworks derived nanocarbon arrays for substrate enhancement in flexible supercapacitors. Small **14**(43), 1–7 (2018)
41. Liu, Y., Li, G., Guo, Y., Ying, Y., Peng, X.: Flexible and binder-free hierarchical porous carbon film for supercapacitor electrodes derived from MOFs/CNT. ACS Appl. Mater. Interfaces **9**(16), 14043–14050 (2017)
42. Salunkhe, R.R., Tang, J., Kamachi, Y., Nakato, T., Kim, J.H., Yamauchi, Y.: Asymmetric supercapacitors using 3D nanoporous carbon and cobalt oxide electrodes synthesized from a single metal-organic framework. ACS Nano **9**(6), 6288–6296 (2015)
43. Sharma, N., Szunerits, S., Boukherroub, R., Ye, R., Melinte, S., Thotiyl, M.O., Ogale, S.: Dual-ligand Fe–metal–organic framework-based robust high-capacity Li ion battery anode and its use in a flexible battery format for electro-thermal heating. ACS Appl. Energy Mater. **2**(6), 4450–4457 (2019)
44. An, T., Wang, Y., Tang, J., Wang, Y., Zhang, L., Zheng, G.: A flexible ligand-based wavy layered metal-organic framework for lithium-ion storage. J. Colloid Interface Sci. **445**, 320–325 (2015)
45. Jiang, S., Li, X.L., Fang, D., Lieu, W.Y., Chen, C., Khan, M.S., Li, D.S., Tian, B., Shi, Y., Yang, H.Y.: Metal–organic-framework-derived 3D hierarchical matrixes for high-performance flexible Li–S batteries. ACS Appl. Mater. Interfaces **15**(16), 20064–20074 (2023)
46. Luo, B., Liu, S., Zhi, L.: Chemical approaches toward graphene-based nanomaterials and their applications in energy-related areas. Small **8**(5), 630–646 (2012)
47. Dikin, D.A., Stankovich, S., Zimney, E.J., Piner, R.D., Dommett, G.H.B., Evmenenko, G., Nguyen, S.T., Ruoff, R.S.: Preparation and characterization of graphene oxide paper. Nature **448**(7152), 457–460 (2007)
48. Zeng, L., Pan, F., Li, W., Jiang, Y., Zhong, X., Yu, Y.: Free-standing porous carbon nanofibers-sulfur composite for flexible Li–S battery cathode. Nanoscale **6**(16), 9579–9587 (2014)
49. Hu, C., Wei, F., Liang, Q., Peng, Q., Yang, Y., Taylor Isimjan, T., Yang, X.: Electronically modulated d-band centers of MOF-derived carbon-supported Ru/HfO_2 for oxygen reduction and aqueous/flexible zinc-air batteries. J. Energy Chem. **80**, 247–255 (2023)
50. Li, Z., Yang, J., Ge, X., Deng, Y.P., Jiang, G., Li, H., Sun, G., Liu, W., Zheng, Y., Dou, H., Jiao, H., Zhu, J., Li, N., Hu, Y., Feng, M., Chen, Z.: Self-assembly of colloidal MOFs derived yolk-shelled microcages as flexible air cathode for rechargeable Zn–air batteries. Nano Energy **89**, 106314 (2021)
51. Zhong, Y., Pan, Z., Wang, X., Yang, J., Qiu, Y., Xu, S., Lu, Y., Huang, Q., Li, W.: Hierarchical Co_3O_4 nano-micro arrays featuring superior activity as cathode in a flexible and rechargeable zinc-air battery. Adv. Sci. **6**(11), 1900575 (2019)
52. Liu, L., Hu, Z., Wang, M., Ma, J., Chen, Z., Ning, X., Yuan, D.: Ultrathin NiFe–LDH nanosheets strongly coupled with MOF-derived hybrid carbon nanoflake arrays as a self-supporting bifunctional electrocatalyst for flexible solid Zn–air batteries. J. Alloys Compd. **925**, 166665 (2022)
53. Ma, J., Liu, L., Chen, Z., Wang, M., Wu, H., Wang, H., Yuan, D., Ning, X.: Interfacial engineering of leaf-like bimetallic MOF-based Co@NC nanoarrays coupled with ultrathin CoFe–LDH nanosheets for rechargeable and flexible Zn–air batteries. Polymers **15**(3), 734 (2023)
54. Tan, Q., Li, X., Zhang, B., Chen, X., Tian, Y., Wan, H., Zhang, L., Miao, L., Wang, C., Gan, Y., Jiang, J., Wang, Y., Wang, H.: Valence engineering via in situ carbon reduction on octahedron sites Mn_3O_4 for ultra-long cycle life aqueous Zn–ion battery. Adv. Energy Mater. **10**, 2001050 (2020). https://doi.org/10.1002/aenm.202001050
55. Ji, D., Fan, L., Li, L., Mao, N., Qin, X., Peng, S., Ramakrishna, S.: Hierarchical catalytic electrodes of cobalt-embedded carbon nanotube/carbon flakes array for flexible solid-state zinc-air batteries. Carbon **142**, 379–387 (2019). https://doi.org/10.1016/j.carbon.2018.10.064

56. Li, T., Xu, Q., Waqar, M., Yang, H., Gong, W., Yang, J., Zhong, J., Liu, Z.: Millisecond-induced defect chemistry realizes high-rate fiber-shaped zinc-ion battery as a magnetically soft robot. Energy Storage Mater. **55**, 64–72 (2023). https://doi.org/10.1016/j.ensm.2022.11.041
57. Guan, C., Liu, X., Ren, W., Li, X., Cheng, C., Wang, J.: Rational design of metal-organic framework derived hollow $NiCo_2O_4$ arrays for flexible supercapacitor and electrocatalysis. Adv. Energy Mater. **7**, 1602391 (2017). https://doi.org/10.1002/aenm.201602391
58. Liu, X., Guan, C., Hu, Y., Zhang, L., Elshahawy, A.M., Wang, J.: 2D metal-organic frameworks derived nanocarbon arrays for substrate enhancement in flexible supercapacitors. Small **14**, 1702641 (2018). https://doi.org/10.1002/smll.201702641

Graphene-Based Multifunctional Hybrid Nanomaterial for Energy Storage

Suman A. Sawant, Maqsood R. Waikar, Gayatri R. Chodankar, Rakesh K. Sonker, and Rajendra G. Sonkawade

Abstract Graphene has drawn a lot of interest in Energy Storage Devices (ESDs) because of their excellent surface area, thermal and electrical conductivities, high mechanical strength, optical transmittance, biocompatibility, and it forms structures that vary from 0 to 3D. Despite this, the zero band gap of pure graphene constrains its applicability. One has to design its hybrid to utilize it in high-performance flexible and wearable ESDs. Superior electrochemical performances have been achieved by integrating the capabilities of other electrochemical materials, including metal oxides, conducting polymers, metal nanoparticles, and doped heteroatoms with the remarkable properties of graphene. This chapter gives the highlights of the different graphene-based hybrids in ESDs based on an extensive literature survey. Herein, diverse aspects and challenges involved in the real-world uses of graphene-based hybrid for batteries and supercapacitors were reviewed rigorously. It will also serve as a catalyst for the fabrication/designing of graphene hybrids to meet the demand of future generations of ESDs.

Keywords Graphene · Graphene-based materials · Energy storage device · Supercapacitors · Batteries · Fuel cells

S. A. Sawant · M. R. Waikar · G. R. Chodankar · R. G. Sonkawade (✉)
Radiation and Materials Research Laboratory, Department of Physics, Shivaji University, Kolhapur, Maharashtra, India
e-mail: rgssuk@gmail.com

S. A. Sawant
Department of Physics, Karmveer Bhaurao Patil College, Urun-Islampur, Sangli, India

R. K. Sonker
Department of Physics, Motilal Nehru College, University of Delhi, New Delhi, India

R. K. Sonker et al. (eds.), *Low-Dimensional Multifunctional Hybrid Nanostructured Materials*, Engineering Materials, https://doi.org/10.1007/978-981-96-7952-2_8

1 Introduction

The rising use of renewable energy sources and the popularity of electric vehicles (EVs) are driving a sharp increase in necessity for energy storage. Projections suggest that by 2050, China alone could achieve an energy storage capacity of 200 GW, with a market size surpassing 2 trillion RMB. This growth is supported by declining battery costs, which enhance the competitiveness of EVs and stationary storage systems. While advancements in technologies like liquid air energy storage (LAES) and lithium batteries are improving efficiency and reducing costs, challenges such as high initial investment costs and the need for better market structures to enable commercial applications remain significant [1]. Today, among the most important societal and scientific issues is securing a reliable energy future. Electrochemical energy storage devices (ESDs) play a pivotal role in modern life due to their increasing use in mobile electronic devices, electric vehicles, and large-scale power grids. Among the key ESD systems are metal-ion batteries (MIBs) and supercapacitors (SCs). MIBs offer high energy density by storing charge and metal ions (such as Li^+ and Na^+) within the electrodes through relatively slow redox activities, though SCs deliver high power density at the cost of energy density as a result of rapid surface or near-surface energy storage mechanisms [2]. An essential objective is to develop ESDs with improved energy density, power density, and cyclic stability to satisfy the expanding industrial and societal needs.

Graphene has gained a considerable attention in recent times for its extensive range of applications in ESDs with its remarkable mechanical strength, electrical conductivity, and huge theoretical specific surface area. According to theoretical research, graphene might be used as an EDLC electrode with an electric double-layer capacitance of 550 F/g and as a lithium–ion battery (LIB) anode with a Li^+ storage capacity of 744 mAh/g [3]. Nevertheless, graphene layers aggregate during its synthesis and numerous charge–discharge cycles, experimental results rarely match mentioned theoretical values. Consequently, there are less useful active sites available for charge storage or electrochemical processes. Several efforts have been made to overcome this problem by synthesizing graphene-based composites that increase surface area and active sites by avoiding graphene layer aggregation. The composites like graphene-based metal oxide, conducting polymer, and heteroatom hybrid nanomaterials play a vital role in the advancement in field of ESDs as they possess high surface area, high electrical conductivity, and tuneable electronic structures. The combination of these elements results in improved electrochemical performance that effectively addresses the growing need for efficient energy storage. Transition metal oxides and hydroxides are widely utilized as electrode materials due to their notable charge storage capacity and stability. Furthermore, various conducting polymers also increase the conductivity of hybrid nanocomposites, thereby enhancing total electrochemical performance of the device. Additionally, heteroatom-doped graphene nanomaterials display synergistic effects, improving electroactive sites and charge transfer [4].

Present chapter explores the advancements made in graphene-based multifunctional hybrid nanomaterials produced for ESDs, with a primary focus on batteries and SCs. First section provides thorough overview of the fundamentals and significance of batteries and SCs in contemporary energy systems along with a background on graphene. The state-of-the-art advancements in the production of graphene-based hybrid nanomaterials are highlighted in the consequent sections wherein conducting polymer–graphene hybrids, metal oxide-based graphene composites, and heteroatom-doped graphene are discussed. These developments highlight how revolutionary hybrid materials may be in advancing sustainable energy. The need for efficient, scalable, and environmentally friendly techniques to advance ESD technology and graphene-based hybrid nanomaterial research is emphasized in this chapter.

2 Background of Graphene

A discovery of graphene by Geim and Konstantin Novoselov (2004) marked a revolution in the field of material research. The preliminary research on graphene (or two dimensional (2D) graphite) dates back to 1947, when Wallace investigated the electronic energy bands in crystalline graphite using the 'tight-binding' approximation. Since it was demonstrated that the semi-metallic phase of monolayer graphene is not stable in 2D, it has long been considered a theoretical substance. However, there have been several attempts to separate monolayer graphene. For instance, a scanning tunnelling microscope (STM) was used in the early 1990s to view monolayer graphene structures made through hydrocarbon breakdown on the surface of Pt(111). In 1997, Japanese researchers who split kish graphite to investigate the effect of thickness on the electrical characteristics of graphite crystals [5]. As a result of these endeavours, the thickness of graphite films was successfully reduced to 30 nm. Building upon this basis, Novoselov and Geim created a reliable and efficient technique for peeling layers from highly orientated pyrolytic graphite (HOPG) to produce monolayer graphene. Employing a clever yet straightforward technique known as mechanical exfoliation (often referred to as the 'Scotch tape method'), where layers of bulk graphite were peeled off using a piece of sticky tape. The tape was repeatedly adhered and peeled to obtain gradually thinner layers. Eventually, graphene, a single sheet of carbon atoms, was created. The further examination of graphene establishes it as one-atom-thick hexagonal lattice of carbon atoms which could exist on its own and can be stably preserved in ambient environments. This paved the way for substantial growth of 2D material research.

The graphene-based industries are now getting some success in innovation and utilization. This is partly because the term 'graphene' now covers a wide range of materials, which is more affordable layered carbon. The versatility of graphene has extended its use case from consumer electronics to concrete, trucks, and even brain implants. But what about the iconic single-layer graphene? It, too, appears to be emerging from its challenges. Industries are improving themselves towards cost-effective and reliable production of graphene. For instance, the use of graphene

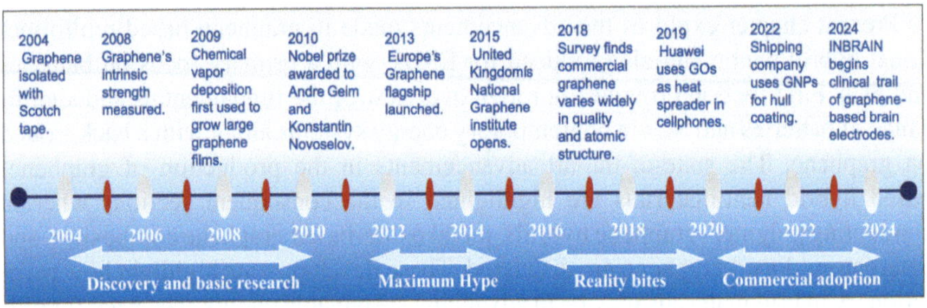

Fig. 1 The journey of graphene from discovery to commercial adoption

with one atom thickness to create magnetic sensors, optical communication chips, and medical diagnostic tools is a good example of innovations emerging in this field. Figure 1 depicts the journey of graphene from its discovery to the commercial adoption.

Carbon is economical and corrosion resistant; hence, it is widely used to manufacture electrodes in industrial-scale SCs and batteries. Carbon-based electrodes maintain their cycle stability along with longer lifespan because their charging and discharging processes involve no chemical reactions. Though their energy storage is constrained by the number and pore size in electrode material. With scientists unveiling increasingly remarkable applications in recent years, graphene has enabled breakthroughs in electronics. Further, to synthesize high-performance electrodes, researchers are exploring quantum dots, nanotubes and nanowires, nanosheets and nanodiscs, and furthermore three-dimensional (3D) structures for SC production. Figure 2 illustrates the journey of discoveries in graphene.

2.1 Properties of Graphene

Graphene possesses exceptional physical, chemical, and mechanical properties. Graphene-based materials are used as electrodes in most industrial SCs and batteries because they are inexpensive and corrosion resistant. Graphene-based electrodes maintain their cycle life because their charging and discharging processes involve no chemical reactions. The electronic structure of graphene reveals unique properties that stands out from other materials. Its structural holes allow phonons to travel freely, resulting in high thermal conductivity, graphene is often treated as a semimetal due to its metallic layers and extremely low bandgap. This unique classification is supported by its exceptional theoretical background and distinct characteristics not found in other non-metallic materials or ideal semimetals. Unlike conventional semiconductors graphene has no bandgap. Even minimal energy is enough to enable its electrons to conduct electricity. Moreover, unusual energy state arrangement of graphene causes its electrons to behave as massless entities, allowing them to pass throughout the material with minimum resistance [7].

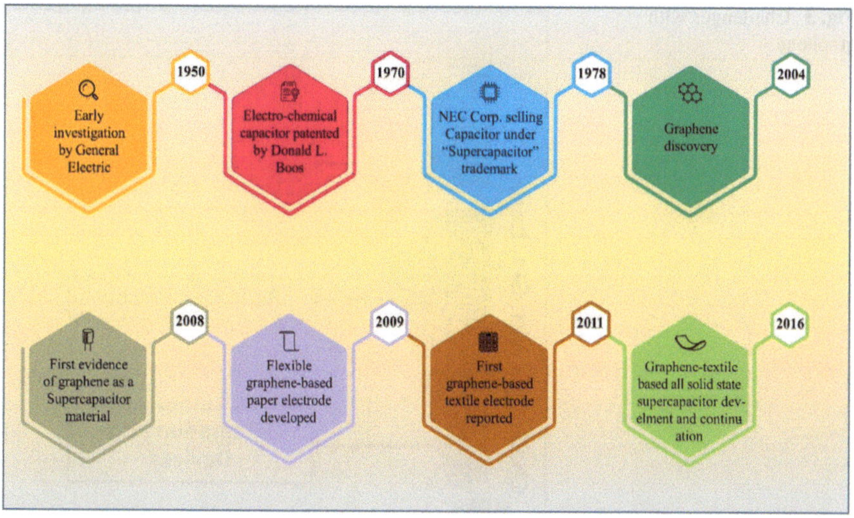

Fig. 2 The discovery model of graphene [6]. Reproduced with permission Copyright @ (2023) Elsevier

Because of its unusual electronic structure, graphene has extraordinarily high electrical conductivity. The delocalized π-electrons in its carbon atoms allow for ballistic transport, making it an excellent conductor. Its electron mobility exceeds 15,000 cm^2/V s at room temperature, surpassing silicon, making it ideal for applications in electronics. In addition to the distinct electronic properties, graphene has a tensile strength of about 130 GPa, which is over 100 times stronger than steel of the same thickness. Despite its strength, it is incredibly lightweight and flexible. The exceptional thermal conductivity of graphene, ranging between 4840 and 5300 W/m K, makes it suitable for heat dissipation applications in electronics and energy systems [8].

Graphene is nearly transparent, absorbing only about 2.3% of visible light. Its transparency, coupled with its conductivity, makes it a promising material for use in transparent conductive films. Graphene is chemically stable but can be functionalized to form graphene oxide or reduced graphene oxide, allowing it to be customized to diverse uses in energy storage, sensors, and catalysis. A single layer of graphene has a theoretical surface area of 2,630 m^2/g, which is beneficial for applications like energy storage (batteries and SCs) and catalysis [9]. Being the first truly two-dimensional material, graphene exhibits quantum mechanical properties that are not observed in bulk materials, such as the quantum Hall effect. These extraordinary properties have spurred significant interest in graphene for a wide range of applications, including in electronics, energy storage, composites, sensors, and biomedical devices. However, challenges remain in large-scale production and preventing issues such as re-stacking, which can diminish its performance as depicted in Fig. 3.

Although carbon materials theoretically possess high surface area and electrical double-layer (EDL) capacitance, practical performance often falls short due

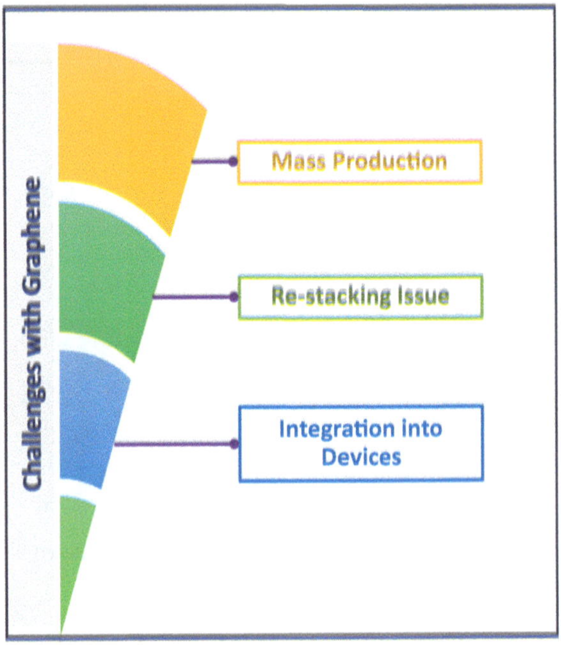

Fig. 3 Challenges with graphene

to low quantum capacitance and limited surface accessibility. Defect engineering in graphene has emerged as a promising approach to overcome these limitations. Defects in graphene play an important role in shaping its capacitance and energy storage capacity. These imperfections modify bond arrangements, introduce dangling bonds, and create localized states that enhance charge accumulation, thereby improving capacitance. Wrinkled structures and surface functionalities further influence the adsorption process by adding energy levels and altering polarities. A comprehensive grasp of graphene's structure and its inherent defects is crucial for achieving tailored performance in various applications, as described in Fig. 4.

2.2 *Synthesis of Graphene*

Graphene synthesis methods involve several approaches, each with unique advantages and challenges. Micromechanical cleavage, first demonstrated by Geim and Novoselov in 2004, produces high-quality graphene using scotch tape, but it is labour-intensive, low-yield, and prone to tape residue impurities, which require heat treatment or three-roll mill exfoliation for improvement [10]. Ball milling exfoliates graphite through high-energy ball collisions, yielding scalable graphene with good properties; however, it introduces defects due to the collisions. Combined techniques like resonance ball milling enhance exfoliation efficiency [11]. Sonication utilizes ultrasonic waves in solvents for scalable, defect-free graphene production,

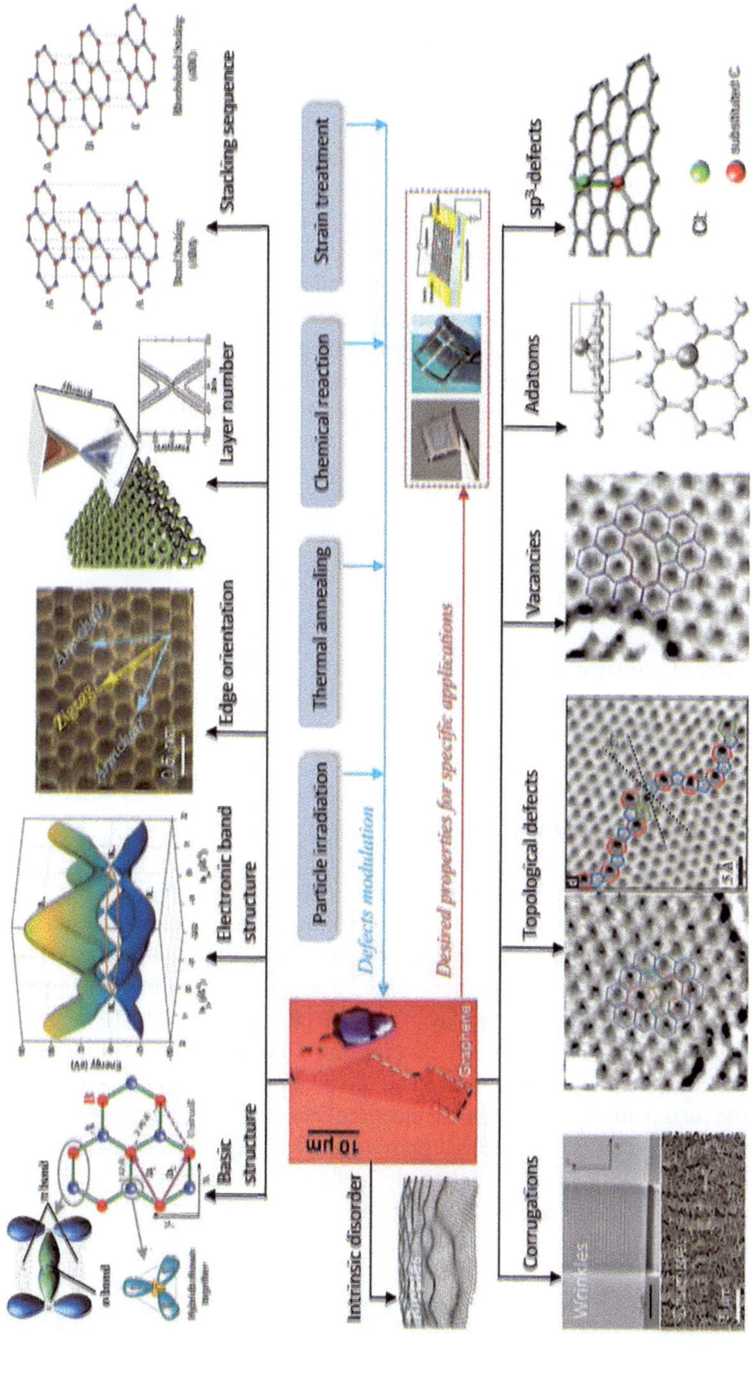

Fig. 4 Structure of graphene and its disorders [5]. Reproduced with permission Copyright @ (2018) Taylor & Francis

though cavitation-induced oxygen functional groups require optimization of parameters such as time, power, and additives [10]. Chemical vapour deposition (CVD) grows graphene on metal substrates like nickel and copper, producing high-quality graphene with challenges such as grain boundaries, costly transfer processes, and limited scalability. Innovations like substrate annealing improve grain size and quality [12]. Epitaxial growth decomposes silicon carbide at high temperatures to produce large-area graphene without substrate transfer, suitable for electronics but costly and environmentally taxing [13]. Unzipping of carbon nanotubes (CNTs) converts CNTs into graphene nanoribbons through cutting or oxidation, yielding conductive nanoribbons but involving expensive and environmentally concerning chemicals. Emerging methods aim to reduce acid use and enhance scalability [14]. Another widely used method is the chemical synthesis of graphene which occurs in two steps. First, graphite is oxidized and exfoliated to generate graphene oxide (GO), which is then reduced by different methods such as chemical reduction, thermal reduction, and hydrothermal reduction. Despite advancements, most methods face trade-offs in cost, scalability, defect control, and environmental impact. It is necessary to develop eco-friendly and effective techniques for industrial applications.

2.3 Applications of Graphene

A combination of physical, chemical, and electronic properties of graphene makes it suitable for a various applications across different fields. Some of the applications are discussed here.

2.3.1 Electronics and Optoelectronics

High electron mobility of graphene positions it as an ideal candidate for next-generation transistors, potentially replacing silicon in ultra-fast electronic devices. Additionally, its thinness and excellent conductivity enable the development of flexible and transparent electronics. Graphene-based transparent films are being utilized more and more in liquid crystal displays (LCDs), OLEDs, and touchscreens as a flexible and long-lasting substitute for indium tin oxide (ITO). In photodetectors, graphene's wide optical absorption range and quick response times improve performance, while its sensitivity to changes in electrical conductivity enables gas sensors to detect trace gases at parts-per-million (ppm) levels. Furthermore, graphene's superior electrical properties and resistance to electromigration make it appropriate as a high-speed interconnected material in integrated circuits [15].

2.3.2 Energy Storage and Conversion

The unique features of graphene-based materials make them strikingly suitable for energy conversion and storage technologies. LIBs are considerably improved by graphene-based anodes as they prevent graphene layers from re-stacking which improves the active sites for Li^+ storage and improves capacity and cycling stability. Graphene composites also improve performance in sodium-ion batteries (SIBs) by addressing problems such as sodium's higher ionic radius [16]. The high surface area and conductivity of graphene favours fast charge/discharge cycles and effective energy storage with a high power density for supercapacitors. The performance of graphene is further improved by combining it with conductive polymers or metal oxides. RGO is widely used in membrane electrode assemblies, leading to greater efficiency. Furthermore, metal oxide and graphene play crucial role in solar cells production as an electrode material that encourages improved charge transfer. Due to its large conductivity and transparency it helps to increase the overall efficiency of solar cell [17].

2.3.3 Composites and Structural Materials

Graphene in form of composites and structural materials has extensive range of applications as it tends to improve strength, thermal stability, and electrical conductivity when added to metals, ceramics, and polymers. These graphene-based composites are used in construction, automotive, and aerospace where robust and lightweight materials are critical [18]. Also, printed electronics use graphene-based conductive inks to produce RFID tags, sensors, and flexible circuits. Graphene's outstanding characteristics make it perfect for anticorrosion coatings that provide outstanding protection in challenging conditions.

2.3.4 Environmental Applications

Considering their nanoscale channels and selective permeability, GO membranes are exceptional for desalinating and filtering water, which makes them beneficial for treating wastewater and eliminating heavy metals. Graphene's high surface area and tuneable porosity allow for the efficient storage of gases such as methane and hydrogen, as well as the capture of CO_2 for clean energy applications. Furthermore, functionalized graphene is excellent in removing contaminants from air and water, including oils and dyes, aiding in environmental clean-up initiatives [19].

2.3.5 Biomedical Applications

Graphene is known as a good drug carrier due to its high surface area and functionalization capabilities. These features permit for regulated release and targeted drug delivery. The high electrical conductivity and sensitivity to the conductivity variations positions it as reliable material for biosensors for distinguishing proteins, glucose, and DNA. Moreover, graphene-based scaffolds enable tissue engineering and recreating medicine by encouraging cell proliferation and differentiation. In addition, functionalized graphene materials have antibacterial properties that make them preferred choice for protective coatings on medical devices and as wound dressings [20].

2.3.6 Thermal Management

Graphene, due to its high thermal conductivity, is used as heat spreaders and thermal interface materials for electrical equipment like CPUs and LEDs. It is well suited for cooling systems due to its efficient heat dissipation, which help sustaining peak performance in batteries and sophisticated electronics [21].

2.3.7 Energy Harvesting

Graphene's use as nanogenerators in piezoelectric and triboelectric devices, which transform mechanical energy into electrical energy, widens it range of applications in wearable technology and Internet of Things (IoT) [22]. Furthermore, graphene composites improve thermoelectric performance by effectively transforming waste heat into electrical power that may be used, thereby promoting energy sustainability.

2.3.8 Aerospace and Automotive Applications

The exceptional physical and electrochemical properties of graphene make it appropriate in the field of aerospace where graphene-reinforced materials are used to make aeroplanes lighter while improving their strength and minimizing impact of environmental factors [23]. Graphene is used in high-performance batteries, lightweight body panels, and thermal management systems in the automobile sector to increase the overall performance and efficiency of vehicles.

2.3.9 Sensors

Its sensitivity to vary the surface charge in chemical sensors allows accurate detection of gases, including CO_2 and NO_2, along with other compounds. Furthermore, graphene is an appropriate material for strain gauges and pressure sensors due to its mechanical flexibility and conductivity, which are critical for wearable technology as well as industrial applications [24].

2.3.10 Catalysis

Graphene-based catalysts, including graphene-supported nanoparticles, are used in chemical reactions, such as hydrogen evolution, oxygen reduction, and CO_2 reduction [24].

2.3.11 Military and Defense

Graphene-based materials are discovered for lightweight armour, electromagnetic shielding, and stealth applications due to their mechanical strength and electrical properties. The mechanical strength and electrical qualities of graphene-based materials are being investigated for applications such as electromagnetic shielding, stealth, and lightweight armour.

2.3.12 Quantum and Spintronics

Graphene's two-dimensional structure and quantum properties are used in spintronics for developing future quantum devices [25]. Spintronics uses graphene's two-dimensional structure and quantum characteristics to create future quantum devices.

Multifunctionality of graphene comes from blend of its strength, flexibility, conductivity, and versatility. By tuning these properties through various synthesis techniques, graphene can be modified to meet the particular requirements of numerous industries such as electronics, energy, healthcare, and environmental sustainability. While remarkable progress has been made, challenges in mass production and integration remain. The ongoing research and invention in this particular area are expected to tackle these challenges in coming years. As more is learnt about its potential, its wide range of applications keeps growing. Graphene-based materials have applications in electronics, energy, the environment, biomedicine, and industry. Even though there has been a lot of development, there are still issues with large-scale production and integration as shown in Fig. 5. However, in the upcoming years, further research and innovation should be able to fully realize their promise.

3 Introduction to Energy Storage Devices (ESDs)

Energy can be stored using a range of fields, including solid-state batteries, flow batteries, flywheels, compressed air energy storage, thermal storage, and pumped hydro-power systems. Among these, batteries and SCs are effective solutions for electrical energy storage, making them particularly suitable for use in electric vehicles (EVs) and plug-in hybrid electric vehicles (PHEVs), both of which are well-established and commercially accessible technologies.

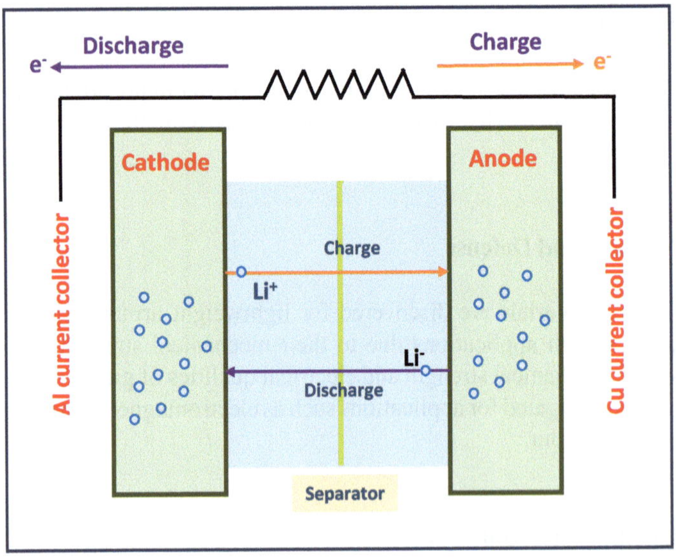

Fig. 5 Schematic of the charge storage in batteries

3.1 Batteries

Batteries are one type of ESD which converts stored chemical energy into electrical energy. Positive anode, negative cathode, and ion-containing electrolytes are the main parts of a battery. To avoid electrical short circuits, a permeable membrane separator is placed between anode–cathode. Additionally, it enables ionic charge carrier's movement, which are necessary for closing the circuit when current flows. In order to conduct the charge via the linked device outside of the battery, a collector is used. After circuit completion, through electrochemical reactions electricity is produced by battery. During this process oxidation occurs at the anode, i.e. electrolyte ions react with anode and releases electron. Whereas the reduction process occurs at the cathode where ions and free electrons react with cathode. That means electricity is produced throughout the cyclic process of donation of electrons from anode with gaining electrons by cathode until the electrodes are depleted of the material required for reaction generation [26].

The two basic categories of batteries are primary and secondary batteries. Since the electrode materials in primary batteries undergo irreversible change while charging, they are only useful once and then become unusable. Typical examples are the alkaline batteries found in toys, flashlights, and a wide range of portable gadgets as well as the zinc–carbon battery. Due to the ability of the electrodes' original composition to restore functioning, repeated charging–discharging is possible in secondary batteries (rechargeable). Lead–acid battery, nickel-based batteries, lithium-metal batteries, and sodium metal-based batteries are examples of rechargeable batteries. Primary and secondary batteries differ primarily in their rechargeability, chemical processes, energy density, cost, environmental impact, and applications. Primary batteries are non-rechargeable, designed for single use, and rely on irreversible chemical reactions to generate electricity. Once depleted, they must be discarded.

Similar to alkaline and lithium primary types, these batteries are valued not only for their high energy density, low initial cost, but also for long shelf-life which make them perfect for low-drain devices. Nevertheless, they generate additional waste, contributing to environmental concerns if not disposed of properly. Whereas secondary batteries are rechargeable, depend on reversible chemical reactions to permit energy to be restored multiple times. Examples such as lithium–ion, lead–acid, and nickel–metal hydride batteries. Their reusability over several charge–discharge cycles makes them more economical and ecologically benign over time, in spite of their slightly lower initial energy density and greater initial cost. High-drain gadgets like computers, smartphones, electric cars, and other portable electronics frequently use secondary batteries. Their versatility and durability make them essential for frequent-use scenarios, while their lower waste output reduces their environmental impact. Overall, the choice between primary and secondary batteries depends on the application. Primary batteries are ideal for long shelf-life and low-power requirements, while secondary batteries are preferred for frequent-use and high-energy applications, offering a more sustainable solution [27].

Different battery types exhibit various benefits and drawbacks. To prevent short circuits during the chemical interaction between the battery's electrodes and electrolyte, essential safety protocols must be followed when handling any rechargeable battery. Environmentally friendly batteries that operate safely and require minimal maintenance should thus be developed to overcome this hurdle. In order to address the aforementioned problems, new advanced battery materials and their combinations need to be developed. Recently, modification of traditional battery materials with graphene is a very effective way to remarkably improve the electrochemical properties of traditional battery materials.

A graphene battery is lightweight, robust, and appropriate for storing large amounts of energy. Addition of graphene not only reduces the charging period but also improves its cycle life. Graphene offers high electrical conductivity and can improve different battery features as energy density and form in various ways [28]. The efficiency of Li–ion batteries (and other rechargeable battery types) can be improved by adding graphene to the battery as the anode. In spite of that as per above discussion, in order to prevent the aggregation of graphene layers and improve the active sites and surface area, much effort has been put into investigating graphene-based hybrids.

- **Key Differences Between Primary and Secondary Batteries**

Feature	Primary batteries	Secondary batteries
Rechargeability	Non-rechargeable	Rechargeable
Lifespan (uses)	Single use	Multiple cycles
Energy density	Higher initially	Lower, but rechargeable

(continued)

(continued)

Feature	Primary batteries	Secondary batteries
Cost	Lower initial cost	Higher upfront cost
Environmental impact	Disposable, more waste	Reusable, less waste
Applications	Emergency, long-term storage	Daily-use, high-drain devices

Graphene-based hybrid materials are shown to be a promising and reliable material for rechargeable Li–ion batteries. Theoretically, graphene exhibits strong Li-storage potential, as lithium ions can bind to both sides along with the sheet edges and various defective positions as given in Fig. 5. Graphene can substantially improve electron conduction in combination with pseudocapacitive material and efficiently avoid material agglomeration while lithiation and delithiation cycles. As a result, graphene-based nanostructures present an innovative approach to deal with problems related to gravimetric energy density. The high surface area of graphene enables key advantages over traditional flat electrode designs, such as (i) enhanced lithium uptake, which increases storage capability and (ii) improved charge transfer kinetics [29].

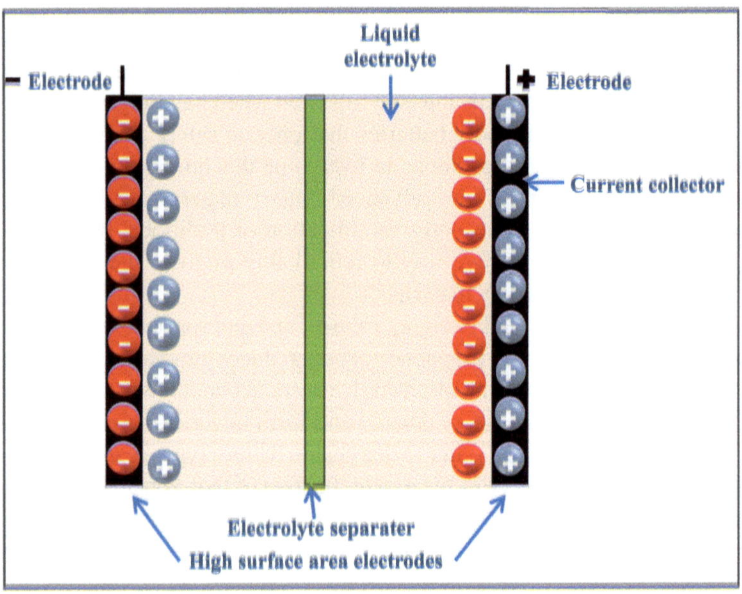

Fig. 6 Schematic of the charge storage in supercapacitor

3.2 Supercapacitors

At the forefront of energy storage technology, due to unique properties, SCs often referred to as ultracapacitors or EDLCs. The illustration of the energy storage mechanism of a SCs is illustrated in Fig. 6. By separating charges conventional capacitors store energy in an electric field, in the case of SCs, energy storage occurs via electrostatic and electrochemical ways, enabling much higher E_d than conventional capacitors.

This is accomplished by the collection of cations/anions at the interface between electrode and electrolyte, where an EDL forms. Furthermore, some SCs can store more energy due to reversible faradaic reactions at the electrode surface known as pseudocapacitors. SCs are distinguished by their high P_d, which allows them to charge and discharge energy quickly. For uses requiring rapid power spikes, including peak shaving in electrical grids and reducing power oscillations in renewable energy systems, this makes them especially well suited. SCs also have an amazing cycle life; they can withstand thousands to millions of charge–discharge cycles without suffering any noticeable damage. Because of their long lifetime, they are a desirable choice for applications that need to be reliable and durable over time. Notwithstanding these benefits, SCs still have drawbacks because of their comparatively lower E_d than batteries. Even though they are excellent at producing high power outputs, they frequently have lower energy storage capacities than batteries. Thus, to make use of the advantages of both technologies, they are often employed in conjunction with batteries in hybrid energy storage systems. By combining the high E_d of batteries with the quick charge and discharge times of SCs, these hybrid systems provide a well-rounded answer for a range of energy storage needs [30]. SC technology is constantly evolving, with research efforts concentrated on increasing E_d, optimizing performance at high temperatures, and minimizing production costs. With their versatility and potential for further improvement, SCs are poised to be crucial in shaping the future of ESDs.

SCs are categorized as EDLCs and pseudocapacitors. These classifications are according to their charge storage system that involves non-faradic, faradic, and combination of these two processes. EDLCs produce an EDL at the electrode–electrolyte interface to store electric energy. Two electrodes, usually composed of materials with a larger SA, such as carbon. A larger SA is readily accessible for ion adsorption on carbon electrodes. The electrolyte solution is inserted between the electrodes which contains ions (cations and anions) that are easily mobile in the solution. Cations and anions are drawn to the anode and the cathode, respectively, when a voltage is applied across the electrodes and by ion accumulation on the electrode's surface EDL forms. In EDL, an inner layer of ions is bound directly to the surface of electrode called the inner Helmholtz layer, whereas an outer layer of loosely bound ions is called the outer Helmholtz layer. Ions move through the electrolyte to the electrode's surface when voltage is applied across the electrodes, storing energy in the EDL. This process is comparatively fast owing to the quick transfer of ions within an electrolyte. During discharging, the stored energy is released since the ions return to the electrolyte, generating an electric current. In EDLCs within EDL, energy is stored electrostatically. Because there are no chemical reactions involved, the charging and discharging operations are extremely reversible and occur quickly. The SA and spacing between the electrodes are the main factors that effect on an EDLC's capacitance. Higher capacitance is the result of thinner electrolyte layers and electrodes with a larger SA. EDLCs are appropriate for applications requiring quick energy bursts because of their rapid delivery and absorption of huge amounts of electricity. EDLCs are highly valued for their high P_d, extended cyclic stability, and quick charge and discharge times.

Reversible faradaic redox processes at the interface of electrode and electrolyte are the source of energy stored by pseudocapacitors, a form of electrochemical ESDs. In EDLCs charge stored by electrostatic way, whereas in pseudocapacitor charge is stored by surface-redox processes. Pseudocapacitors utilize electrode materials capable of undergoing fast and reversible redox reactions. These contain transition metal oxides, metal–organic framework (MOFs) conducting polymers (CPs). Similar to EDLCs, pseudocapacitors use electrolyte solution which permits unrestricted ion movement within it. Ions are accumulated onto the electrode material when a pseudocapacitor is charged. During the charging–discharging processes, reversible oxidation–reduction reactions occur in the CPs or transition metal oxides at the electrode surface. Both electrostatic and redox-based charge storage occur since ions are adsorbed on electrode surface during charging. As ions return to the electrolyte and the redox reactions at the electrode surface are reversed, the stored energy is released during discharging. The combination of electrostatic, textile, and redox-based charge storage mechanisms enables pseudocapacitors to attain higher energy densities compared to EDLCs [31]. Compared to others, pseudocapacitors can have faster charge and discharge rates because they can go through quick oxidation/reduction. Charge storage mechanism of pseudocapacitors is further sub-categorized depending upon the way of charge storage. Firstly, surface-redox reactions involve charge storage through redox processes occurring on electrode's surface. Secondly, intercalation-type reactions encompass the reversible insertion or extraction of ions into or from the crystal lattice of the electrode material. Lastly, battery-type faradic reactions mimic the redox processes observed in conventional batteries, involving the reversible movement of ions and electrons at the electrode–electrolyte boundary [32].

High porosity and surface area in graphene-based hybrid for efficient transportation of hydrated ions, leading to high double-layer capacitance in various electrolytes as shown in Fig. 7. The porosity enables effective transfer of ions and electrons, increases the active surface area, reduces ion diffusion lengths, and ensures better electrolyte access to the electrode. These factors collectively contribute to remarkable improvements in overall performance of SCs.

The previously mentioned conversation indicates that existing industrial energy storage solutions possess specific drawbacks, and creating a device that combines the benefits of EDLCs and batteries remains a significant challenge. Lithium–ion capacitors (LICs), also known as lithium–ion hybrid SCs, successfully reduce the differences between LIBs and EDLCs. It combines high energy density, high power

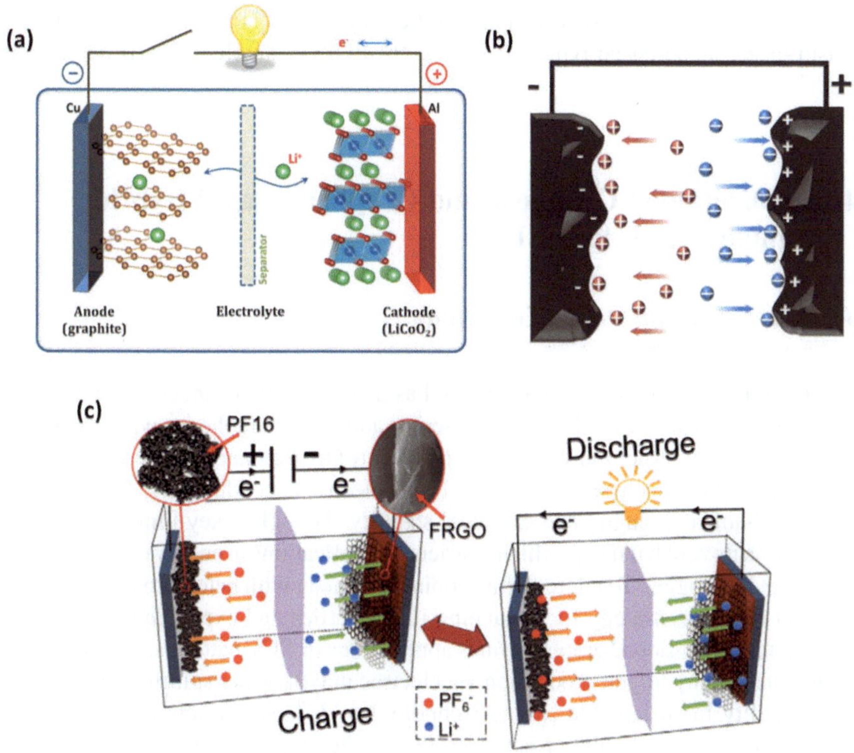

Fig. 7 Illustration of the charge/discharge mechanisms of **a** LIBs, **b** EDLCs, and **c** LICs [33]. Reproduced with Open access @ (2021) MDPI

output, wide operating voltage, and more extended cycle stability within a single device. Generally, LICs consist of a battery-type electrode as the anode, which provides high energy, and a capacitor-type electrode as the cathode, which ensures high power and cycling stability. Unlike LIBs and EDLCs, LICs exhibit a hybrid energy storage mechanism. While charging, the absorption of anions at the capacitor-type electrode's interface is carried out as its potential increases, lithium ions from the electrolyte simultaneously intercalate into the battery-type electrode, as shown in Fig. 7c. In the process of discharging, the lithium ions and anions return to the electrolyte and release the stored energy after exiting the cathode and anode, respectively. Additionally, since the cathode and anode operate in different potential regions, LICs achieve a significantly higher working voltage than EDLCs, resulting in improved energy density. Overall, LICs employ a hybrid charge storage mechanism enabled by their innovative device configuration. The combination of a battery-type anode from LIBs and a capacitor-type cathode from EDLCs provides high energy and power densities along with extended cycling life in LICs, and thus LICs are a desirable alternative for energy storage systems of the future [33]. The charge storage mechanism of graphene-based typical (a) LIBs, (b) EDLCs, and (c) LICs is depicted in Fig. 7.

4 Application of Graphene and Graphene-Based Hybrid Nanomaterials for ESDs

4.1 Heteroatom-Doped Graphene

Heteroatom-doped graphene has emerged as a significant advancement in the field of energy storage, particularly for batteries and SCs. Graphene, with its unique two-dimensional structure, has emerged as a promising material due to its exceptional properties, such as high thermal and electrical conductivity, large surface area, mechanical strength, and chemical stability. Though, a key limitation of pure graphene is its zero band gap, which restricts its applications in energy conversion and storage technologies. To address this, doping graphene with heteroatoms has proven to be an effective strategy for enhancing its properties and expanding its functionality. Heteroatom-doped graphene finds applications in SCs, batteries, dye-sensitized solar cells, and hydrogen production via electrocatalytic water splitting, showcasing its versatility in sustainable energy technologies [34]. When heteroatoms such as nitrogen, oxygen, and sulphur get inserted into the graphene structure, it increases its electrochemical properties which again lead to improved performance in energy storage applications. Processing methods, electrochemical performance, and potential applications of heteroatom-doped graphene will all be addressed in it. Nowadays, functionalized nanostructured carbon materials gained a considerable interest for high-performance ESDs. Remarkably, the combination of three-dimensional (3D) and heteroatom-doped graphene materials gives exceptional electrochemical performance. The 3D architecture allows effective ion transport during the electrochemical activities, the incorporated heteroatoms add a large number of active sites that improve electrochemical performance [35]. There are two ways to dope herotoatoms in graphene: first is adsorbing dopants onto its surface and another is replacement of carbon atoms with heteroatoms. Since covalent bonds are formed, substitutional doping is more stable; nonetheless, it must be carefully managed to avoid flaws that could damage the graphene structure or decrease conductivity. Elements including phosphorus, sulphur, nitrogen, and boron are favoured dopants because they are well suitable with carbon. As GO is perfect starting material for doping, while replacing oxygen directly these oxygen functionalities damage the graphene structure which promoted the decrement in conductivity. Nevertheless, active sites created by concurrent doping and reduction of GO make it easier to incorporate heteroatoms and enhance the electrochemical properties of rGO for uses such as energy conversion and storage. Latif et al. [36] used a single-step hydrothermal process to prepare heteroatom-doped GO electrodes, which have outstanding performance for SC applications and expand the possibilities of GO-based electrodes in energy storage technologies. They incorporated heteroatoms nitrogen (NGO) and boron (BGO) in graphene oxide (GO) separately using urea and boric acid, respectively, as sources. Also, they synthesized a co-doped version of BNGO. From the X-ray photoelectron spectroscopy (XPS), it was observed that NGO contained 6.13% nitrogen, BGO contained 17.94% boron, and BNGO contained 23.76% boron and 3.64% nitrogen. In 3 M KOH NGO and BNGO electrodes achieved specific capacitances of 855 F/g and 811 F/g, respectively. The charge storage mechanism was mainly diffusion controlled, signifying that along with structure of electrode, doping has a major impact on their electrochemical performance. Liang et al. [37] presented easy way to prepare nitrogen-doped graphene and fabricated nitrogen-doped graphene/SnO_2 composite as a cathode material for lithium battery anode. The collaborative interaction of SnO_2 nanoparticles with nitrogen-doped graphene noticeably improved the electroconductivity and structural stability of the anode material, which facilitate its ability to resist the structural deterioration frequently observed in SnO_2-based electrodes throughout prolonged cycling.

In situ gas reduction has been acknowledged as a competent approach, as noted by Li et al. for the uniform distribution of SnS-SnO_2 nanoparticles on N-doped graphene, thus facilitating enhanced charge transfer kinetics and reducing aggregation. The resultant material achieved high initial Coulombic efficiency with 630 mAh/g at 2.0 A/g over 1000 cycles and 85% of capacity retention over 500 cycles. This work validates the potential of SnS-SnO_2/N-doped graphene as durable, high-performance anodes for future lithium–ion batteries. Li et al. [38] reported nitrogen and phosphorus dual doping improves sodium storage by expanding interlayer spacing and introducing active sites such as pyridinic-N, which facilitate ion transport and charge storage. The 3D interconnected networks of GAs further contribute to structural stability and high conductivity. Studies have demonstrated that N, P dual-doped GAs achieve high reversible capacities, excellent rate capabilities, and ultra-stable cycling performance, delivering capacities of 195 mAh/g after 1000 cycles at 1 A/g and 132 mAh/g after 2000 cycles at 5 A/g. Despite the practical capacity being halved due to potential range limitations (0.005–1.5 V), these materials show remarkable promise for SIBs and electrochemical capacitors. This simple phosphidation-based synthesis approach underscores the feasibility of scalable, high-performance carbonaceous anode materials for next-generation energy storage technologies. Three-dimensional nitrogen-doped graphene aerogel/sulphur (NGA/S) composites have been developed by Kang et al. [39]. In this study composites were synthesized using ethylenediamine (EDA) as a nitrogen source and graphene structural modifier via a one-pot hydrothermal self-assembly followed by freeze-drying and a heat melt-diffusion process. This process is used to produce a porous 3D structure with a 77.3 wt% of sulphur loading. The nitrogen-doped graphene aerogel improves charge transfer by stabilizing sulphur particles and contributing a conductive framework which also avoids lithium polysulphides from dissolving. The NGA/S composite showed a high initial discharge capacity of 1258 mAh/g and maintains 1077 mAh/g after 75 cycles. This outstanding performance and stability is owed to the conductive nature and polysulphide-immobilization of N-doped graphene, with its high sulphur loading and flexible porous structure. Uesugi et al. [40] demonstrated an innovative approach for the stabilization of sulphur cathodes in lithium–sulphur batteries. A straightforward acid–base reaction at room temperature is used to incorporate sulphur between graphene sheets that are attached with flexible polypropylene glycol (PPG) which ultimately yielded enhanced sulphur utilization, cycling stability, and overall battery efficiency. The strong interfacial attraction between sulphur and highly conductive graphene is provided by this design, which avoids polysulphide dissolution and improves the reversibility of redox reactions. The PPG anchor also bestows structural flexibility to support the volume expansion of sulphur during charge–discharge cycles. Relative to traditional carbon-based electrodes, the resultant sulphur–carbon composite exhibits enhanced cyclability and a high initial discharge capacity of 1154 mAh/g. The confined layered structure helps quick electron and lithium–ion transport, further enhancing electrochemical performance.

Recently, carbon nano-onions (CNOs) have gained a lot of attention for high-performance electrode materials for ESDs, because of their excellent electrochemical characteristics, huge surface area, and distinctive onion-like structure. Prior research has been concerned with improving the performance of CNOs by doping them with heteroatoms which increase active sites for electrochemical processes, surface wettability, and electrical conductivity. Traditional techniques to produce doped CNOs usually entail difficult steps, costly precursors, or severe reaction conditions, which restricts their scalability and usefulness. In this regard, Mohapatra et al. [41] utilized simple and low-cost flame-pyrolysis techniques for the synthesis of CNO allowing for the concurrent doping of several heteroatoms while maintaining the inherent structural characteristics. Traditional techniques used to synthesize boron/nitrogen co-doped carbon materials need complex procedures or multiple steps, making them less effective and scalable. Recent developments have focused on simplifying the process and improving the performance of these materials. Utilizing a single-step carbonization process using a ferric catalyst and boric acid treatment is one such approach that has been demonstrated to improve the materials' specific surface area and electrochemical characteristics. It is well known that doping carbon materials with boron and nitrogen increases their conductivity, stability, and electrochemical reactivity, which makes them ideal for energy storage devices. In this regard, Li et al. [42] revealed that the use of ferric catalysts is beneficial for the prevention of inert B-N bond creation which is responsible for the lowering of performance. Significant electrochemical characteristics, like high specific capacitance, superior rate performance, and long cycle life, are displayed by the resultant boron/nitrogen co-doped carbon materials. This material achieved 330 F/g of capacitance with 89.9% of their capacitance retention over 5000 charge–discharge cycles. Moreover, the assembled symmetric SCs showed an energy density of 7.9 W h/kg at a power density of 505 W/kg. In order to enhance the charge storage and conductivity, Govindarasu et al. [43] synthesized these phosphorus-doped graphitic hollow carbon spheres (P-GHCS) through P-GHCS via catalytic chemical vapour deposition (CVD) on a Fe-KIT-6 substrate. This doping of P introduced additional active sites which promoted the supercapacitive properties of the material. This study has shown that the best performing sample, P-GHCS-800, exhibits a specific capacitance

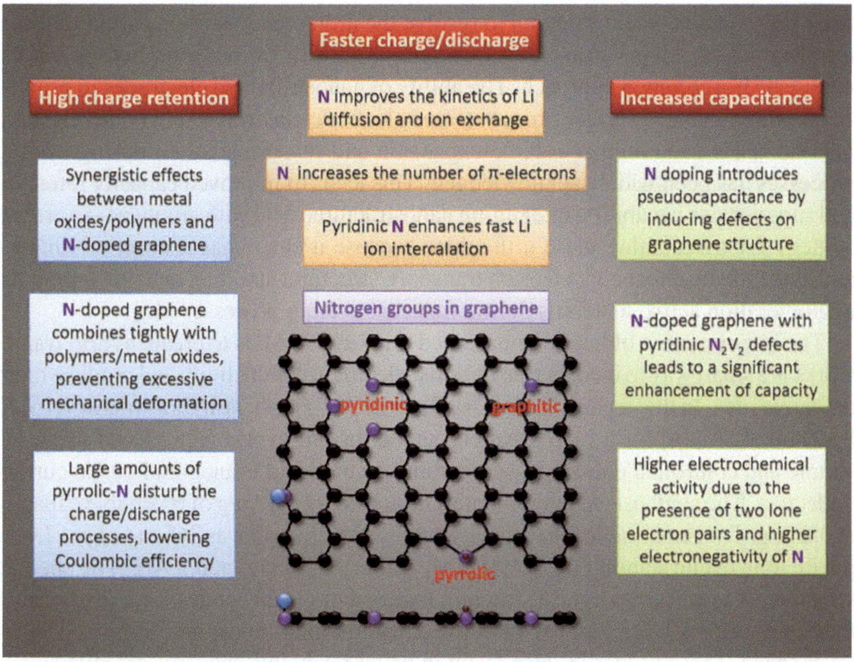

Fig. 8 Key implications of N-doping for doped graphene-based SCs and LIBs [35]. Reproduced with permission Copyright @ (2020) Elsevier

of 321 F/g with capacitance retention 97.1% of its capacitance after 2000 charge–discharge cycles. Moreover, it operates at a high voltage of 2.0 V, delivering an energy density of 10.83 Wh/kg, which surpasses the performance of many commercial SCs. The successful demonstration of a P-GHCS-based SC powering a red LED further highlights the practical potential of these materials in real-world applications. This research contributes to the ongoing efforts to develop efficient and durable materials for energy storage, offering a promising solution for advancing SC technology. Fig. 8 identifies some of the most significant effects of N-doping on doped graphene-based SCs and LIBs. As a visual assistance, the locations and nomenclature of the most prevalent N groups in graphene materials are also displayed.

4.2 *Metal Oxide-Based Graphene*

The high specific capacity of transitional metal oxides (TMOs), which is often twice or even three times more than that of carbon-based materials, makes them extremely promising in ESDs. But face several difficulties such as intrinsically insufficient electrical conductivity, short cyclic stability, fast capacitance loss, and extreme volume changes during cycling [44]. There are two primary approaches that have been suggested in recent research to resolve the aforementioned issues and improve the electrochemical performances. Designing MO with various morphologies at the nanoscale is a common approach. The ability of nanostructured electrodes to provide easily available large surface areas between electrode/electrolyte and relatively large internal space to minimize volume expansion and releasable strain during cycling processes has been widely acknowledged. This leads to improved capacity retention and superior rate capability. Integrating nanostructured MO with graphene is a crucial strategy for overcoming these difficulties because it can enhance reaction kinetics and cycle life by enhancing electron transport, structural stability, and preventing the agglomeration active materials during repeated cycling processes [45].

The incorporation of heteroatom-doped graphene in MOs to create hybrid frameworks is a smart way to increase the hybrids' electrical conductivity and produce more defective sites. In this regard, MnO/Mn_3O_4/N-doped-graphene hybrid was developed by Zhou et al. [46] for a Li–ion battery anode. In situ hybrid is produced using the simple and controlled one-step laser scribing method and is anchored on Cu current collectors. The synthesized hybrid material showed 992 mAh/g of reversible capacity at 0.2 A/g. The remarkable rate capacity achieved by hybrid was 365 mAh/g at 2.0 A/g with long cyclic stability up to 699 mAh/g after 400 cycles at 0.2 A/g. The specific structure of synthesized hybrid structure makes them ideal anodes for LIBs because they may offer a large number of active sites, internal voids, and improved interlayer space. These properties also result in better reversible capacity, extended cycle stability, and superior rate capability. Additionally, these distinctive architectures may effectively prevent the MnO/Mn_3O_4 nanoparticles' volume growth during the charge–discharge process. Moreover, introducing evenly spread and controllable in-plane nanoholes to graphene sheets is not only useful to prevent aggregation and increasing surface area, but also offers an abundance of ion/mass transport channels. With more uncovered edges and active sites in the graphene nanosheets due to these homogenously dispersed nanoholes, more surface modification, and even catalysis is possible. Additionally, despite maintaining nearly unchanged mechanical properties, holey graphene can exhibit impressive hole-dependent semiconducting bandgap and optical properties in comparison to pure graphene. Thus, the use of graphene materials for energy storage may be expanded by holey graphene with distinct structures and physicochemical features. In another study, Gao et al. [47] combined SnO_2 with metal–organic framework (MOF) and graphene to improve the efficiency of SnO_2. By serving as a barrier, the MOF layer increases the lithium storage capacity of the SnO_2. SnO_2 nanoparticles were integrated into Al-MOF at an ideal mass ratio and then wrapped in graphene to form the SnO_2@MOF/graphene composite. The composite demonstrated noticeably improved cycling stability, reaching a specific capacity of 450 mAh/g after 1,000 charge/discharge cycles at a current density of 1,000 mA/g. This enhancement is attributed to the composite's unique structural design, which offers numerous accessible electroactive sites, shorter ion transport pathways, and enhanced electronic conductivity.

Instead of fact that LIB is strong candidate in the field of ESDs, lack of resources and high prices of lithium limits its development. However, sodium-ion batteries

(SIBs) are seen as a possible choice for large-scale energy storage systems and potential substitute for LIBs due to the availability of sodium resources and comparatively good safety. However, due to SIBs' poor conductivity and significant volumetric change during cycling, the material frequently exhibits inferior cycle stability and capacity rate. The development of hybrid material by adding graphene to MOs with specific nanostructure and composition has shown great promise in SIBs [48]. Ren et al. [49] fabricated interstitial Mn-incorporated Fe_3O_4/graphene nanocomposite as anode material for SIBs and demonstrated that expansion of Fe_3O_4 lattice occurred due to the interstitial Mn, which inevitably increases the electrical conductivity of material. Meanwhile, it is discovered that the graphene substrate enables the in situ growth of Fe_3O_4 particles, greatly lowering the particle size. These benefits make it possible to reduce the Na ion diffusion path, increase the specific surface area for active sites on the surface, and offer a high storage capacity. With a current density of 100 mA/g and a high reversible capacity of 439.8 mAh/g, the hybrid material demonstrated 94% capacity retention after 200 cycles. By acting as a flexible skeleton to adapt the volume change, lamellar graphene successfully prevents particle aggregation during long-term cycling when structural pulverization is absent. The inclusion of the lamellar grapheme substrate has also been shown to improve electrical conductivity, which enhances the composite's electrochemical performance. B. Wang et al. (2019) developed Mn_3O_4@pGS hybrid material which contains Mn_3O_4 nanotubes implanted in porous graphene sheets as an anode for both LIBs and SIBs. This specially constructed framework achieved 770 mAh/g of specific capacity after 200 cycles for LIBs. Whereas it showed charge capacity of 233 mAh/g after 55 cycles for SIBs.

Barmi et al. [50] prepared electrochemically deposited iron oxide nanoparticles on the N-doped porous graphene (N-PG) nanosheets for ESD application. This study explores the development of advanced Fe_3O_4-based hybrid electrodes for SCs by combining Fe_3O_4 with N-doped porous graphene (N-PG). The hybrid electrode, Fe_3O_4@N-PG/Ni foam, exhibited significantly enhanced electrochemical performance, with specific capacitances of 822 F/g at 0.5 A/g and 631 F/g at 10 A/g, surpassing the pristine Fe_3O_4/Ni foam electrode, which achieved only 279 and 131 F/g at the same current densities. This improvement is attributed to the synergistic interactions between N-doped graphene sheets and Fe_3O_4 nanoparticles, which enhance electrical conductivity, reduce structural degradation, and maximize the pseudocapacitance of Fe_3O_4. Using a one-step co-deposition method, fabricated hybrid electrode with mesoporous structure and high surface area of 138 m^2 g^{-1} enhanced its energy density, rate capability, and cycling stability. The study offers a viable solution for high-performance SC electrodes by addressing the issues of Fe_3O_4's low electrical conductivity and structural instability. Whereas, by preparing Bi_2O_3-Fe_3O_4/graphene composite utilizing a chronoamperometric deposition process, Pattananuwat et al. [51] combined solar energy conversion and storage in a single electrode. Significant photo-assisted charging capabilities are demonstrated by the BiO_3-FeO_4/graphene electrode, which achieves a high specific capacitance of 306.67 F/g under UV light, which is 39.29% higher than under dark conditions. The synergistic effects of graphene improved charge separation and transport efficiency to improve photoelectrochemical properties. After 1,000 cycles, the electrode still has more than 60% of its capacitance, demonstrating longer cycling stability. This work offers a viable strategy for enhanced SC technology and next-generation energy devices by permitting simultaneous solar energy conversion and storage. In this context, El-Hallag et al. [52] demonstrated a novel method for creating high-performance SC electrodes via a one-pot in situ electrochemical reduction and deposition method to create palladium nanoparticle (Pd NP)-embedded electrochemically reduced GO nanosheets (Pd-E-ErGO) on nickel sheets. This method forms a hybrid structure with Pd NPs uniformly embedded within ErGO nanosheets by concurrently reducing GO to ErGO and Pd^{2+} ions to Pd NPs.

With a high specific capacitance of 1524.7 F/g at a scan rate of 20 mV/s, remarkable rate capability, and 92.1% cycling stability after 2000 cycles, the Pd-E-ErGO electrode revealed outstanding electrochemical performance. These improvements are owed to the strong interaction between Pd NPs and ErGO as well as the hybrid structure's effective charge transport. Figure 9 gives the advantages of graphene/metal-oxide composite in various structures.

4.3 Conducting Polymer-Based Graphene

CPs showed potential candidature for environmentally friendly adaptable ESDs due to their unique properties. Polypyrrole (PPy), polyaniline (PANI), polythiophene (PTh), and poly (3,4-ethylenedioxythiophene) (PEDOT) are popular CPs for ESD applications. Their exceptional flexibility, ease of processing, and different morphologies, along with their excellent electrochemical performance, provide them a distinct advantage in a range of battery systems, including lithium ion, sodium ion, metal air, lithium sulphur, all-polymer batteries, and SCs. The electroactivity and conductivity of CPs are the key features utilized in ESDs. The polymer backbone of CPs is made up of alternate single and double bonds. The configuration of delocalized electrons is produced by their overlapping p-orbitals, which results in certain fascinating and advantageous optical and electrical features. Only doping offers electrical conductivity to CPs, and doping amount has a significant impact on it [54]. The doping/de-doping process and mobility of ions are both implicated in the ability of CPs to be charged or discharged. Because of their electrochemical stability in p-doped states, CPs are frequently utilized as cathodes in batteries. The use of CPs electrodes in rechargeable batteries and SCs is inhibited by a significant problem with cycle stability. When using pristine CPs for electrochemical applications, they deteriorate because of volume changes during charging and discharging [54]. To address this issue, introduction of graphene sheets helps to reduce this volume change during cycling and improves the stability at higher voltages [55].

Currently, CP-based zinc rechargeable batteries have made substantial progress. Zn serves as a suitable anode for aqueous batteries due to its larger theoretical capacity

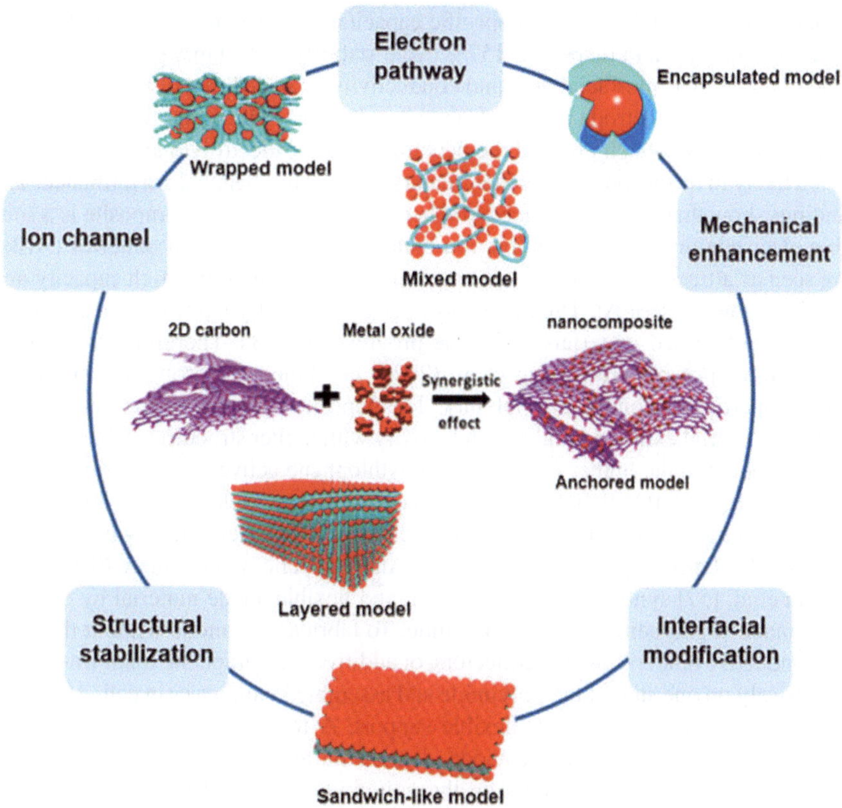

Fig. 9 Advantages of graphene/metal-oxide composite in various structures [53]. Reproduced with Open access @ (2019) MDPI

with lower redox potential and good compatibility with water. The potential for large-scale manufacture of Zn batteries is facilitated by their plentiful resources, lack of toxicity, and earthly air stability [44]. In comparison to conventional batteries, Zn/PANI batteries possess high specific capacity and an inexpensive, simple production. Despite this, it has a reduced real specific capacity as a result of the substantial PANI aggregation that occurs during the synthesis and the mechanical degradation of the electrodes caused on by deformation, which reduces the electrode's capacity and cycle performance. Hence, in order to enhance the performance of Zn/PANI batteries Wang et al. [44] reported PANI/graphene composites made by a new in situ chemical oxidative polymerization process combining two oxidants. The prepared composite showed higher discharge specific capacity (192 mAh/g) than pristine PANI (100 mAh/g) at 0.2 C constant current. Also, composition with graphene increased cyclic stability of PANI from 80.4 to 95.05% after 100 cycles. Cycle performance prior to modification was significantly enhanced. Furthermore, Xu et al. [56] synthesized 3D ZnS/RGO/polypyrrole ternary composites for the LIBs which showed 1446.9 mAh/g of specific capacitance at 100 mA/g with 955.6 Wh/kg energy density and displayed 157% cycle stability at 100 mA/g for 200 cycles. Due to the higher interfacial area and conductivity provided by polypyrrole, charge transfer between electrode and electrolyte was boosted. Moreover, the 3D structure of the composite, which enabled the penetration of the electrolyte and the insertion Li^+, a likely to be responsible for their remarkable electrochemical performance. The findings show that the ZnS/reduced graphene oxide/polypyrrole composite is a suitable electrode material for lithium–ion batteries. Aluminium–ion batteries (AIBs) are seen as attractive alternatives as upcoming ESD owing to the high capacity and relatively low price of Al. Though the low performance of pristine carbon and metal chalcogenide-based materials limits the progress of AIBs. Therefore, to improve the performance of AIBs Kong et al. (2020) combined the properties of carbon with properties of conducting polymer. The deposited polythiophene on GO as a cathode material exhibited higher conductivity with higher structural stability which was attributed to the higher exposure of polythiophene active sites. Storage mechanism utilizing polythiophene was in accordance with strong electrostatic interactions between charge carriers and the Cβ-H sites of the polythiophene chain which showed 130 mAh/g of discharge capacity and 100 mAh/g of cyclic stability after 4000 cycles. Aslam et al. [57] synthesized a lightweight and flexible anode material by creating laser-induced graphene foam on polyimide. To fabricate an anode without the need for conductive binders, current collectors, or additives, the graphene foam was grown exclusively on one side of the polyimide and assembled within a coin cell. The polyimide served as a lightweight, flexible support. A folded laser-induced graphene foam sample was utilized as the anode, demonstrating the successful movement of electrons across the foam grown on the polyimide surface. The fabricated anode achieved an areal capacity of 203 $\mu Ah/cm^2$, exhibiting stable rate performance over 100 cycles at a current density of 0.1 mA/cm^2, with an average Coulombic efficiency of 99%. This work paved the way for lightweight Li–ion batteries based on laser-induced graphene foam and innovative flexible wearable high energy density storage devices.

The performance of ESDs has been significantly improved by recent developments in PPy- and PANI-based graphene materials. Via in situ polymerization PANI/graphene hybrid fibres (PANI/GF) were prepared by Zhang et al. [58] on graphene fibres which are prepared by wet spinning and chemical reduction. This combination improved the structural stability even dispersibility of PANI resulted in a specific area capacitance of 87.8 mF/cm^2 at 0.22 mA/cm^2, energy density of 12.2 $\mu Wh/cm^2$, and excellent cycling stability over 10,000 cycles. The wrinkled morphology with networked architecture of the composite fibres enhanced conductivity and ion diffusion promoted such high electrochemical performance. A novel technique to develop self-standing, interconnected 3D nanostructured CP/graphene hydrogels (CPHs) was presented by Li et al. [59]. The use of GO and PANI as model materials for 3D PANI/RGO hybrid hydrogel exhibited decreased aggregation, improved structural homogeneity, and strong intermolecular interactions. The hybrid hydrogel delivered a volumetric energy density of 8.80 mWh/cm^3 at a power density of 30.77 mW/cm^3. These results enlightened the potential of 3D hybrid hydrogels in advancing flexible, lightweight, and high-performance energy storage solutions for future wearable and portable electronics. Furthermore, in another study, Moussa et al. [60] revealed an innovative and additive-free approach for synthesis of CP hydrogels (CPHs) with enhanced electronic conductivity, high surface areas, and superior mechanical properties. In this technique, there is no need of external cross linkers since it makes use of intrinsic intermolecular interactions such as ionic contacts, hydrogen bonds, and π–π stacking. The synthesis of several CPHs, such as PANi, PPy, and PEDOT, which are biocompatible and appropriate for bioactive applications, such as tissue regeneration and stimulation, demonstrates the adaptability of this technique. Among PANI nanofiber (PANi-NF) hydrogels achieved a high specific capacitance of 492 F/g at a current density of 1 A/g in flexible SCs. Additionally, PANi-NF hydrogels were employed as both gelling and reducing agents to fabricate graphene composite hydrogels, demonstrating enhanced gravimetric and areal capacitances of up to 927 F/g and 578 mF/cm, respectively. This shows the candidature of additive-free CPHs as advanced materials for high-performance supercapacitors and multifunctional applications in energy storage and bioelectronics.

Yazar et al. [61] explored the enhancement of PANI composites by doping graphene oxide (GO) with heteroatoms—specifically sulphur (S), nitrogen (N), and chlorine (Cl)—to improve their electrochemical performance in supercapacitors. The composites were synthesized using a two-step process: first, S-GO, N-GO, and Cl-GO were prepared through chronoamperometry, followed by hydrothermal synthesis of PANI composites. Among these, the PANI/Cl-GO electrode exhibited the highest specific capacitance of 1217 mF/cm (608.3 F/g) at a current density of 4 A/g, surpassing other hybrid electrodes in performance. PANI/Cl-GO//PANI/Cl-GO exhibited the energy density of 42.8 Wh/kg at a power density of 250 W/kg with 96.9% capacitance retention over 5000 cycles. The work emphasizes improvement in the composites' capacitance and cycling stability by increasing their surface area, electronic conductivity, and redox-active sites after Cl doping. The substantial promise of Cl-GO-doped PANI composites for long-lasting, high-performance ESD is demonstrated by this work.

Cymann-Sachajdak et al. [62] carried out a systematic study on electrochemical behaviour of composite films consisting of poly(3,4-ethylenedioxythiophene) (PEDOT), graphene oxide (GOx), and oxidized multiwalled carbon nanotubes (oxMWCNTs) across various aqueous electrolytes, including different sulphates (Li_2SO_4, Na_2SO_4, K_2SO_4, $MgSO_4$) and H_2SO_4 solutions. The study examines influence of the properties of electrolytes on the capacitance, rate capability, and cycling stability of composite electrodes. The study suggests that acidic electrolytes, in this case H_2SO_4, exhibit the highest capacitance values (98.6 mF/cm at 1 mA/cm), while neutral electrolytes like $MgSO_4$ contribute to the enhancement of the cycling stability, where 88.4% capacitance retention after 10,000 cycles was observed. Furthermore, the study emphasizes the notable role of electrolyte properties in the improved capacitance, highlighting that smaller cations such as Li^+ and Na^+ foster higher capacitance than larger ions like Mg^{2+}. The enhanced capacitance in acidic electrolytes is ascribed to additional redox reactions promoted by the oxygen functionalities on GOx and oxMWCNTs. These findings reiterate the critical role of electrolyte selection to retrieve optimized performance of PEDOT-based supercapacitors, with enhanced energy storage efficiency and cycling stability.

Zhao et al. [63] presented the vapour–liquid interfacial polymerization method which has been an effective approach for synthesizing aforementioned composites. This method provides better dispersion of conducting polymer nanoparticles within the GO sheets, resulting in effective ion transport and charge storage in a three-dimensional network structure. Studies have shown GO/PPy/PANI composites exhibit superior capacitance retention and rate capability, with the GO/PPy/PANI composite hydrogel maintaining 71.5% of its capacitance at high current densities, compared to 32.8% for the PPy/PANI composite. Furthermore, the composite hydrogel exhibits improved cyclic stability, retaining 91% of its initial capacitance after 2,000 cycles, which is a noteworthy improvement over the 62% retention observed for PPy/PANI films. The interaction between GO, PPy, and PANI through hydrogen bonding and π–π stacking contributes to the stability and enhanced electrochemical performance of the hydrogel. Furthermore, the resulting flexibility of hydrogel makes it possible to use it in solid-state supercapacitors, which reveal the potential for flexible electronic devices with their exceptional electrochemical properties under various bending states and longer cycling stability (81% retention after 10,000 cycles).

In recent research, aqueous redox-active electrolytes have come up as promising materials with their potential to significantly enhance energy storage in electrochemical capacitors. Moyseowicz et al. [64] demonstrated promising approach of hybrid capacitors consisting of porous polypyrrole/reduced graphene oxide (PPy/rGO) composite electrodes and redox-active electrolytes, such as hydroquinone. The PPy/rGO composites synthesized using hydrothermal methods has been shown to produce a porous material with a high specific surface area (597 m^2 g^{-1}), which further improves electrochemical performance. The reversible Faradaic reactions due to redox-active species, such as hydroquinone, leads to the increased charge storage capacity. Studies have showcased that these composites outperform the standard rGO electrodes, particularly in terms of specific capacity and cycling stability. For instance, a symmetric hybrid capacitor consisting of PPy/rGO-HT composite electrodes accompanied with hydroquinone-based aqueous electrolyte achieved a specific energy of 6.5 Wh/kg and retained 104% of its initial capacitance after 12,000 cycles, showing excellent long-term stability. The enhanced mesoporosity and nitrogen content of the PPy/rGO-HT facilitate the efficient interaction with the quinone–hydroquinone redox couple resulting in better electrochemical performance. Additionally, the composite demonstrated good rate capability, maintaining 3.1 Wh/kg at a high specific power of 7.7 kW/kg. The binder-free design of the electrodes simplifies the fabrication process, further enhancing the device performance. These results highlight environmentally friendly and cost-effective approach of combining porous polymer–graphene composites with redox-active electrolytes towards developing high-performance, sustainable, and stable aqueous electrochemical capacitors. Chang et al. [65] demonstrated the covalent grafting of TANI onto 3D graphene networks (GN) using perfluorophenylazide coupling chemistry, significantly improving both the electrochemical performance and cycling stability of the composite material. The resulting ATgGN composite has shown excellent cycling stability, retaining over 85% of its capacitance over 30,000 charge–discharge cycles. Furthermore, the addition of redox-active electrolytes further boosted the specific capacitance of the supercapacitors while maintaining its cycling stability, with 82% capacitance retention after 100,000 cycles. This study opens the door for the production of high-performance, long-cycle-life ESDs by offering a practical method to get over one of the main weaknesses of conducting polymer-based SCs.

5 Challenges and Future Research Directions

For extensive utilization of potential of graphene in ESDs, particularly batteries and SCs, several challenges and limitations need to be addressed. A well-organized electrode structure is essential and it is achievable through streamlined preparation and synthesis techniques. In these processes, reducing the production cost of graphene-based materials with high surface area and optimal composition for large-scale applications remains critical. A well-conductive network of graphene materials must be developed which can serve as active electrode components to mitigate electrode imbalances that reduces electrical performance. Enhancing the packing density of graphene-based materials while preserving their inherent capacitive and flexible properties can further enable high volumetric capacitance and gravimetric energy density, particularly for wearable technologies. Moreover, along with the electrodes, the performance of ESDs depends also on the synergy of all components, such as current collectors, electrolytes, substrates, electrodes, separators, and insulating layers. Designing these components helps in advancing the understanding of reaction mechanisms and electrochemical behaviour of graphene electrodes in different electrolytes. Future efforts should emphasize development of innovative strategies to improve energy density, reduce costs, and enhance the durability and stability of graphene-based electrodes.

In general, the key challenges in developing graphene-based hybrid materials for ESDs include complex and expensive fabrication processes, difficulties in scalability, and uniformity, along with interfacial compatibility between graphene and other materials. Further, decay in the performance during cycling, limited stability, and mismatched properties between hybrid components further complicate their practical application. Performance improvement is further complicated by material-specific concerns such as polymer degradation, low conductivity in transition metal oxides, and stability issues in hydroxides, as well as a lack of knowledge about charge storage mechanisms and ion dynamics. Economic and environmental issues also need to be taken into account, such as the high cost of materials, energy-intensive procedures, difficulties with recycling, and toxicity control. The major challenges in developing graphene-based hybrid materials for SCs include complex and costly fabrication processes, difficulties in scalability, and ensuring uniformity, alongside interfacial compatibility between graphene and other materials. Further, degradation in the performance during cycling, limited stability, and mismatched properties between hybrid components complicate their use case. Material-specific issues like polymer degradation, low conductivity in transition metal oxides, and stability problems in hydroxides combined with a limited understanding of charge storage mechanisms and ion dynamics create additional challenges in the improvement of performance. There are also environmental and economic concerns, which include high material costs, energy-intensive processes, and challenges in recycling and toxicity management that need to be considered.

Future studies need to aim to address current limitations and explore innovative approaches to enhance performance of graphene-based materials for SC application. Combining graphene with transition metal oxides, hydroxides, or CPs to improve pseudocapacitance, energy density, and flexibility is one of the key directions to proceed with designing novel hybrid architectures. Additionally, developing three-dimensional porous structures and incorporating surface engineering techniques, such as defect introduction, heteroatom doping, and functionalization, can further increase active sites, conductivity, and ion accessibility. Scalable and sustainable synthesis methods, using low-cost precursors, are necessary for large-scale production without compromising material quality, structural and electrolyte optimizations, such as redox-active or solid electrolytes, which can help in improving energy density, cycling stability, and rate capability. Advanced characterization techniques will enable deeper insights into ion transport mechanisms and material behaviour during cycling. Further, integration into flexible and wearable devices opens new possibilities for portable energy storage. In this direction, computational modelling and machine learning will assist in material design and optimization, while environmental considerations emphasize the need for recyclable and sustainable materials. These efforts aim to overcome current challenges and enable the development of high-performance, sustainable graphene-based multifunctional hybrids for energy storage applications.

6 Summary and Conclusions

The advancements in energy storage technologies heavily rely on the strategic integration of nanostructured materials with enhanced properties. Graphene holds immense potential in the development of ESDs due to some of its exceptional properties such as high surface area, excellent thermal and electrical conductivities, superior mechanical strength, optical transmittance, and biocompatibility. However, its zero band gap significantly limits its wider applicability. This enables the development of graphene-based hybrids for high-performance, flexible, and wearable ESDs. By integrating graphene with other electrochemical materials, such as MOs, CPs, metal nanoparticles, and doped heteroatoms, researchers have achieved superior electrochemical performance. Heteroatom-doped graphene has demonstrated its potential to improve electrochemical properties and performance in ESDs through increased reaction kinetics and structural stability. Similarly, nanoscale-designed MO, when combined with graphene, address challenges such as volume expansion and active material agglomeration, leading to improved capacity retention and rate capability. CPs, with their unique optical and electrical properties, face limitations in cycle stability and structural changes during charging/discharging. However, incorporating graphene sheets into CP systems significantly enhances their stability and performance under high-voltage conditions. Together, these innovations underscore the critical role of nanostructured materials and their composites in overcoming traditional limitations, paving the way for next-generation high-performance ESDs. This chapter reviews extensive literature on graphene-based hybrids used in batteries and SCs, discussing the challenges and practical considerations in their application. The insights aim to catalyse the design and fabrication of advanced graphene hybrids to meet the growing demands of future ESDs, highlighting the critical role of continued innovation in this research area.

References

1. Ali, A., et al.: Latest energy storage trends in multi-energy standalone electric vehicle charging stations: a comprehensive study. Energies **15**(13), 1–19 (2022). https://doi.org/10.3390/en15134727
2. Devendran, A., Nagai, A.: Recent advancements in 3D porous graphene-based electrode materials for electrochemical energy storage applications. Mater. Adv. **4**(12), 2524–2543 (2023). https://doi.org/10.1039/d3ma00022b
3. Dey, R.S., Hjuler, H.A., Chi, Q.: Approaching the theoretical capacitance of graphene through copper foam integrated three-dimensional graphene networks. J. Mater. Chem. A **3**(12), 6324–6329 (2015). https://doi.org/10.1039/c5ta01112d
4. Islam, M.S., Mubarak, M., Lee, H.J.: Hybrid nanostructured materials as electrodes in energy storage devices. Inorganics **11**(5), 1–21 (2023). https://doi.org/10.3390/inorganics11050183
5. Yang, G., Li, L., Lee, W.B., Ng, M.C.: Structure of graphene and its disorders: a review. Sci. Technol. Adv. Mater. **19**(1), 613–648 (2018). https://doi.org/10.1080/14686996.2018.1494493
6. Ahmad, F., et al.: Advances in graphene-based electrode materials for high-performance supercapacitors: a review. J. Energy Storage **72**, 108731 (2023). https://doi.org/10.1016/j.est.2023.108731
7. Tiwari, S.K., Sahoo, S., Wang, N., Huczko, A.: Graphene research and their outputs: status and prospect. J. Sci. Adv. Mater. Dev. **5**(1), 10–29 (2020). https://doi.org/10.1016/j.jsamd.2020.01.006
8. Senturk, A.E., Oktem, A.S., Konukman, A.E.S.: Influence of defect locations and nitrogen doping configurations on the mechanical properties of armchair graphene nanoribbons. J. Mol. Model. **24**(2), 9 (2018). https://doi.org/10.1007/s00894-018-3581-3
9. Mathew, E.E., Balachandran, M.: Crumpled and porous graphene for supercapacitor applications: a short review. Carbon Lett. **31**(4), 537–555 (2021). https://doi.org/10.1007/s42823-021-00229-2
10. Yi, M., Shen, Z.: A review on mechanical exfoliation for the scalable production of graphene. J. Mater. Chem. A **3**, 11700–11715 (2015). https://doi.org/10.1039/c5ta00252d
11. Zhang, Z., Sun, J., Lai, C., Wang, Q., Hu, C.: High-yield ball-milling synthesis of extremely concentrated and highly conductive graphene nanoplatelet inks for rapid surface coating of diverse substrates. Carbon **120**, 411–418 (2017). https://doi.org/10.1016/j.carbon.2017.05.065
12. Yakubu, A.: Graphene synthesis by chemical vapour deposition (CVD): a review on growth mechanism and techniques. Int. J. Sci. Eng. Res. **8**(5), 15–26 (2019)
13. Jiat, X., et al.: Review on graphene and its derivatives: synthesis methods and potential industrial implementation. J. Taiwan Inst. Chem. Eng. (2018). https://doi.org/10.1016/j.jtice.2018.10.028

14. Dimiev, A.M., Khannanov, A., Vakhitov, I., Kiiamov, A., Shukhina, K., Tour, J.M.: Revisiting the mechanism of oxidative unzipping of multiwall carbon nanotubes to graphene nanoribbons. ACS Nano (2018). https://doi.org/10.1021/acsnano.8b01617
15. Guo, S., Chen, J., Zhang, Y.: Graphene-based films: fabrication, interfacial modification, and applications. J. Mater. Chem. A **11** (2021)
16. Ma, X., et al.: S-doped mesoporous graphene microspheres: a high performance reservoir material for Li–S batteries. Electrochim. Acta **269**, 83–92 (2018). https://doi.org/10.1016/j.electacta.2018.02.163
17. Sonker, R.K., Shastri, R., Johari, R.: Superficial synthesis of CdS quantum dots for an efficient perovskite-sensitized solar cell. Energy Fuels **35**(9), 8430–8435 (2021). https://doi.org/10.1021/acs.energyfuels.1c00608
18. Kirchner, H.P.: Potential applications. In: Strengthening of Ceramic Materials, pp. 229–236. CRC Press (2020). https://doi.org/10.1201/9781498710466-9
19. Lanças, F.M.: Food and environmental water samples. In: Chromatographic Analysis of Environmental and Food Samples (2023)
20. Ye, T., et al.: The mechanical, optical, and thermal properties of graphene influencing its pre-clinical use in treating neurological diseases. Front. Neurosci. **17**, Article 1162493 (2023). https://doi.org/10.3389/fnins.2023.1162493
21. Sun, H., et al.: Graphene and 2D hexagonal boron nitride heterostructure for thermal management in actively tunable manner. Nanomaterials **12**(22), 1–10 (2022). https://doi.org/10.3390/nano12224057
22. Nurmakanov, Y., Kalimuldina, G., Nauryzbayev, G., Adair, D., Bakenov, Z.: Structural and chemical modifications towards high-performance of triboelectric nanogenerators. Nanoscale Res. Lett. **16**(1), 1–10 (2021). https://doi.org/10.1186/s11671-021-03578-z
23. Sadia, S.I., et al.: Effect of graphene for ablation study of advanced composite materials for aerospace applications. Key Eng. Mater. **778**, 118–125 (2018). https://doi.org/10.4028/www.scientific.net/KEM.778.118
24. Ananda Murthy, H.C., Kelele, K.G., Ravikumar, C.R., Nagaswarupa, H.P., Tadesse, A., Desalegn, T.: Graphene-supported nanomaterials as electrochemical sensors: a mini review. Results Chem. **3**, Article 100131 (2021). https://doi.org/10.1016/j.rechem.2021.100131
25. Droth, M., Burkard, G.: Spintronics with graphene quantum dots. Phys. Status Solidi Rapid Res. Lett. **10**(1), 75–90 (2016). https://doi.org/10.1002/pssr.201510182
26. Borah, R., Hughson, F.R., Johnston, J., Nann, T.: On battery materials and methods. Mater. Today Adv. **6**, 100046 (2020). https://doi.org/10.1016/j.mtadv.2019.100046
27. Ullah, K., Shah, N., Wadood, R., Khan, B.M., Oh, W.C.: Recent trends in graphene based transition metal oxides as anode materials for rechargeable lithium-ion batteries. Nano Trends **1**, Article 100004 (2023). https://doi.org/10.1016/j.nwnano.2023.100004
28. Hossain, S., et al.: Nanostructured graphene materials utilization in fuel cells and batteries: a review. J. Energy Storage **29**, 101386 (2020). https://doi.org/10.1016/j.est.2020.101386
29. Ji, L., Meduri, P., Agubra, V., Xiao, X., Alcoutlabi, M.: Graphene-based nanocomposites for energy storage. Adv. Energy Mater. **6**(16), 1602159 (2016). https://doi.org/10.1002/aenm.201502159
30. Shao, Y., et al.: Design and mechanisms of asymmetric supercapacitors. Chem. Rev. **118**(18), 9233–9280 (2018). https://doi.org/10.1021/acs.chemrev.8b00252
31. Gurav, S.R., Waikar, M.R., Rasal, A.S., Sonker, R.K., Sonkawade, R.G.: Current development and challenges in textile-based flexible supercapacitors. In: Smart and Flexible Energy Devices, pp. 539–550. CRC Press (2022)
32. Chodankar, N.R., et al.: True meaning of pseudocapacitors and their performance metrics: asymmetric versus hybrid supercapacitors. Small **16**(37), 2002806 (2020). https://doi.org/10.1002/smll.202002806
33. Sui, D., et al.: Graphene-based cathode materials for lithium-ion capacitors: a review. Energy Storage Mater. **35**, 1–28 (2021)
34. Kumar, U., Sikarwar, S., Sonker, R.K., Yadav, B.C.: Carbon nanotube: synthesis and application in solar cell. J. Inorg. Organomet. Polym. Mater. **26**, 1231–1242 (2016). https://doi.org/10.1007/s10904-016-0402-8

35. Kumar, R., et al.: Heteroatom doped graphene engineering for energy storage and conversion. Mater. Today **39**, 47–65 (2020). https://doi.org/10.1016/j.mattod.2020.04.010
36. Latif, U., et al.: Binder free heteroatom-doped graphene oxide as high energy density electrodes for supercapacitor applications. Int. J. Energy Res. **46**(7), 9643–9666 (2022). https://doi.org/10.1002/er.7835
37. Liang, K., Zhao, Z.W., Zhou, X., Xu, A.W.: A novel route to prepare N-graphene/SnO_2 composite as a high-performance anode for lithium batteries. Dalton Trans. **47**(30), 10206–10212 (2018). https://doi.org/10.1039/c8dt01944d
38. Li, M., et al.: In-situ gas reduction in reversible SnS–SnO_2@N-doped graphene anodes for high-rate and lasting lithium storage. J. Power Sources **387**, 20–29 (2018)
39. Kang, J.H., Chen, J.S.: Using ethylenediamine to prepare three dimensional nitrogen-doped graphene aerogel/sulfur composite for lithium-sulfur batteries. Diam. Relat. Mater. **88**, 222–229 (2018). https://doi.org/10.1016/j.diamond.2018.07.015
40. Uesugi, N., et al.: Sulfur-inserted polymer-anchored edge exfoliated graphite for durable positive electrodes for lithium-sulfur batteries. RSC Adv. **11**(29), 18093–18102 (2021). https://doi.org/10.1039/d1ra01225h
41. Mohapatra, D., Sayed, M.S., Shim, J.J.: Co-doped in-situ engineered carbon nano-onions enabled high-performance supercapacitors. Nanomaterials **13**(1), 1–15 (2023). https://doi.org/10.3390/nano13010019
42. Li, R., Qin, C., Zhang, X., Lin, Z., Lv, S., Jiang, X.: Boron/nitrogen Co-doped carbon synthesized from waterborne polyurethane and graphene oxide composite for supercapacitors. RSC Adv. **9**(3), 1679–1689 (2019). https://doi.org/10.1039/C8RA09043B
43. Govindarasu, K.G., Ganesan, B., Venkatesan, R., Arumugam, P.: Simple CVD growth of P-doped graphitic hollow carbon spheres for high-voltage (2.0 V) aqueous symmetric supercapacitor. J. Mater. Sci. Mater. Electron. **32**(7), 8475–8490 (2021). https://doi.org/10.1007/s10854-021-05466-9
44. Wang, Z., Han, J., Zhang, N., Sun, D., Han, T.: Synthesis of polyaniline/graphene composite and its application in zinc-rechargeable batteries. Ionics **25**, 3373–3382 (2019)
45. Mei, J., Liao, T., Sun, Z.: Two-dimensional metal oxide nanosheets for rechargeable batteries. J. Energy Chem. **26**(5), 970–986 (2017). https://doi.org/10.1016/j.jechem.2017.10.012
46. Zhou, C., et al.: Laser-induced MnO/Mn_3O_4/N-doped-graphene hybrid as binder-free electrode. Chem. Eng. J. **375**, 123720 (2019). https://doi.org/10.1016/j.cej.2019.123720
47. Gao, C., Jiang, Z., Wang, P., Rosgaard, L., Zhang, Y., Yue, Y.: Optimized assembling of MOF/SnO_2/graphene leads to superior anode for lithium-ion batteries. Nano Energy **74**, 104868 (2020). https://doi.org/10.1016/j.nanoen.2020.104868
48. Wang, L., Wei, Z., Mao, M.: Metal oxide/graphene composite anode materials for sodium-ion batteries. Energy Storage Mater. **14**, 22–48 (2018). https://doi.org/10.1016/j.ensm.2018.06.027
49. Ren, H., et al.: High-capacity interstitial Mn-incorporated $Mn_xFe_{3-x}O_4$/graphene nanocomposite for sodium-ion battery anodes. ACS Appl. Mater. Interfaces **11**(45), 42387–42395 (2019). https://doi.org/10.1021/acsami.9b14003
50. Barmi, A.A.M., Aghazadeh, M., Moosavian, M.A., Golikand, A.N.: Binder-free high-performance Fe_3O_4 fine particles in situ grown onto N-doped porous graphene layers Co-embedded into porous substrate as supercapacitor electrode. J. Mater. Sci. Mater. Electron. **31**(18), 15198–15217 (2020). https://doi.org/10.1007/s10854-020-04085-0
51. Pattananuwat, P., Khampuanbut, A., Haromae, H.: Novel electrode composites of mixed bismuth-iron oxide/graphene utilizing for photo assisted supercapacitors. Electrochim. Acta **370**, 137741 (2021). https://doi.org/10.1016/j.electacta.2021.137741
52. El-Hallag, I.S., El-Nahass, M.N., Youssry, S.M., Kumar, R., Abdel-Galeil, M.M., Matsuda, A.: Facile in-situ simultaneous electrochemical reduction and deposition of reduced graphene oxide embedded palladium nanoparticles as high performance electrode materials for supercapacitor with excellent rate capability. Electrochim. Acta **314**, 124–134 (2019). https://doi.org/10.1016/j.electacta.2019.05.065
53. Dohyeong, S., Jeong, Y., Han, K.H., Yoon, D.Y., Seok, H.S.: Recent progress of electrochemical energy devices: metal oxide-carbon nanocomposites as materials for next-generation chemical

storage for renewable energy. Sustainability **11**(13), 3694 (2019). https://doi.org/10.3390/su11133694
54. Li, Y., et al.: Remarkably enhanced performances of novel polythiophene-grafting-graphene oxide composite via long alkoxy linkage for supercapacitor application. Carbon **147**, 519–531 (2019). https://doi.org/10.1016/j.carbon.2019.03.030
55. Jia, X., Ge, Y., Shao, L., Wang, C., Wallace, G.G.: Tunable conducting polymers: toward sustainable and versatile batteries. ACS Sustain. Chem. Eng. **7**(24), 19630–19648 (2019). https://doi.org/10.1021/acssuschemeng.9b02315
56. Xu, Z., et al.: Three-dimensional ZnS/reduced graphene oxide/polypyrrole composite for high-performance supercapacitors and lithium-ion battery electrode material. J. Solid State Electrochem. **23**(12), 3419–3428 (2019). https://doi.org/10.1007/s10008-019-04434-y
57. Aslam, S., et al.: Graphene decorated polymeric flexible materials for lightweight high areal energy lithium-ion batteries. Appl. Mater. Today **17**, 123–129 (2019). https://doi.org/10.1016/j.apmt.2019.08.003
58. Zhang, M., et al.: Polyaniline/graphene hybrid fibers as electrodes for flexible supercapacitors. Synth. Met. **268**, 116484 (2020). https://doi.org/10.1016/j.synthmet.2020.116484
59. Li, P., et al.: Stretchable all-gel-state fiber-shaped supercapacitors enabled by macromolecularly interconnected 3D graphene/nanostructured conductive polymer hydrogels. Adv. Mater. **30**(18), 1800124 (2018). https://doi.org/10.1002/adma.201800124
60. Moussa, M., et al.: Self-assembly and cross-linking of conducting polymers into 3D hydrogel electrodes for supercapacitor applications. ACS Appl. Energy Mater. **3**(1), 923–932 (2020). https://doi.org/10.1021/acsaem.9b02007
61. Yazar, S.: S, N and Cl separately doped graphene oxide/polyaniline composites for hybrid supercapacitor electrode. J. Electrochem. Soc. **169**(11), 110502 (2022). https://doi.org/10.1149/1945-7111/acadb1
62. Cymann-Sachajdak, A., Graczyk-Zajac, M., Trykowski, G., Wilamowska-Zawłocka, M.: Understanding the capacitance of thin composite films based on conducting polymer and carbon nanostructures in aqueous electrolytes. Electrochim. Acta **383**, 138356 (2021). https://doi.org/10.1016/j.electacta.2021.138356
63. Zhao, Z., Wang, H., Huang, H., Li, L., Yu, X.: Graphene oxide/polypyrrole/polyaniline composite hydrogel synthesized by vapor-liquid interfacial method for supercapacitors. Colloids Surf. A Physicochem. Eng. Asp. **626**, 127125 (2021). https://doi.org/10.1016/j.colsurfa.2021.127125
64. Moyseowicz, A., Gryglewicz, G.: High-performance hybrid capacitor based on a porous polypyrrole/reduced graphene oxide composite and a redox-active electrolyte. Electrochim. Acta **354**, 136661 (2020). https://doi.org/10.1016/j.electacta.2020.136661
65. Chang, X., et al.: 3D graphene network with covalently grafted aniline tetramer for ultralong-life supercapacitors. Adv. Funct. Mater. **31**(32), 2102397 (2021). https://doi.org/10.1002/adfm.202102397

Two-Dimensional Nanomaterials for Energy Storage

Deepika Gupta, Shweta Sharma, Aniket Bharti, and Rajesh Kumar

Abstract Two-dimensional (2D) nanomaterials are gaining attention for energy storage applications on account of their exceptional characteristics and versatility. This chapter offers an overview of latest research on 2D nanomaterials in energy storage device applications, focusing on the various synthesis methods and electrochemical behavior. It also examines the recent progress in the development of flexible, free-standing supercapacitors utilizing 2D materials. 2D materials gained significant interest on account of their large surface area, making them suitable candidate for energy storage applications. The extensive need for energy generation applications substantially enhances the advanced research in the synthesis of 2D nanomaterials specifically from the discovery of graphene. Key factors in designing high-performance energy storage devices include electronic conductivity, available intercalation sites, and cycling stability. The chapter emphasizes the properties of 2D materials and provides description of electrodes based on 2D heterostructured materials that demonstrate unique efficiency, resulting in improved electrochemical performance.

1 Introduction

Supercapacitors and secondary batteries contain an important role in modern energy storage systems. They exhibit significant role in addressing global energy, climate, and environmental challenges by supplying and storing the clean energy, such as solar and wind power, to operate electronic and electrical devices without generating greenhouse gases [1, 2]. Beyond their growing importance in electric vehicles [3, 4], various energy storage technologies are widely utilized in wearable electronic devices like smartphones. Despite these developments, many modern high-demand

D. Gupta · A. Bharti · R. Kumar (✉)
University School of Basic and Applied Sciences, Guru Gobind Singh Indraprastha University, New Delhi, India
e-mail: rajeshkumar@ipu.ac.in; kumarrrpi@gmail.com

S. Sharma
Department of Botany, Sri Venkateswara College, University of Delhi, New Delhi, Delhi, India

R. K. Sonker et al. (eds.), *Low-Dimensional Multifunctional Hybrid Nanostructured Materials*, Engineering Materials, https://doi.org/10.1007/978-981-96-7952-2_9

power devices and vehicles continue to rely heavily on non-renewable fossil fuels, on account of efficiency limitations of current energy storage applications [5, 6]. For instance, secondary batteries typically contain low power density (less than 1 kW kg^{-1}), while supercapacitors contain limited energy density (ranging from ~1 to 10 Wh kg^{-1}), making them inefficient for long-term applications [7, 8]. In smaller electronic devices, there is also a need to improve energy storage systems by altering their power performance while incorporating features like light weight, thin form, and flexibility [9–11].

The major difference between supercapacitors and batteries is supercapacitors are suitable for circumstances where fast bursts of energy are required, like in the field of acceleration of vehicle or give backup of power during sudden drops of power supply. Their efficiency to charge and discharge very quickly also exhibits long cycle life, and makes them significant for short time energy demands. Batteries are more suitable for applications that need continuous energy over longer time. They exhibit higher energy density, and they are generally utilized in devices that required to work for extended periods of time without recharging, like portable electronics, electric vehicles, and energy storage systems for the applications in grid. A significant challenge in this domain is development of electrode materials that meet these criteria and efficiently deliver and store energy. Supercapacitors (SCs), like other electrochemical energy storage devices, contain the electrode materials and electrolytes. The performance of SCs, specifically their energy density (E), is largely dependent on electrode materials, as illustrated by the equation $E = 1/2CV^2$, where C represents specific capacitance and V represents the voltage. SCs contain two main energy storage mechanisms: one involves charge storage through double-layer formation at electrolyte/electrode interface, known as electric double-layer capacitance (EDLC), and the other depends on fast redox reactions near the electrode surface, referred to as pseudo-capacitance. Pseudo-capacitors can store more charge than EDLCs, though EDLCs offer superior electrochemical stability, faster charge/discharge rates, and better charge-transfer kinetics. 2D nanomaterials contain different properties in contrast to other nanomaterial structures, such as one-dimensional (1D) nanowires, and three-dimensional (3D) bulk counterparts, and zero-dimensional (0D) quantum dots (QDs).

In 2D nanomaterials, charge carriers are confined along the thickness, and they can move freely in the plane. These characteristics make them highly suitable for fundamental electronic device applications. The electronic properties of 2D nanomaterials can be tuned by adjusting their thickness, a capability not significantly acquired with 0D, 1D, or 3D nanomaterials. Additionally, their ultra-thin nature contains mechanical flexibility and optical transparency, making them suitable candidate for highly transparent and flexible optoelectronic devices. On account of their large surface area, 2D nanomaterials contain high sensitivity to external factors, including chemical modifications, chemical doping, molecular adsorption, and mechanical deformation [12]. Recent research material science shows that 2D transition metal dichalcogenides (TMDs) like $MoSe_2$, MoS_2, WS_2, TiS_2, NbS_2, and VS_2 may hold the role to remove the gap between current energy storage efficiency and the evolving demands of device applications [13–17]. These materials offer different characteristics, such as reactive sites over the traditional electrode materials. Transition metal dichalcogenides (TMDs) is considered to be important domain of materials has general chemical formula T, TM refers to transition metal and X represent chalcogen. TMDs exhibit structure in layered form, where the layers are weakly bonded by van der Waals forces, although some TMDs are non-layered [13]. Bulk TMDs typically contain three types of polymorphs: 1T, 2H, and 3R, numeric value contains number of TM-X-TM units which are present in the structure of unit cell.

When TMDs reduced to monolayer or few-layer forms, they exhibit significantly altered properties on account of confinement of charge carriers in only two dimensions, which cause the changes in their electronic structure. The efficiency of transportation of charge between electrode and the monolayer of TMD is enhanced by eliminating the need for electron hopping between layers, which occurs in bulk materials [18–23]. Additionally, the increased edge-to-surface ratio in 2D TMDs allows for greater tunability, such as through chemical doping, and the formation of heterostructures provides a range of physical and chemical properties with varying band gap energies and types of charge carrier [24, 25]. These unique characteristics, along with the materials, give 2D TMDs significant chemical, electrical, and mechanical properties, making them ideal for next-generation electrochemical technologies, such as flexible and energy storage devices. Recent advancements in 2D TMDs hold great potential which differentiate them from bulk materials and other 2D substances. The chart in Fig. 1a displays the number of papers published between the year 2013 and 2023 on the electrochemistry of materials for pollutant remediation in environment. Figure 1b displays the distribution of types of publication, like book chapters, articles, and conference papers. Figure 1c shows the data associated to electrochemistry of materials for energy conversion and storage. Figure 1d shows the publication types across different sources, such as review articles conference proceedings and research papers.

This chapter deals with development which takes place in 2D TMD materials mainly in MoS_2 for energy storage device.

2 Synthesis Techniques of 2D Nanostructured Materials and Electrochemical Performance

Synthesis route of 2D nanostructured materials is pivotal as it directly affects the efficiency, capacity, stability, and charge–discharge rates of devices like batteries and supercapacitors. The synthesis method impacts the various aspects crucial for energy storage applications, including surface area, conductivity, ion diffusion pathways, and mechanical stability. Synthesis of 2D nanostructured materials for energy storage devices involves a variety of techniques, each suited to achieving specific structural, compositional, or functional properties. The chemical or physical techniques are generally used for fabricating 2D nanostructured materials.

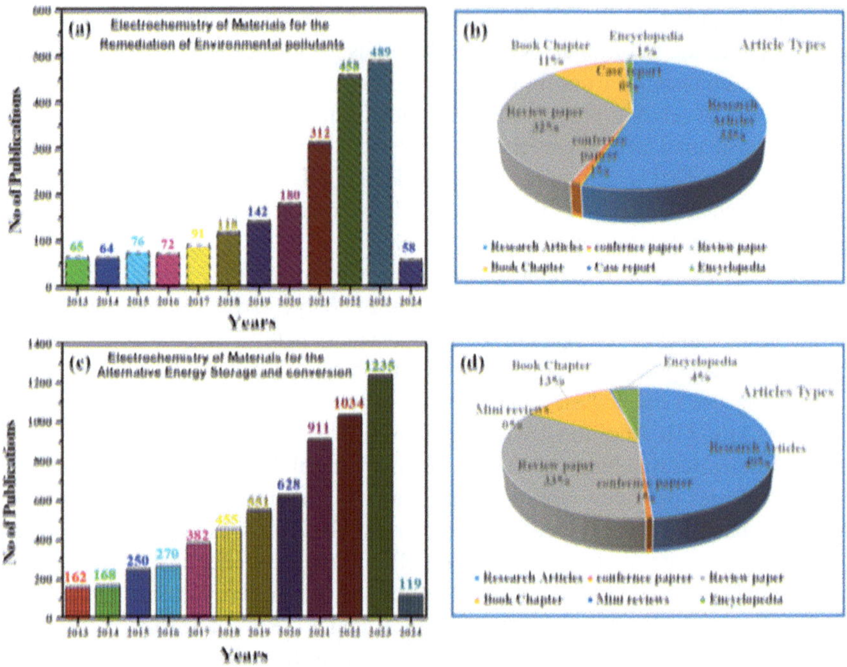

Fig. 1 Graphical representation of articles published from 2013 to 2023, every year on topic of 2D-materials' electrochemistry for the remediation of environmental pollutant applications, **a** articles published per year, **b** article types. The data published on the topic of electrochemistry of 2D-materials for the alternative energy storage/conversion materials applications, **c** article published per year, from 2013 to 2024, and **d** article types [25]. Reproduced with permission Copyright @ (2024) Elsevier

2.1 *Physical Methods*

The synthesis of 2D nanostructured materials can be synthesized through various physical methods, each offering distinct advantages and capabilities. Here are some commonly used physical methods.

2.1.1 Thermal Evaporation Technique

Thermal evaporation is considered to be a physical method for the thin films' fabrication explored for synthesis of 2D nanostructured materials. The thermal evaporation equipment schematic's representation is given in Fig. 2. In this technique, the material of the source which is in solid form is evaporated in a high vacuum conditions. The target bulk material, or material to be deposited, is put into alumina crucible within the high vacuum chamber of the deposition system. The material that gets evaporated travels directly to the substrate material via gaseous phase, on reaching the substrate,

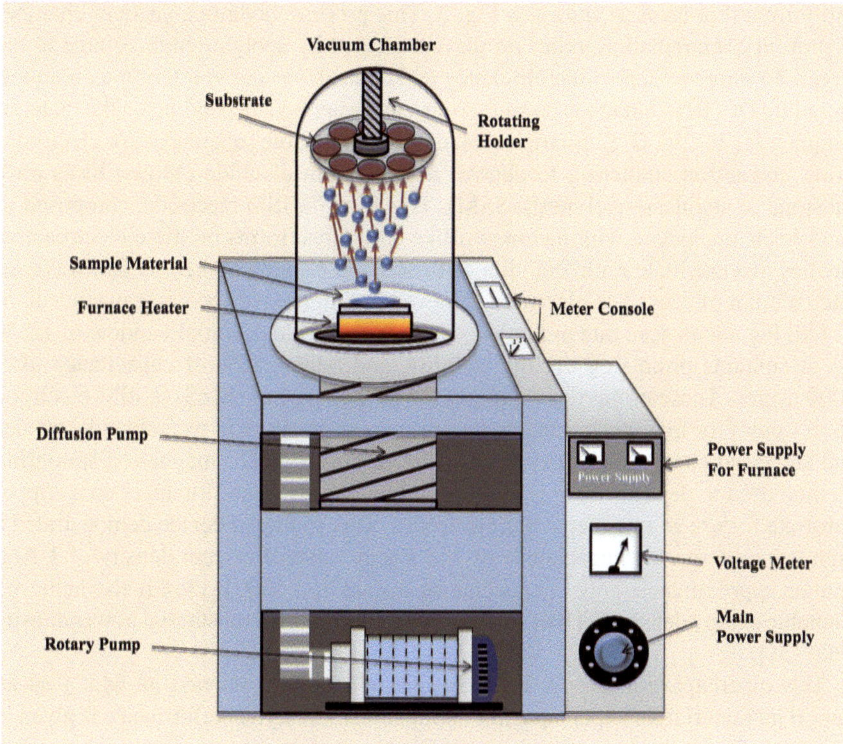

Fig. 2 A block diagram of thermal evaporation system

the vapor particles condense into the solid state and fabricate a thin film as depicted in Fig. 2 [26]. Shahbazi et al. synthesized molybdenum disulfide (MoS_2) thin films using thermal evaporation deposition on quartz substrates, aiming to improve energy storage performance. By reducing the film thickness to approximately 20 nm, they observed a significant shift in the band gap from 1.2 to 1.78 eV, suggesting a transition from an indirect to a direct band gap semiconductor. This modification enhances the material's optical and electronic properties, which is advantageous for energy storage applications. Electrochemical measurements indicated a value of specific capacitance of 24.42 $mFcm^{-2}$ (~12.21 F g^{-1}) at a 10 mV s^{-1} scan rate for MoS_2 synthesized via thermal evaporation deposition display of its potential in energy storage device applications [27].

2.1.2 Sputtering Technique

Sputtering technique is mainly utilized for the deposition of thin film, in which bombardment of a target material occurred by high-energy ions of inert gas such as argon, which eject the material's atoms. The atoms then move to a nearby substrate and form a thin layer as shown in Fig. 3. This process, done in a vacuum chamber to prevent contamination, relies on plasma created by applying high voltage to the target. To improve deposition efficiency and control, various sputtering techniques, including DC, RF, or magnetron sputtering, can be applied, depending on the material parameters [28, 29]. G. K. Sharma et al. developed flexible supercapacitor electrodes using magnetron sputtering to deposit molybdenum disulfide (MoS_2) films on a substrate of stainless-steel mesh (SSM). This MoS_2@SSM electrode, composed of nanosheets embedded with nanoworm-like structures, forms multiple electroactive cavities that facilitate rapid Na^+ ion diffusion and provide numerous pathways for intercalation of ions. The electrode achieved an impressive specific capacitance of 214.90 F/g a with scan rate of 5 mV/s, along with a wide potential window of 1.2 V. So, it contains promising cycling stability, and retains 88% of capacitance after 3000 cycles. These properties highlight the potential of the MoS_2@SSM electrode (1×1 cm^2) for high-performance, flexible supercapacitors in wearable electronics and advanced energy storage systems [30]. Additionally, DC magnetron sputtering method used to deposit MoS_2 thin films with nanoworm-like structures on a copper substrate for use as supercapacitor electrodes. This sputtered device demonstrated a high value of specific capacitance of 138 F/g at value of current density of 1 A/g, contain approximately 86% of capacitance sustain over 5000 cycles. It also achieved a notable energy density (ED) of 12.26 Wh/kg, along with a substantial power density (PD) [31].

The supercapacitor uses a 1 M Na_2SO_4 electrolyte, showed an MoS_2 as an important material for supercapacitor applications. Catalytic performance is given in Fig. 4a–d. Figure 4a shows the electrode CV curve of sweep integral area in contrast to current collector value at 200 mV/s, illustrating that MoS_2 nanoworms is the significant reason for the electrochemical efficiency. Figure 4b shows the nanoworm's CV curves of MoS_2 electrode for 0.2–0.6 V potentials value with the Ag/AgCl reference electrode at distinct scan rates. The values of specific capacitance (C_s) at various scan rates (i.e., 5, 10, 20, 50, 100, and 200 mV/s) were studied to be 589, 511, 407, 298, 222, and 182 F/g, respectively, given in Fig. 4c. Figure 4d is used to study the Nyquist plots took for the nanoworm's MoS_2 electrode in 1 M Na_2SO_4. The results suggested that significant ion conductivity for the device applications. Figure 5a, b displays the FE-SEM images of electrode surface and the cross-sectional image corresponding to the nanoworms of MoS_2 with 2 mm thickness. The synthesized electrode MoS_2 material contained densely grown nanosheets which exhibit the great application in mechanism of charge storage. Figure 5c is 2D AFM of synthesized active material. The value of root mean square (RMS) roughness is studied to be 19.62 nm. These promising electrochemical characteristics of the MoS_2 nanoworm electrodes showed they could be highly suitable for high-performance supercapacitors in energy storage applications, marking them as effective candidates for advancing energy storage technologies [31]. Magnetron sputtering technique for the deposition MoS_2 thin films on a substrate of copper and used as binder-free supercapacitor electrodes. MoS_2 nanoworm's FE-SEM images, with a 2 mm active material thickness, reveal well-distributed and densely grown nanosheets.

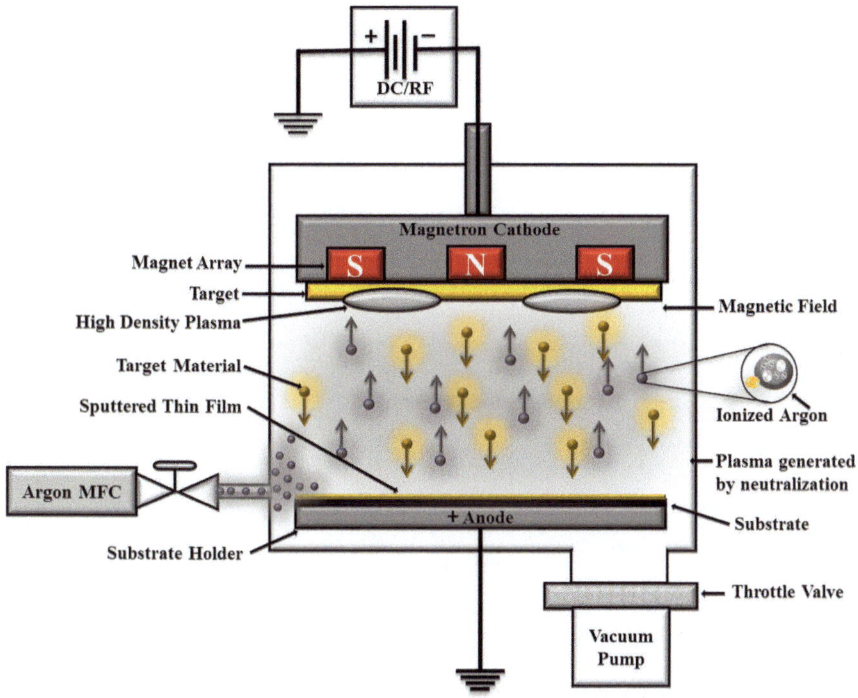

Fig. 3 A block diagram of sputtering system

2.1.3 Pulsed Laser Deposition (PLD)

Pulsed laser deposition (PLD) technique is utilized to synthesize a good quality films with control on their composition, thickness, and morphology [32]. In PLD, a high-energy pulsed laser targets a material in vacuum chamber, causes the vaporization, and forms a plasma plume. The generated plume exhibits atoms, ions, and molecules from the desired target, then on a substrate it gets deposited, and thin film fabrication occurs [33]. By regulating the parameters like gas pressure, laser energy, and temperature of the substrate, modification in thin film properties such as crystallinity and morphology is possible [34, 35]. The schematic illustration of the PLD setup is given in Fig. 6. Siegel et al. used pulsed laser deposition (PLD) to form MoS_2 thin films on sapphire substrates, acquiring precise control over the count of layer on a 2-cm scale. They studied a direct linear relationship between the resulting MoS_2 layer thickness and number of laser pulses. Also, by changing the number of pulses, they selectively grow monolayer, bilayer, and trilayer films. They observed that monolayer 2D-MoS_2 was fabricated at 700 °C in 5 min. This method provides a scalable approach to grow controlled MoS_2 layers, significant for numerous potential applications [36]. Serna et al. observed the influence of reducing the growth temperature from 700 to 300 °C

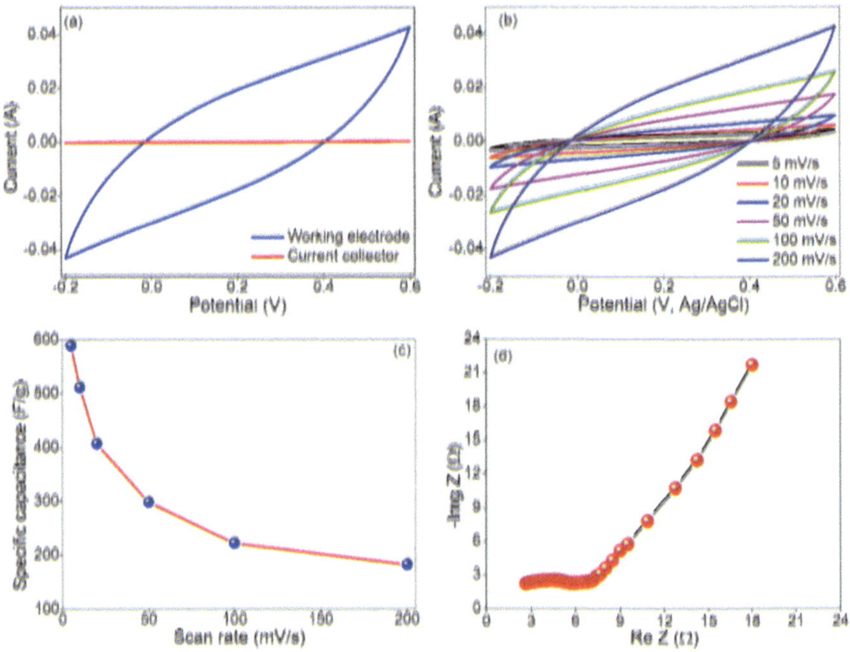

Fig. 4 **a** MoS_2 device CV curves at various rates of scan range from 5 to 200 mV/s in 1 M Na_2SO_4, **b** value of the MoS_2 symmetric device-specific capacitances at various rates of scan from 5 to 200 mV/s in 1 M Na_2SO_4 electrolyte, **c** GCD curves of device at different current densities in the electrolyte 1 M Na_2SO_4, and **d** MoS_2-specific capacitance-based electrode in various current densities in 1 M Na_2SO_4 [31]. Reproduced with permission Copyright @ (2018) Elsevier

on MoS_2 films' stoichiometry. They determined that at the low temperature, the stoichiometric ratio between molybdenum (Mo) and sulfur (S) was altered, with sulfur deficiency becoming significant [37].

2.1.4 Lithography Processes

Lithography is important technique for the growth of 2D nanostructures on various types of substrates, significantly used for patterning of surface in the domains like optics and electronics. It exhibits the deposition of a design of photomask on a desired substrate, allowing various similar copies to be generated rapidly from a single exposure. This technique is compatible with numerous materials, making it appropriate for different applications. Lithography's adaptability makes it a significant method for generating the precise and reproducible patterns [38]. M.-A. Kang et al. used a technique based on soft-lithographic patterning for the fabrication of stacked MoS_2/graphene on polyethylene terephthalate (PET) substrates. This method involves a transferring and patterning of chemical vapor deposition (CVD)-grown MoS_2 and graphene layers by utilizing the stamps of polymer coated with the polar

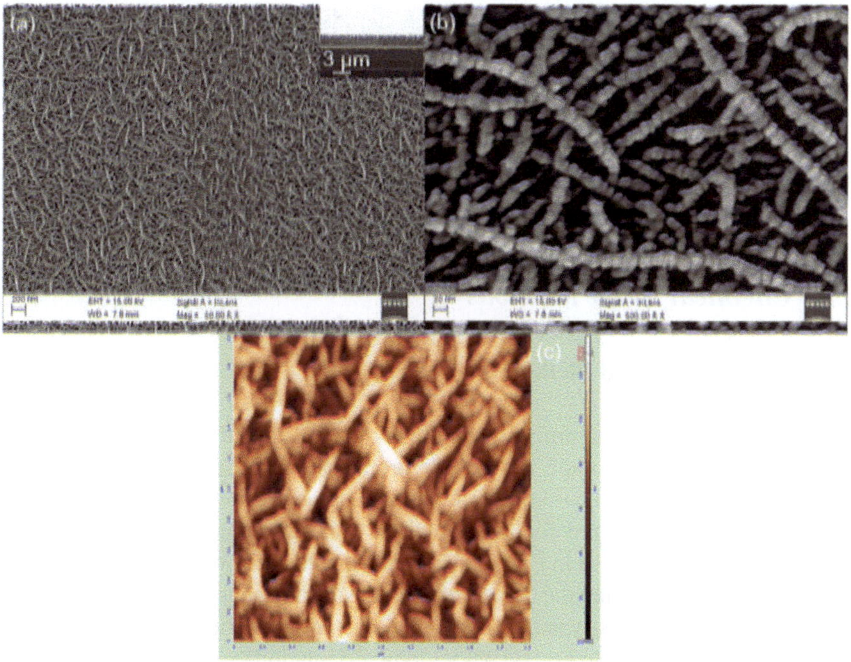

Fig. 5 **a, b** Working electrode based on MoS_2 nanoworms observed in FE-SEM images at distinct scales, and **c** MoS_2 nanoworms' AFM image in 2 × 2 mm area of active material [31]. Reproduced with permission Copyright @ (2018) Elsevier

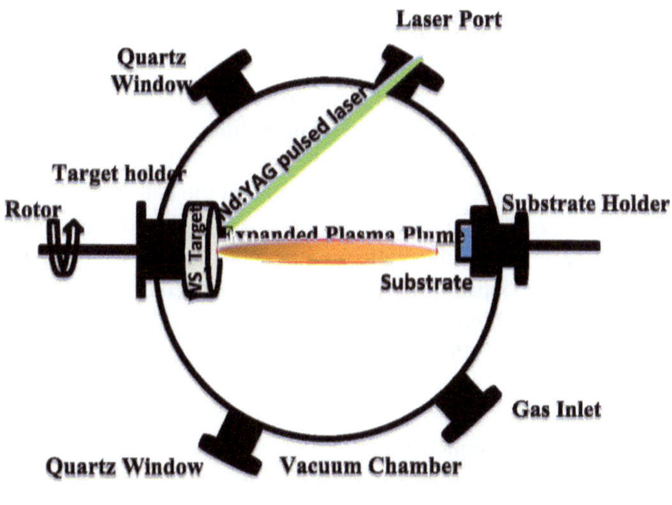

Fig. 6 Schematic diagram of pulsed laser deposition (PLD) setup [32]. Reproduced with permission Copyright @ (2023) Elsevier

molecules. The arrangement of these 2D materials in cross-stacked form on flexible substrates provides a robust platform to understand the physical properties of ultra-thin electronic devices. The 2D materials' cross-stacking arrangement forms an interconnecting network that increases the electrochemical efficiency by offering the rapid pathways of electron and caused the enhancement in active sites, making it significant choice for energy storage device applications [39].

2.2 Chemical Route Methods

2.2.1 Electrochemical Method

Electrodeposition is a significant technique for the growth of 2D materials at nanoscale for energy storage devices. It produces controlled nanoparticles with particular properties, which involve composition, structure, and size. In this method, electrical current is utilized for the growth of nanostructured material layer on a substrate material as shown in Fig. 7. In contrast to other methods, this technique has the importance of attaching the nanoparticle to the desired substrate. Additionally, by changing the bath chemistry and operating parameters, the nanostructured materials' thickness, mass, morphology, crystallographic orientation, and particle size can all be altered [40–44]. Bamidele D. Falola et al. formed MoS_2 doped with copper (Cu) thin films by using a one-step electrodeposition technique. This process contains an electrolyte solution with 10.0 mM ammonium tetrathiomolybdate ($(NH_4)_2MoS_4$), 5.0 mM copper sulfate ($CuSO_4$), 0.10 M potassium chloride (KCl), and 0.50 M potassium thiocyanate (KSCN) at a pH of 6.95. After growth of the films, they were annealed in atmosphere of argon at 500 °C temperature for 3 h. The films contain copper at concentrations of 1–4 atomic percent and contain a sulfur deficiency compared to the ideal composition of MoS_2. The addition of copper reduced the films' resistivity, achieving approximately ten times lower resistivity than undoped MoS_2. Additionally, capacitance measurements showed that the Cu-doped films contained specific capacitances 2.5–3.5 times high in contrast to undoped films over the cyclic stability of 1000 cycles of cyclic voltammetry (CV) testing in the 1.0 M sodium sulfate (Na_2SO_4). These doped films also showed enhanced retention of the capacity and superior efficiency at high rate of scans, underscoring their potential for energy storage device applications [45].

2.2.2 Chemical Vapor Deposition

Chemical vapor deposition (CVD) involves a chemical method in which the transformation of molecules of gas takes place in a solid material, in 2D nanostructured materials' form. It is widely used for synthesizing oxide-based 2D nanostructured materials for energy storage devices on account of its ability to fabricate thin films of high quality and precise control over the thickness of film, composition, and

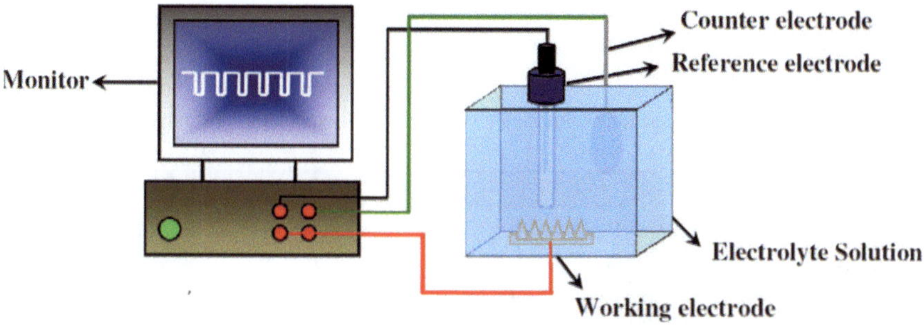

Fig. 7 A schematic of an electrodeposition system [44]. Reproduced with permission Copyright @ (2012) Elsevier

morphology. First, the substrate is properly cleaned to make sure that all contaminants that may impact the quality of the film should be removed. Thereafter volatile precursor gases are introduced into a heated chamber of reaction, decompose on the substrate to form a uniform, high-purity oxide layer (Fig. 8). This uniformity, along with high crystallinity, benefits supercapacitor electrodes by enhancing conductivity and stability. By adjusting certain parameters, such as temperature, precursor flow rates, and pressure, one can achieve the desired film thickness, deposition rate, and other essential properties [46–48]. CVD is utilized extensively in a variety of industries, including the manufacturing of semiconductors, thin film coatings, supercapacitors, and material science research [49]. A. Raza et al. successfully synthesized MoS_2 and heterostructured ZnS/MoS_2 electrode materials (E-Ms) on the foam of nickel (NiF) using a CVD method process for energy storage device fabrication. The $ZnS/MoS_2/NiF$ heterostructure consisted of interconnected nanofibers and nanorods, which greatly enhanced the electrochemical performance, containing specific capacitance (Csp) of 3540 F/g as compared to MoS_2/NiF which exhibited a Csp of 1666 F/g at 1 A/g. The interconnected nanostructures reduced charge-transfer resistance compared to MoS_2/NiF electrodes, which shows a resistance of 0.51 Ω cm^2. In addition to high capacitance, the $ZnS/MoS_2/NiF$ electrodes demonstrated impressive energy densities ranging from 72 to 122 Wh/kg and power densities between 250 and 2500 W/kg. These electrodes also contained exceptional cyclic stability, over 95% of their Csp after 20,000 cycles. Due to their unique nanostructures and outstanding electrochemical properties, ZnS/MoS_2 electrodes show strong potential as high-performance pseudo-capacitors for next-generation energy storage solutions [50].

2.2.3 Hydrothermal Method

Hydrothermal technique is an effective way to synthesize two-dimensional nanostructured materials, with different high edge states exposed to environments with higher chemical activities and morphologies, considered to be important in various

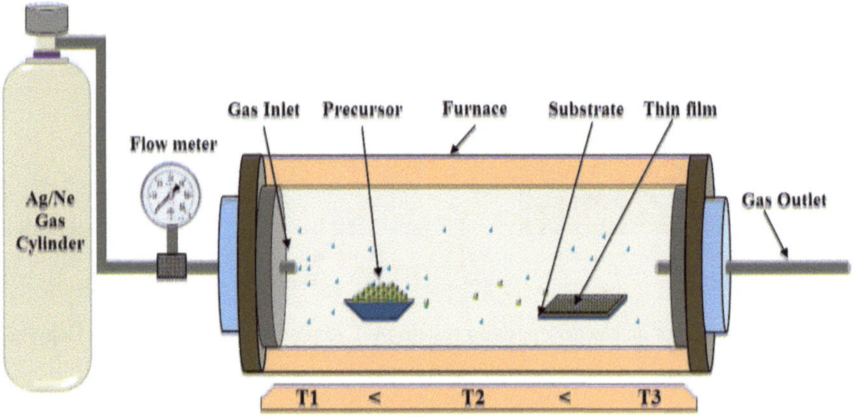

Fig. 8 Schematic representation of chemical vapor deposition (CVD) method

applications including energy storage, gas sensing, and fabrication of electronic device. Basic hydrothermal process is shown in Fig. 9. Another significance is that functional nanocomposites with other 2D materials like graphene, tellurium (Te), TiO_2, and carbon nanotubes (CNTs) and according to various demand which is not easily possible in the other synthesis techniques. Radhakrishnan et al. synthesized a MoS_2/multiwalled carbon nanotube (MWCNT) nanohybrid using a straightforward hydrothermal method, showcasing its effectiveness as a positive electrode in both symmetric and asymmetric supercapacitor systems. The nanohybrid electrode achieved capacitance of approximately 274.63 F/g, with energy density of 20.70 Wh/kg. The inclusion of MWCNT enhanced the cyclic stability, retaining nearly 99.8% of the capacitance even after 5,000 cycles. In comparison to other MoS_2-based asymmetric cells and its symmetric equivalent, the synthesized asymmetric supercapacitor exhibited a greater energy density (32.18 Wh/kg). The MWCNT nanohybrids' porous structure is responsible for this enhancement since it allowed for effective ion and charge transport from the pseudo-capacitive MoS_2 materials to the MWCNT electrodes, leading to high capacitance. These configurations make the MoS_2/MWCNT nanohybrid a significant candidate for advanced energy storage device applications [51]. Additionally, G. Muthulakshmi et al. grow a SnO_2/MoS_2 nanocomposite via the hydrothermal technique, varying the MoS_2 composition in the composite at 1, 2, and 5% weight. The electrochemical efficiency of these electrodes displays that among these composites, the 5% SnO_2/MoS_2 electrode contained high specific capacitance value of 146 F/g in 2 M KOH electrolyte at value of current density of 1 A/g. This result shows that the 5% SnO_2/MoS_2 composite contains a strong potential as material for future energy storage applications [52].

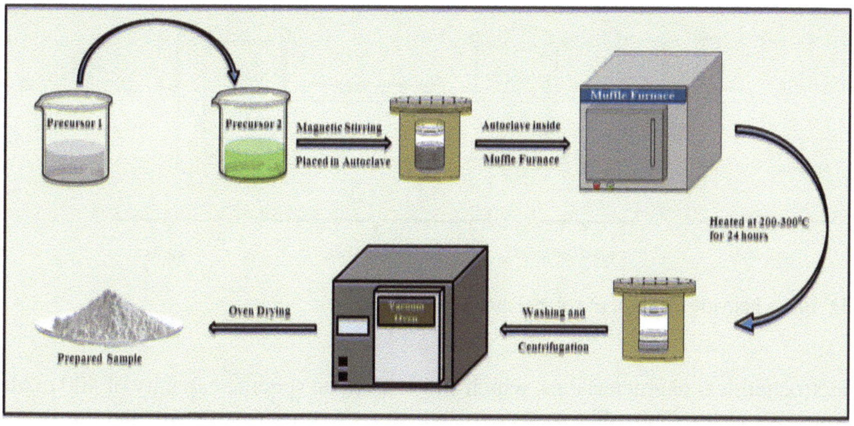

Fig. 9 Schematic diagram of hydrothermal method

2.2.4 Sol–Gel Technique

This method is mainly used to create various 2D nanostructured materials, specifically metal oxide nanoparticles [53]. The low-temperature condition of this process is beneficial, which might be specifically used for the production of ceramic and nanomaterials [54]. This method also used the simultaneous growth of two or more nanoparticles by using the various metal precursors in the proper ratios [55]. In the sol–gel technique, a homogenous sol is first obtained from numerous precursors and is eventually changed into a gel, as shown in Fig. 10. This transformation occurs in two primary steps: in the first stage, the precursors react to form solid colloidal particles at nanoscale, obtain a stable sol. In the second stage, these colloidal particles, with the solvent, begin to connect and an interconnected network obtained shows transition from a liquid sol to a semi-solid gel [56]. This technique is used in the formation of materials such as coatings, thin films, and powders because it gives the fine control over the material's structure and composition. M. V. Nardi et al. studied an innovative and scalable sol–gel technique for fabrication of MoS_2 films through a spin coating of the aqueous solution on various types of flexible substrates. This approach contains low-temperature annealing temperature of 300 °C and does not need sulfurization or hydrogen supplementation. This cost-effective method gives a pathway for production of good quality MoS_2 films appropriate for various applications, specifically in energy storage devices. The results showed the potential of this sol–gel process for MoS_2's use in industrial applications [57]. Xingzhong Guo et al. synthesized porous MoS_2/C nanocomposites by using a sol–gel technique facilitated by tetramethyloxysilane (TMOS). In this synthesis method, ammonium molybdatetetrahydrate ($(NH_4)_6Mo_7O_{24}{\cdot}4H_2O$) served as the precursor for molybdenum (Mo) source, glucose as carbon precursor, and TMOS as gelation accelerator and silica template. After high-temperature sulfuration, NaOH etching was used to remove the silica derived from TMOS. The MoS_2/C nanocomposite showed excellent

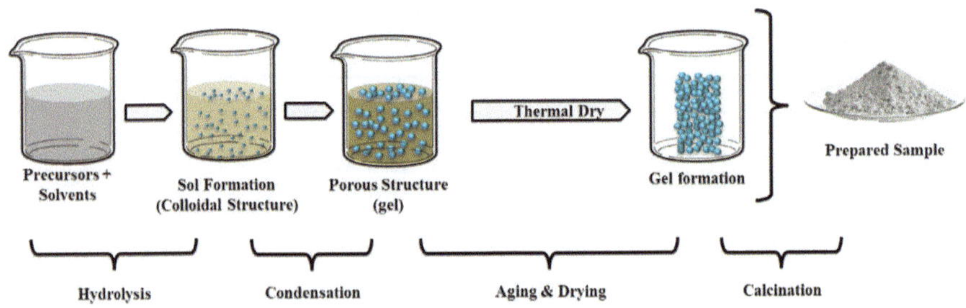

Fig. 10 Schematic diagram of sol–gel method

electrochemical characteristics, which shows a stable specific capacity of 400 mAh/g over 60 cycles. This efficiency is attributed to the porous structure and the synergistic effect between MoS_2 and amorphous carbon, which increases the electronic conductivity. These characteristics make the MoS_2/C nanocomposite a significant candidate for lithium-ion battery applications on account of its cycling stability [58].

3 Energy Storage Applications of MoS_2 and Its Various Composites

Supercapacitors and batteries are the significant devices for energy storage, where MoS_2 utilized as an anode material in lithium-ion and sodium-ion batteries. Sodium-ion batteries contain high weight but low cost in contrast to lithium-ion batteries, which make it efficient for stationary applications. Lithium-Ion Batteries (LIB) contain significant amount of capacity and are considered to be recyclable. LIB considered to be portable devices but their capacity is found to be low in various electrical vehicles. Metallic form of MoS_2 acts as anode in LIB on account of its specific capacity and conductivity which offers appreciable intercalation of ions and increase the stability of battery and performance efficiency. Research in this domain aims on two important directions, enhancing the stability of metallic phase of MoS_2-based batteries or increasing the conductive behavior of the 2H form of MoS_2 to act as an anode, as it provides the superior stability. Supercapacitors are important of storing valuable amounts of energy at low limits of voltage. In contrast to conventional batteries, they exhibit more storage energy per unit volume. Supercapacitors offer high value of power density but lower energy density value in contrast to batteries, making them appropriate as complementary devices in electric vehicles along with batteries. They contain high value of including high power densities, high specific capacitances (SCs), wide operating temperature range and stability over many cycles. While they contain the higher power densities than batteries and fuel cells, their main disadvantage is low energy density. The mechanism transfer of ions between the electrolyte and electrolyte and the supercapacitors' energy storage mechanism is generally found in various types. MoS_2 has also gained significant attention from researchers for its potential applicability in supercapacitor applications. In the domain of energy generation applications, Hydrogen Evolution Reactions (HER) used as the substitution of fuel as an energy source. MoS_2 is considered to be a cheap catalyst in [59] water splitting reactions and electrochemical HER [60]. The various electrostatic active defects establish MoS_2 as a catalyst. 1T phase of MoS_2 is considered to be a significant catalyst in HER reactions than 2H MoS_2 on account of basal planes, which acquires activity from the affinity of binding with hydrogen at the S sites [61]. Studies done to increase the activity of MoS_2-based catalyst so it becomes replacement of noble metals. The catalytic activity of MoS_2 relies on the active edges. A study by [62] proposed different types of defects in layers of MoS_2 layers, such as disulfur vacancies (VS_2), molybdenum vacancies (VMo), sulfur tri-vacancies (VS_3), vacancy complexes involving molybdenum and adjacent tri-sulfur atoms ($VMoS_3$), and combined molybdenum and sulfur vacancies ($VMoS_2$). The results indicated that vacancy of Mo and vacancy of MoS_2 significantly activate otherwise inert basal planes, contributing to water splitting in HER. The influence of strain on Mo vacancies in the single-layer MoS_2 studied in literature report [63]. Carbon dioxide (CO_2) is a primary reason to climate change, ozone layer depletion, global warming, and it contains severe health problems, including life-threatening situations [64]. Modern solutions are focused on reduction of CO_2 into carbon monoxide (CO) or converting it into fuel. This process, known as e-fuel (electrofuels), contains using hydrogen and CO_2 to form fuel through electrocatalytic or photocatalytic reactions. Significant research has been going on CO_2 photoreduction using the semiconductors [65], including conversion of solar-driven CO_2 where semiconductors present in solar cells act as catalysts to directly change CO_2 into fuels [66]. While production of synthetic fuel may show additional costs and require environmental considerations [67]. Metal chalcogenides showed catalytic characteristics in CO_2 reduction and HER [68, 69], and are specifically valuable due to their significant atomic structures, significant electrical transport, and scalability. Molybdenum disulfide (MoS_2) is considered to be a prominent metal chalcogenide, and is an important candidate for reduction of CO_2 because of its high light absorption capabilities. The MoS_2 catalytic performance in CO_2 reduction was studied in [70], where sharp edges of demonstrated the capacity to adsorb CO_2 and convert it into CO [71]. This highlights the importance of focusing more on MoS_2 composites for energy applications. Various combinations of MoS_2 with noble metals, carbon, or metal oxides need additional exploration for further commercial use. Also, increasing the crystalline structure, purity, and particle size of MoS_2 through further engineering is important. Chemical exfoliation technique is considered to be a cost-effective technique for the synthesis of MoS_2 nanosheets. The chemically exfoliated fabricated MoS_2 nanosheets illustrate the significant electrochemical efficiency when it is explored as electrodes in the applications of supercapacitors. There are various parameters that decide the efficiency of energy storage capacity of the electrode materials. Energy density is the important parameter for the energy storage efficiency study. The energy density is the energy stored per unit volume or per unit mass of the energy storage medium. Cyclic voltammetry (CV) is an important electrochemical method used to study the oxidation and reduction processes of species of various molecules. CV is explored to understand the transfer of electron behavior of various chemical reactions, which involves catalysis. Table 1 contains the details of distinct energy storage devices, their fabrication information, and key observations.

4 Conclusion

Due to the increasing requirement for higher power densities, with noteworthy factors such as extended lifetime, enhanced safety, and cost-effectiveness, both secondary batteries and supercapacitors require significant changes in their energy storage methods. Various approaches have been used to increase their efficiency, such as changing the traditional materials of electrode and using the alternative energy storage methods other than traditional lithium-ion systems. The recent evolution of two-dimensional (2D) thin transition metal dichalcogenides (TMDs) has become as a more significant research domain for increasing the performance of energy storage devices. This is primarily associated with the unique properties of TMDs, which give several applications in terms of energy storage efficiency and stability. Optimization of the phase structure and crystalline purity are significant approaches for enhancing the electrical transport characteristics of MoS_2. As research on nanostructured MoS_2 keeps growing, its limitations will be addressed. This progress will facilitate the energy storage device application domain. In conclusion, the advancements in the domain of materials science, particularly with the exploration of 2D materials like TMDs, give significant path for the advancement in energy storage device applications. These innovations could generate the sustainable, cost-effective, and efficient ways to meet the global demands for energy storage device applications.

Table 1 Details of distinct energy storage devices, their fabrication information, and key observations

Energy storage applications	Structure of materials	Key observations	References
MoS_2-based lithium-ion batteries (LIBs)	Metallic form of MoS_2, placed it over graphene with an interlayer spacing of 0.98 nm	High capacity value of 1700 mAh g^{-1}, value of current density 70 mA g^{-1}, Coulombic efficiency (70%). Battery also contained a capacity value of 666 mAh g^{-1} value at density of current value of 3500 mA g^{-1}	[72]
	MoS_2 on the carbon fiber	Specific capacity, i.e., 1789 mAh g^{-1} at 0.1 A g^{-1} and managing a capacity of 853 mAh g^{-1} at 140 cycles at 1 A g^{-1}	[73]
	Composite form of 1T-phase MoS_2 and conductive molybdate ($NiMoO_4$)	Coulombic efficiency of 99.5% and exhibit stable nature over 750 cycles	[70]
	Pure metallic form of MoS_2 structure	Specific-capacity value of 935 mAh g^{-1} for cyclic stability of 200 cycles at 5 A g^{-1}, which could be enhanced to 1150 mAh g^{-1}, efficient rate performance across a current density value range of 0.2–20 A g^{-1}	[74]
Sodium-ion batteries (NIBs)	Theoretical study observed that monolayer MoS_2 contains high sodium adsorption	Theoretical capacity value of 335 mAh g^{-1}. The monolayer form work at a low value of voltage (1.0 V) in contrast to the bulk counterpart (1.7–2.0 V). The sodium ions' low mobility is addressed by the monolayer form, as the decrease in dimensions low barrier of diffusion value from 0.7 to 0.11 eV	[75]
	Sandwich arrangement composed of MoS_2-graphene-MoS_2	Capacity of 175 mAh g^{-1} at a value of current density of 2 A g^{-1} and a reversible capacity value of 313 mAh g^{-1} at a low value of current density of 50 mA g^{-1}, stabilized at 313 mAh g^{-1} after value of 200 cycles	[76]

(continued)

Table 1 (continued)

Energy storage applications	Structure of materials	Key observations	References
	Dual-phase form of MoS_2 structure (2H and 1T phases)	Capacity value of 300 mAh g^{-1} after 200 cyclic readings with value of high Coulombic efficiency (99%). The increase interlayer spacing advanced high reversibility for sodium-ion intercalation	[77]
	MoS_2 and carbon microtubes (C-MTs) are combined	Capacity value of 563.5 mAh g^{-1} at 0.2 A g^{-1}, with 86.6% value of Coulombic efficiency and stable cycling efficiency of 484.9 mAh g^{-1} at 2.0 A g^{-1}	[78]
	Progress in alkali metals' multilayer intercalation between bilayer of graphene give higher storage capacities value than bulk structures. Intercalation of MoS_2 and bilayer graphene for different alkali metals such as Li, Rb, Cs, Na, and K. The weak van der Waals forces existing between the MoS_2 layers offer efficient lithium-ion intercalation without volume expansion, exhibit a storage capacity	The storage value of up to 700 mAh g^{-1}. Although value of storage capacity of MoS_2 is low in contrast to graphene, it can be enhanced by vertical van der Waals interactions in heterostructures of graphene-MoS_2. These structures are significant due to graphene's lightweight characteristics and low formation energy of MoS_2	[79, 80]
Supercapacitors	Charge storage in 1T MoS_2 was studied for interlayer spacings between in ionic liquids	The highest gravimetric and volumetric capacitances were recorded to be 42 F g^{-1} and 118 F cm^{-3}, MoS_2 can contain remarkable capacitance ranges from 400 to 700 F cm^{-3}. Figure 7 shows the MoS_2 CV curves at different scan rates in 1 M Na_2SO_4 solution. Figure 8 shows the MoS_2 morphology (FE-SEM and AFM) used for the electrode material	[81–84]

(continued)

Table 1 (continued)

Energy storage applications	Structure of materials	Key observations	References
	A micro-supercapacitor fabricated using MoS_2 nanosheets sprayed on Si/ SiO_2 substrate and laser-patterned showed excellent electrochemical and cyclic performance in contrast to graphene-based micro-supercapacitors	It acquired a significant capacitance area which shows capacitance of 8 mF cm^{-2} and a value of volumetric capacitance is 178 F cm^{-3}. This innovation provides possibilities for flexible microelectronic devices	[85]
	Nano-MoS_2 structures have displayed a superior energy storage performance. Metallic 1T phase of MoS_2 nanosheets effectively shows intercalation of ions like H^+, Li^+, Na^+, and K^+	High value of capacitance of approximately 400–700 F cm^{-3} in various aqueous electrolytes	[86]
	A MoS_2 flower-shaped nanostructure composite with 3D graphene	Supercapacitor contained a high value of specific capacitance (Csp) of 58 F g^{-1}, value of energy density is noted to be 24.59 W h kg^{-1}, at a power density value of 8.8 W kg, it showed application in affordable supercapacitor without the need for ionic media based on liquid	[87]
	DC magnetron sputtering technique for the deposition MoS_2 thin films on a substrate of copper and used as binder-free supercapacitor electrodes	MoS_2 nanoworm FE-SEM images, with a 2 mm active material thickness, high densely grown and well-distributed nanosheets. The supercapacitor uses a 1 M Na_2SO_4 electrolyte which acquired a capacitance value of 138 F g^{-1} at 1 A g^{-1}, energy density of 12.26 Wh kg^{-1}, and a 0.4 kW kg^{-1} power density. This shows an MoS_2 as important candidate for supercapacitor applications. Active material contained densely grown nanosheets which exhibit a great application in charge storage process	[31]

(continued)

Table 1 (continued)

Energy storage applications	Structure of materials	Key observations	References
	rGO-MoS_2 shows promising path for applications in the field of wearable electronics and printed electronics	The fibers contained significant mechanical characteristics with MoS_2 (10 wt%), beyond which mechanical efficiency began to reduce. H_2SO_4-based polymer gel electrolyte is used for the cyclic voltammetry (CV) and galvanostatic charge–discharge (GCD) testing of the samples Well-defined CV curves acquired in 0.0 to −0.8 V of the potential window shown in Fig. 9a–e. Current density enhanced by increasing the scan rate for devices MoS_2 loading up to 20 wt% shows higher current density at 5 mV s^{-1} but reduction is observed at higher loading shown in Fig. 9f. With rGO fiber contained resistive and pseudo-capacitive behavior on account of low conductivity shown in Fig. 8a. Addition of 5 wt% MoS_2 enhanced the conductivity behavior shown in Fig. 8b. Addition of 10–50 wt% MoS_2 increased the capacitive nature shown in Fig. 9c–e	[88]
Hydrogen evolution Reactions (HER)	First-principle density functional theory showed distinct cell vacancies of MoS_2 and observed that the significant catalytic activity	Mo defects in inert basal plane of MoS_2 displayed its significant efficiency in the HER and reactions based on water splitting applications	[89]
	A hybrid catalyst containing MoS_2 grown on cobalt diselenide (MoS_2/$CoSe_2$) showed catalytic behavior equivalent to commercial platinum catalysts	This hybrid contained a 36 mV dec^{-1} Tafel slope, and value of exchange current density is 7.3 × 10^{-2} mA cm^{-2} which is observed in acidic electrolytes	[81]

(continued)

Table 1 (continued)

Energy storage applications	Structure of materials	Key observations	References
	1T MoS_2 was grown through a hydrothermal method. The process contains ultrasonication of 2H MoS_2 with lithium to enhance intercalation, followed by exfoliation to grow 1T MoS_2. This material contained excellent HER efficiency	Low value of overpotential of 156 mV at a current density of 10 mA cm^{-2} in an acidic medium	[82]
CO_2 reduction	A composite of Bi_2S_3/MoS_2 in literature report contained a superior light absorption capability	CO_2 adsorption in contrast to individual catalysts. This improvement was associated to the high value of catalytic activity in n-type Bi_2S_3 and strong light absorption characteristics of p-type MoS_2	[90]
	Nanoflower-shaped powder of MoS_2 grown through chemical vapor deposition (CVD) technique	Displayed improved photocatalytic behavior, the production rate of CO remained under 0.01 μmol-gcat^{-1} h^{-1} at room temperature (25 °C). Treatment with hydrogen doubled the production rate of CO after 30 min, illustrating the potential for enhancing MoS_2-based systems	[91]

References

1. Braff, W.A., Mueller, J.M., Francik, J.E.: Value of storage technologies for wind and solar energy. Nat. Clim. Change **6**, 964–969 (2016). https://doi.org/10.1038/nclimate3045
2. Rahul, Singh, P.K., Parvaz, M., Ahmed, S., Sonker, R.K., Bhattacharya, B., Khan, Z.H.: Less toxic tin incorporated perovskite solar cell using polymer electrolyte processed in the air. Optik **169**, 166–171 (2018)
3. Marzougui, H., Amari, M., Kadri, A., Bacha, F., Ghouili, J.: Energy management of fuel cell/battery/ultracapacitor in electrical hybrid vehicle. Int. J. Hydrog. Energy **65**, 6592–6601 (2017). https://doi.org/10.1016/j.ijhydene.2016.09.190
4. Farmann, A., Waag, W., Sauer, D.U.: Application-specific electrical characterization of high power batteries with lithium titanate anodes for electric vehicles. Energy **112**, 294–306 (2016)
5. Saxena, S., Sanchez, G., Pecht, M.: Batteries in portable electronic devices: a user's perspective. IEEE Ind. Electron. Mag. **11**, 35–44 (2017). https://doi.org/10.1109/MIE.2017.2688483
6. Sonker, R.K., Shastri, R., Johari, R.: Superficial synthesis of CdS quantum dots for an efficient perovskite-sensitized solar cell. Energy Fuels **35**(9), 8430–8435 (2021)
7. Zhao, C., Zheng, W.: A review for aqueous electrochemical supercapacitors. Front. Energy Res. **3**, 23 (2015). https://doi.org/10.3389/fenrg.2015.00023

8. Faraji, S., Ani, F.N.: The development of supercapacitors from activated carbon by electroless plating—a review. Renew. Sustain. Energy Rev. **42**, 823–834 (2015). https://doi.org/10.1016/j.rser.2014.10.068
9. Gurav, S.R., Waikar, M.R., Rasal, A.S., Sonker, R.K., Sonkawade, R.G.: Current development and challenges in textile-based flexible supercapacitors. In: Smart and Flexible Energy Devices, pp. 539–550. CRC Press (2022)
10. Liu, Y., He, K., Chen, G., Leow, W.R., Chen, X.: Nature-inspired structural materials for flexible electronic devices. Chem. Rev. **117**, 12893–12941 (2017). https://doi.org/10.1021/acs.chemrev.7b00291
11. Li, L., Wu, Z., Yuan, S., Zhang, X.: Advances and challenges for flexible energy storage and conversion devices and systems. Energy Environ. Sci. **7**, 2101–2122 (2014)
12. Ghosh, S., Behera, S.K., Mishra, A., Casari, C.S., Ostrikov, K.K.: Quantum Capacitance of Two-Dimensional-Material-Based Supercapacitor Electrodes. Springer, pp. 17836–17862 (2023). https://doi.org/10.1021/ef3037
13. Chhowalla, M., Shin, H.S., Eda, G., Li, L.-J., Loh, K.P., Zhang, H.: The chemistry of two-dimensional layered transition metal dichalcogenide nanosheets. Nat. Chem. **5**(3), 263–275 (2013). https://doi.org/10.1038/nchem.1589
14. Gupta, D., Chauhan, V., Kumar, R.: A comprehensive review on synthesis and applications of molybdenum disulfide (MoS_2) material: past and recent developments. Inorg. Chem. Commun. **121**, 108200 (2020). https://doi.org/10.1016/j.inoche.2020.108200
15. Ho, S., Nandi, P., Paulraj, I., Ali, D., Sonker, R.K.: Chalcogenides and their nanocomposites: fundamental, properties and applications. In: Metal-Chalcogenide Nanocomposites, pp. 1–27. Woodhead Publishing (2024)
16. Choudhary, N., Islam, M.A., Kim, J.H., Ko, T.-J., Schropp, A., Huntado, L., Weitzman, D., Zhai, L., Jung, Y.: Two-dimensional transition metal dichalcogenide hybrid materials for energy applications. Nano Today **19**, 16–40 (2019). https://doi.org/10.1016/j.nantod.2018.02.007
17. Chhowalla, M., Voiry, D., Yang, J., Shin, H.S., Loh, K.P.: Phase-engineered transition-metal dichalcogenides for energy and electronics. MRS Bull. **40**(7), 585–591 (2015)
18. Lin, L., Miao, N., Huang, J., Zhang, S., Zhu, Y., Horsell, D.W., Ghosez, P., Sun, Z., Allwood, D.A.: A photocatalyst of sulfur-depleted monolayered molybdenum sulfide nanocrystals for dye degradation and hydrogen evolution reaction. Nano Energy **38**, 544–552 (2017). https://doi.org/10.1016/j.nanoen.2017.06.008
19. Lin, L., Xu, Y., Zhang, S., Ross, I.M., Ong, A.C.M., Allwood, D.A.: Fabrication of luminescent monolayered tungsten dichalcogenides quantum dots with giant spin-valley coupling. ACS Nano **7**(10), 8214–8223 (2013). https://doi.org/10.1021/nn403682r
20. Mak, K.F., Lee, C., Hone, J., Shan, J., Heinz, T.F.: Atomically thin MoS_2: a new direct-gap semiconductor. Phys. Rev. Lett. **105**(13), 136805 (2010). https://doi.org/10.1103/PhysRevLett.105.136805
21. Sie, E.J., Lui, C., Lee, Y.-H., Fu, L., Kong, J., Gedik, N.: Large, valley-exclusive Bloch-Siegert shift in monolayer WS_2. Science **355**(6326), 1066–1069 (2017). https://doi.org/10.1126/science.aal2241
22. Yu, Y., Huang, S.-Y., Li, Y., Steinmann, S.N., Yang, W., Cao, L.: Layer-dependent electrocatalysis of MoS_2 for hydrogen evolution. Nano Lett. **14**(2), 553–558 (2014). https://doi.org/10.1021/nl403620g
23. Acerce, M., Voiry, D., Chhowalla, M.: Metallic 1T phase MoS_2 nanosheets as supercapacitor electrode materials. Nat. Nanotechnol. **10**(5), 313–318 (2015). https://doi.org/10.1038/nnano.2015.40
24. Singh, M., Sonker, R.K., Kumar, P., Jain, A., Dagar, M., Singh, N.S.: Bandgap optimization in N-doped Ag-enhanced ZnO-MWCNT nanocomposites for improved absorption. Phys. Solid State **66**(10), 375–386 (2024)
25. Raza, S., Hayat, A., Bashir, T., Chen, C., Shen, L., Orooji, Y., Lin, H.: Electrochemistry of 2D-materials for the remediation of environmental pollutants and alternative energy storage/conversion materials and devices: a comprehensive review. Sustain. Mater. Technol. e00963 (2024)

26. Fouad, O.A., Ismail, A.A., Zaki, Z.I., Mohamed, R.M.: Zinc oxide thin films prepared by thermal evaporation deposition and its photocatalytic activity. Appl. Catal. B **62**(1–2), 144–149 (2006)
27. Shahbazi, M., Khanlary, M.R.: Study of optical, electrochemical, and morphological properties of MoS_2 thin films prepared by thermal evaporation. Braz. J. Phys. **51**(4), 1182–1190 (2021)
28. Gupta, D., Chauhan, V., Upadhyay, S., Koratkar, N., Singh, F., Kumar, S., Mahajan, A., Chandra, R., Kumar, R.: Defects engineering and enhancement in optical and structural properties of 2D-MoS_2 thin films by high energy ion beam irradiation. Mater. Chem. Phys. **276**, 125422 (2022). https://doi.org/10.1016/j.matchemphys.2021.125422
29. Gupta, D., Kumar, R.: Ion implantation as an approach for tuning of electronic structure, optical, morphological and electrical transport properties of sputtered molybdenum disulfide thin films. Mater. Sci. Semicond. Process. **158**, 107326 (2023). https://doi.org/10.1016/j.mssp.2023.107326
30. Sharma, G.K., Ranjan, B., Kaur, D.: Electrochemical kinetics of 2D-MoS_2 sputtered over stainless-steel mesh: insights into the Na^+ ions storage for flexible supercapacitors. Ceram. Int. **48**(16), 23404–23414 (2022)
31. Sanger, A., Malik, V.K., Chandra, R.: One-step sputtered grown MoS_2 nanoworms binder-free electrodes for high-performance supercapacitor application. Int. J. Hydrog. Energy **43**(24), 11141–11149 (2018)
32. Sasi, S., Midhun, P.S., Sharon, A., Reshmi, R.: Growth of few layers of WS_2 thin films by pulsed laser deposition. Mater. Today Proc. (2023). https://doi.org/10.1016/j.matpr.2023.06.179
33. Krebs, H., Ulrich, M., Weisheit, J., Faupel, J., Süske, E., Scharf, T., Fuhse, C., Störmer, M.: Pulsed laser deposition (PLD)—a versatile thin film technique. Adv. Solid State Phys. **43**, 505–518 (2003)
34. Wei, X.Q., Huang, J.Z., Zhang, M.Y., Du, Y., Man, B.Y.: Effects of substrate parameters on structure and optical properties of ZnO thin films fabricated by pulsed laser deposition. Mater. Sci. Eng. B **166**(2), 141–146 (2010)
35. Ojeda, G.P., Döbeli, M., Lippert, T.: Influence of plume properties on thin film composition in pulsed laser deposition. Adv. Mater. Interfaces **5**(18), 1701062 (2018)
36. Siegel, G., Subbaiah, Y.P.V., Prestgard, M.C., et al.: Growth of centimeter-scale atomically thin MoS_2 films by pulsed laser deposition. APL Mater. **3**, 056103 (2015)
37. El Bouanani, L., Serna, M.I., Hasan, S.M.N., et al.: Large-area pulsed laser deposited molybdenum diselenide heterojunction photodiodes. ACS Appl. Mater. Interfaces **12**, 51645–51653 (2020)
38. Baig, M.K., Atiq, S., Bashir, S., Riaz, S., Naseem, S., Soleimani, H., Yahya, N.: Pulsed laser deposition of SmCo thin films for MEMS applications. J. Appl. Res. Technol. **14**(5), 287–292 (2016)
39. Schvartzman, M., Mathur, A., Hone, J., Jahnes, C., Wind, S.J.: Growth of MoS_2 thin films. Appl. Phys. Lett. **93**, 153105 (2008)
40. Kang, M.A., Kim, S.J., Song, W., Chang, S.J., Park, C.Y., Myung, S., Lim, J., Lee, S.S., An, K.S.: Fabrication of flexible optoelectronic devices based on MoS_2/graphene hybrid patterns by a soft lithographic patterning method. Carbon **116**, 167–173 (2017)
41. Lu, D.L., Tanaka, K.I.: Gold particles deposited on electrodes in salt solutions under different potentials. J. Phys. Chem. **100**(5), 1833–1837 (1996). https://doi.org/10.1021/jp952015u
42. Huang, H., Yang, X.: One-step, shape-controlled synthesis of gold nanoparticles stabilized by 3-thiopheneacetic acid. Colloids Surf. A **255**(1–3), 11–17 (2005). https://doi.org/10.1016/j.colsurfa.2004.12.046
43. Finot, M.O., Braybrook, G.D., McDermott, M.T.: Characterization of electrochemically deposited gold nanocrystals on glassy carbon electrodes. J. Electroanal. Chem. **466**(2), 234–241 (1999). https://doi.org/10.1016/S0022-0728(99)00153-7
44. Tiwari, J.N., Tiwari, R.N., Kim, K.S.: Zero-dimensional, one-dimensional, two-dimensional and three-dimensional nanostructured materials for advanced electrochemical energy devices. Prog. Mater. Sci. **57**(4), 724–803 (2012). https://doi.org/10.1016/j.pmatsci.2011.08.003

45. Falola, B.D., Fan, L., Wiltowski, T., Suni, I.I.: Electrodeposition of Cu-doped MoS_2 for charge storage in electrochemical supercapacitors. J. Electrochem. Soc. **164**(9), D674–D678 (2017). https://doi.org/10.1149/2.1111709jes
46. Li, W.Z., Wen, J.G., Ren, Z.F.: Effect of temperature on growth and structure of carbon nanotubes by chemical vapor deposition. Appl. Phys. A **74**, 397–402 (2002). https://doi.org/10.1007/s003390100992
47. Li, W.Z., Wen, J.G., Tu, Y., Ren, Z.F.: Effect of gas pressure on the growth and structure of carbon nanotubes by chemical vapor deposition. Appl. Phys. A **73**(2), 259–264 (2001). https://doi.org/10.1007/s003390100735
48. Sharifi, Y., Achenie, L.E.K.: Effect of substrate geometry on the deposition rate in chemical vapor deposition. J. Cryst. Growth **304**(2), 520–525 (2007). https://doi.org/10.1016/j.jcrysgro.2007.03.043
49. Choy, K.L.: Chemical Vapour Deposition (CVD): Advances, Technology and Applications. CRC Press, Boca Raton, FL (2019)
50. Raza, A., Rasheed, A., Farid, A., Yousaf, M., Ayub, N., Khan, I.A.: Synthesis of binder-free nanofibers ZnS/MoS_2/NiF electrode material for asymmetric supercapacitor applications. J. Energy Storage **84**, 110811 (2024). https://doi.org/10.1016/j.est.2024.110811
51. Radhakrishnan, S., Sree Raj, K.A., Kumar, S.R., Johari, P., Rout, C.S.: Energy storage performance of 2D MoS_2 and carbon nanotube heterojunctions in symmetric and asymmetric configuration. Nanotechnology **32**(15), 155403 (2021). https://doi.org/10.1088/1361-6528/abe5f1
52. Muthulakshmi, G., Ismail, M.M., Ramya, R., Arivanandhan, M., Arjunan, S., Bhaskaran, A.: Facile preparation of SnO_2/MoS_2 nanocomposites with high electrochemical performance for energy storage applications. Inorg. Chem. Commun. **143**, 109802 (2022). https://doi.org/10.1016/j.inoche.2022.109802
53. Danks, A.E., Hall, S.R., Schnepp, Z.J.M.H.: The evolution of 'sol–gel' chemistry as a technique for materials synthesis. Mater. Horiz. **3**(2), 91–112 (2016). https://doi.org/10.1039/C5MH00260E
54. Rahman, A., Padavettan, V.: Synthesis of silica nanoparticles by sol-gel: size-dependent properties, surface modification, and applications in silica–polymer nanocomposites—a review. J. Nanomater. **2012**, Article ID 132424 (2012). https://doi.org/10.1155/2012/132424
55. Verma, R., Mantri, B., Kumar Srivastava, A.: Shape-controlled synthesis, characterizations, mechanisms and optical properties of large-scaled metal oxide nanostructures of ZnO and TiO_2. Adv. Mater. Lett. **6**(4), 324–333 (2015). https://doi.org/10.5185/amlett.2015.5602
56. Bokov, D., Jalil, A.T., Chupradit, S., Suksatan, W., Ansari, M.J., Shewael, I.H., Valiev, G.H., Kianfar, E.: Nanomaterial by sol-gel method: synthesis and application. Adv. Mater. Sci. Eng. **2021**(1), 5102014 (2021). https://doi.org/10.1155/2021/5102014
57. Nardi, M.V., Timpel, M., Ligorio, G., Zorn Morales, N., Chiappini, A., Toccoli, T., Verucchi, R., Ceccato, R., Pasquali, L., List-Kratochvil, E.J., Quaranta, A.: Versatile and scalable strategy to grow sol-gel derived 2H-MoS_2 thin films with superior electronic properties: a memristive case. ACS Appl. Mater. Interfaces **10**(40), 34392–34400 (2018). https://doi.org/10.1021/acsami.8b12468
58. Guo, X., Yin, P., Wang, Z., Yang, H.: Template-assisted sol-gel synthesis of porous MoS_2/C nanocomposites as anode materials for lithium-ion batteries. J. Sol-Gel Sci. Technol. **85**, 140–148 (2018). https://doi.org/10.1007/s10971-017-4574-4
59. Xiang, T., Fang, Q., Xie, H., Wu, C., Wang, C., Zhou, Y., Liu, D., Chen, S., Khalil, A., Tao, S., et al.: Vertical 1T-MoS_2 nanosheets with expanded interlayer spacing edged on a graphene frame for high rate lithium-ion batteries. Nanoscale **9**(14), 6975–6983 (2017). https://doi.org/10.1039/C7NR00137C
60. Wu, M., Zhan, J., Wu, K., Li, Z., Wang, L., Geng, B., Wang, L., Pan, D.: Metallic 1T MoS_2 nanosheet arrays vertically grown on activated carbon fiber cloth for enhanced Li-ion storage performance. J. Mater. Chem. A **5**(30), 14061–14069 (2017). https://doi.org/10.1039/C7TA02869K

61. Li, Z., Zhan, X., Zhu, W., Qi, S., Braun, P.V.: Carbon-free, high-capacity and long cycle life 1D–2D $NiMoO_4$ nanowires/metallic 1T MoS_2 composite lithium-ion battery anodes. ACS Appl. Mater. Interfaces **11**(39), 44593–44600 (2019). https://doi.org/10.1021/acsami.9b14445
62. Jiao, Y., Mukhopadhyay, A., Ma, Y., Yang, L., Hafez, A.M., Zhu, H.: Ion transport nanotube assembled with vertically aligned metallic MoS_2 for high rate lithium-ion batteries. Adv. Energy Mater. **8**(3), 1702779 (2018). https://doi.org/10.1002/aenm.201702779
63. Su, J., Pei, Y., Yang, Z., Wang, X.: Ab initio study of graphene-like monolayer molybdenum disulfide as a promising anode material for rechargeable sodium ion batteries. RSC Adv. **4**(73), 43183–43188 (2014). https://doi.org/10.1039/C4RA05754E
64. Geng, X., Jiao, Y., Han, Y., Mukhopadhyay, A., Yang, L., Zhu, H.: Freestanding metallic 1T MoS_2 with dual ion diffusion paths as high rate anode for sodium-ion batteries. Adv. Funct. Mater. **27**(17), 1702998 (2017). https://doi.org/10.1002/adfm.201702998
65. Wu, J., Liu, J., Cui, J., Yao, S., Ihsan-Ul-Haq, M., Mubarak, N., Quattrocchi, E., Ciucci, F., Kim, J.-K.: Dual-phase MoS_2 as a high-performance sodium-ion battery anode. J. Mater. Chem. A **8**(4), 2114–2122 (2020). https://doi.org/10.1039/C9TA11694A
66. Pan, Q., Zhang, Q., Zheng, F., Liu, Y., Li, Y., Ou, X., Xiong, X., Yang, C., Liu, M.: Construction of MoS_2/C hierarchical tubular heterostructures for high-performance sodium ion batteries. ACS Nano **12**(12), 12578–12586 (2018). https://doi.org/10.1021/acsnano.8b07384
67. Kühne, M., Börrnert, F., Fecher, S., Ghorbani-Asl, M., Biskupek, J., Samuelis, D., Krasheninnikov, A.V., Kaiser, U., Smet, J.H.: Reversible superdense ordering of lithium between two graphene sheets. Nature **564**(7734), 234–239 (2018). https://doi.org/10.1038/s41586-018-0770-z
68. Chepkasov, I.V., Ghorbani-Asl, M., Popov, Z.I., Smet, J.H., Krasheninnikov, A.V.: Alkali metals inside Bi-layer graphene and MoS_2: insights from first-principles calculations. Nano Energy **75**, 104927 (2020). https://doi.org/10.1016/j.nanoen.2020.104927
69. Saha, D., Kruse, P.: Editors' choice—review—conductive forms of MoS_2 and their applications in energy storage and conversion. J. Electrochem. Soc. **167**(12), 126517 (2020). https://doi.org/10.1149/1945-7111/ab9a0a
70. Acerce, M., Voiry, D., Chhowalla, M.: Metallic 1T phase MoS_2 nanosheets as supercapacitor electrode materials. Nat. Nanotechnol. **10**(4), 313–318 (2015). https://doi.org/10.1038/nnano.2015.40
71. Miao, L., Song, Z., Zhu, D., Li, L., Gan, L., Liu, M.: Recent advances in carbon-based supercapacitors. Mater. Adv. **1**(9), 945–966 (2020). https://doi.org/10.1039/D0MA00451G
72. Liang, Z., Zhao, C., Zhao, W., Zhang, Y., Srimuk, P., Presser, V., Feng, G.: Molecular understanding of charge storage in MoS_2 supercapacitors with ionic liquids. Energy Environ. Mater. **5**, eem2.12147 (2021). https://doi.org/10.1002/eem2.12147
73. Cao, L., Yang, S., Gao, W., Liu, Z., Gong, Y., Ma, L., Shi, G., Lei, S., Zhang, Y., Zhang, S., et al.: Direct laser-patterned micro-supercapacitors from paintable MoS_2 films. Small **9**(17), 2905–2910 (2013). https://doi.org/10.1002/smll.201300411
74. Singh, K., Kumar, S., Agarwal, K., Soni, K., Gedela, V.R., Ghosh, K.: Three-dimensional graphene with MoS_2 nanohybrid as potential energy storage/transfer device. Sci. Rep. **7**, 9458 (2017). https://doi.org/10.1038/s41598-017-09848-7
75. Gao, M.-R., Chan, M.K.Y., Sun, Y.: Edge-terminated molybdenum disulfide with a 9.4-Å interlayer spacing for electrochemical hydrogen production. Nat. Commun. **6**, 7493 (2015). https://doi.org/10.1038/ncomms8493
76. Ye, K., Li, M., Luo, J., Wu, B., Lai, L.: The H_2O dissociation and hydrogen evolution performance of monolayer MoS_2 containing single Mo vacancy: a theoretical study. IEEE Trans. Nanotechnol. **19**(1), 163–167 (2020). https://doi.org/10.1109/TNANO.2019.2959363
77. Tang, Q., Jiang, D.: Mechanism of hydrogen evolution reaction on 1T-MoS_2 from first principles. ACS Catal. **6**(8), 4953–4961 (2016). https://doi.org/10.1021/acscatal.6b01068
78. Ye, K., Li, M., Luo, J., Wu, B., Lai, L.: Activating inert basal plane of MoS_2 for H_2O dissociation and HER via formation of vacancy defects: a DFT study. In: 2019 IEEE 19th International Conference on Nanotechnology (IEEE-NANO), Macao, China, July 22–26, 2019, pp. 48–53. IEEE, Macao, China (2019). https://doi.org/10.1109/NANO.2019.8875911

79. Li, J., Joseph, T., Ghorbani-Asl, M., Kolekar, S., Krasheninnikov, A.V., Batzill, M.: Mirror twin boundaries in $MoSe_2$ monolayers as one dimensional nanotemplates for selective water adsorption. Nanoscale **13**(3), 1038–1047 (2021). https://doi.org/10.1039/D0NR07589C
80. Ye, K., Lai, L., Li, M., Luo, J., Wu, B., Ren, Z.: Strain effect on the hydrogen evolution reaction of V Mo-SL MoS_2. IEEE Trans. Nanotechnol. **19**(2), 192–196 (2020). https://doi.org/10.1109/TNANO.2020.2967683
81. Gao, M.-R., Liang, J.-X., Zheng, Y.-R., Xu, Y.-F., Jiang, J., Gao, Q., Li, J., Yu, S.-H.: An efficient molybdenum disulfide/cobalt diselenide hybrid catalyst for electrochemical hydrogen generation. Nat. Commun. **6**, 5982 (2015). https://doi.org/10.1038/ncomms6982
82. Li, Y., Wang, L., Zhang, S., Dong, X., Song, Y., Cai, T., Liu, Y.: Cracked monolayer 1T MoS_2 with abundant active sites for enhanced electrocatalytic hydrogen evolution. Catal. Sci. Technol. **7**(3), 718–724 (2017). https://doi.org/10.1039/C6CY02649D
83. Jacobson, T.A., Kler, J.S., Hernke, M.T., Braun, R.K., Meyer, K.C., Funk, W.E.: Direct human health risks of increased atmospheric carbon dioxide. Nat. Sustain. **2**(8), 691–701 (2019). https://doi.org/10.1038/s41893-019-0323-1
84. He, J., Janáky, C.: Recent advances in solar-driven carbon dioxide conversion: expectations versus reality. ACS Energy Lett. **5**(6), 1996–2014 (2020). https://doi.org/10.1021/acsenergylett.0c00645
85. Tanaka, Y., Hasanuzzaman, M.: A review of global current techniques and evaluation methods of photocatalytic CO_2 reduction. In: Proceedings of the 5th IET International Conference on Clean Energy and Technology (CEAT2018), Kuala Lumpur, Malaysia, Sept 5–6, 2018, p. 6. Institution of Engineering and Technology: Kuala Lumpur, Malaysia (2018). https://doi.org/10.1049/cp.2018.1371
86. Ueckerdt, F., Bauer, C., Dirnaichner, A., Everall, J., Sacchi, R., Luderer, G.: Potential and risks of hydrogen-based e-fuels in climate change mitigation. Nat. Clim. Chang. **11**(5), 384–393 (2021). https://doi.org/10.1038/s41558-021-01032-7
87. Shao, X., Zhang, X., Liu, Y., Qiao, J., Zhou, X.-D., Xu, N., Malcombe, J.L., Yi, J., Zhang, J.: Metal chalcogenide-associated catalysts enabling CO_2 electroreduction to produce low-carbon fuels for energy storage and emission reduction: catalyst structure, morphology, performance, and mechanism. J. Mater. Chem. A **9**(5), 2526–2559 (2021). https://doi.org/10.1039/D0TA09766C
88. Roy, A.K., Faisal, S.N., Spickenheuer, A., Scheffler, C., Wang, J., Harris, A.T., Islam, M.S.: Loading dependency of 2D MoS_2 nanosheets in the capacitance of 3D hybrid microfibre-based energy storage devices. Carbon Trends **5**, 100097 (2021). https://doi.org/10.1016/j.cartre.2021.100097
89. Ye, K., Li, M., Luo, J., Wu, B., Lai, L.: Activating inert basal plane of MoS_2 for H_2O dissociation and HER via formation of vacancy defects: a DFT study. In: Proceedings of the 2019 IEEE 19th International Conference on Nanotechnology (IEEE-NANO), pp. 48–53. IEEE, Macao, China (2019). https://doi.org/10.1109/NANO46743.2019.8993741
90. Wang, J., Zhang, X., Wu, J., Chen, H., Sun, S., Bao, J., Bi, H.: Preparation of Bi_2S_3/carbon quantum dot hybrid materials with enhanced photocatalytic properties under ultraviolet-, visible-, and near infrared-irradiation. Nanoscale **9**(41), 15873–15882 (2017). https://doi.org/10.1039/C7NR04213D
91. Meier, A.J., Garg, A., Sutter, B., Kuhn, J.N., Bhethanabotla, V.R.: MoS_2 nanoflowers as a gateway for solar-driven CO_2 photoreduction. ACS Sustain. Chem. Eng. **7**(1), 265–275 (2019). https://doi.org/10.1021/acssuschemeng.8b03526

Graphene Oxide-Based Hybrid Composite Films for Fuel Cell Applications

Rahul Johari, Rakesh K. Sonker, Rashi Johari, Urmila Samariya, Pawan Kumar, Pramod K. Singh, and Zishan H. Khan

Abstract Recent research around energy is directed primarily towards fuel cells, and graphene-assisted materials have been found to increase efficiency by leaps and bounds in real-world applications. This paper discusses recent developments in graphene-based cathode materials and how this is revolutionizing the pace of fuel cell technology. Graphene is put forward as an extremely promising material to be included in solid oxide fuel cells (SOFCs) and proton-exchange membrane fuel cells (PEMFCs) due to its large surface area, better electrical conductivity, and greater mechanical stability. This paper describes in detail the various forms of graphene, including graphene oxide (GO), reduced graphene oxide (rGO), and doped graphene. It highlights their unique properties and how good they are as catalysts. Apart from this, we discuss how doping, structural modification, and functional

R. Johari (✉)
Department of Electrical and Computer Engineering, Ben-Gurion University of the Negev, Beer-Sheva, Israel
e-mail: rahuljohari.phy@gmail.com

R. Johari · Z. H. Khan
Department of Applied Sciences and Humanities, Jamia Millia Islamia, New Delhi, India

R. Johari · P. K. Singh
Centre of Excellence on Solar Cells and Renewable Energy, Sharda University, Greater Noida, India

R. K. Sonker
Department of Physics, Motilal Nehru College, University of Delhi, New Delhi, Delhi, India

R. Johari
Department of Computer Science and Engineering, Jaypee Institute of Information Technology, Noida, India

U. Samariya
Department of Information Technology, Indira Gandhi Delhi Technical University for Women (IGDTUW), New Delhi, Delhi, India

P. Kumar
Department of Physics, Acharya Narendra Dev College, University of Delhi, New Delhi, Delhi, India

R. K. Sonker et al. (eds.), *Low-Dimensional Multifunctional Hybrid Nanostructured Materials*, Engineering Materials, https://doi.org/10.1007/978-981-96-7952-2_10

group introduction influence the stability and electrochemical behavior of graphene-based cathodes. This study highlights the key challenge of ensuring the thermal stability of graphene derivatives in high-temperature SOFCs and discusses ways of enhancing their longevity. This paper discusses recent advancements as well as present issues to highlight the promising potential of graphene-based materials in fuel cells. These advancements can lead to more efficient, cost-saving, and long-duration energy solutions.

1 Introduction

Development of effective and sustainable energy conversion systems is a global challenge. Among the available technologies, fuel cells have been demonstrated to be effective alternatives for their high efficiency and environmental compatibility [1]. However, performance of fuel cells is often restricted by limitations including catalyst degradation, high cost, and poor durability. Owing to their excellent electrical, mechanical, and chemical properties, graphene oxide (GO) hybrid composite films have been explored as a potential solution to the limitations [2]. Owing to its characteristic structure, comprising oxygen-functional groups, GO can be mixed with other materials to expand its application in fuel cells [3]. Assuming there is a fuel source, a series of energy conversion devices known as fuel cells can generate electricity. Owing to the fact that chemical energy in the fuel is directly transformed into electrical energy, fuel cells provide much higher system efficiency and lower emissions of pollutants than combustion engines. Therefore, fuel cells are anticipated to be among the most crucial technologies to mitigate environmental issues and the world's energy crisis as well as to enable a better and cleaner life.

There is an electrolyte layer between two electrodes in a typical fuel cell (Fig. 1) [4]. In a fuel cell, anode oxidizes fuel, releasing electrons that travel through an external circuit to supply oxygen (O_2) reduction at the cathode. At the same time, charge carriers like protons (H^+), hydroxide ions (OH^-), carbonate ions (CO_3^{2-}), or oxide ions (O_2^-) move through the electrolyte, ensuring continuous circulation of the electrical circuit. Fuel cells are divided on the basis of the kind of electrolyte used in them. The principal types are solid oxide fuel cells (SOFCs), phosphoric acid fuel cells (PAFCs), alkaline fuel cells (AFCs), molten carbonate fuel cells (MCFCs), and polymer electrolyte membrane fuel cells (PEMFCs). Among these, PAFCs, PEMFCs, and AFCs are low-temperature operating (below 300 °C), and thus they are a very good choice for automotive and portable power equipment applications. Fuel cells are categorized on the basis of the nature of the electrolyte in them into types. The most prominent among them are solid oxide fuel cells (SOFCs), phosphoric acid fuel cells (PAFCs), alkaline fuel cells (AFCs), molten carbonate fuel cells (MCFCs), and polymer electrolyte membrane fuel cells (PEMFCs). They include PAFCs, PEMFCs, and AFCs, which have low-temperature operation (below 300 °C) and therefore they are a good choice for automotive and portable power equipment applications. MCFCs and SOFCs are high-temperature operation (above 500 °C) and therefore they can

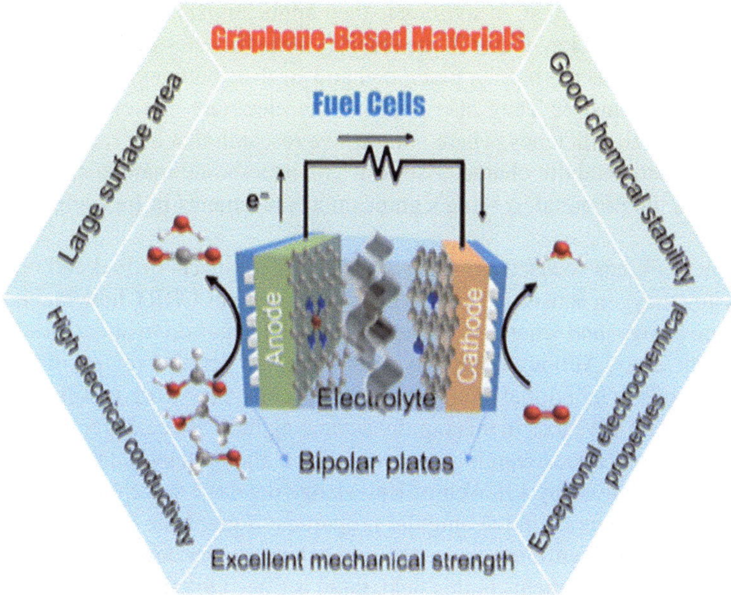

Fig. 1 Fuel cell schematic with graphene-based materials in each part [4]. Reproduced with Open access @ (2020) Wiley

be powered with a large range of fuels at high efficiency. Fuel cell material selection is done on the basis of the most critical parameters of efficiency, durability, and electrochemical properties. These have to be optimized for fuel cell technology to enhance reliability and commercial viability [5, 6].

Graphene and graphene derivatives have been of special interest as a potential fuel cell material due to their high mechanical toughness, chemical stability, and electrical conductivity. In the last decade, tremendous effort has been focused on the application of graphene-based materials, particularly as electrocatalyst supports. Their ability to increase active sites and the efficiency of electron transfer is a contributing factor to increasing fuel oxidation and the oxygen reduction reaction (ORR) and fuel cell efficiency. Metal-free graphene materials, for example, have demonstrated superior ORR activity with high electrocatalytic activity, high resistance to poisoning, and low cost. Researchers have examined in detail how modification of graphene—modification of its electronic structure, doping, introduction of defects, or addition of functional groups—can affect fuel cell efficiency. Such advancements are in progress to propel fuel cell technology, and graphene-based materials are an interesting area for the development of next-generation energy [7, 8].

The incorporation of graphene-based materials into polymer membranes greatly enhances ionic conductivity while simultaneously reducing fuel crossover, thus making them a key element in fuel cell technology. Their enhanced high proton conductivity and water, hydrogen (H_2), and methanol permeation resistance have rendered such materials as potential proton-exchange membranes, which has drawn extensive scholarly interest. In addition to their application in membranes, graphene-based materials are a key factor in the stability and efficiency of important fuel cell components, such as the bipolar plate, fuel and air distribution systems, and current collection mechanisms. They also improve the electrode and electrolyte, further increasing overall efficiency. There is extensive research that has explored the electrochemical and catalytic characteristics of graphene-based materials, and they are seen to have the potential to bring significant advancements in fuel cell technology [9–12].

Recent work has focused primarily on individual aspects of fuel cell operation, such as how oxygen is used (oxygen reduction reaction or ORR), how fuel is treated to produce energy, and what membranes and special plates do to allow the system to operate efficiently. Although previous work has given detailed review of graphene-based materials in individual components of fuel cells, new work is still needed that reflects the latest advances in this rapidly advancing field. Since 2016, the overlap of fuel cell and graphene research has grown exponentially, hundreds of publications per year, as evidenced by ISI Web of Knowledge database searches. This literature records growing interest in the singular properties of graphene to enhance the efficiency of fuel cells, but more is needed as a balanced overview of recent progress. With the rapid developments in this field, there is a need to critically examine recent developments to be abreast with recent progress [13–15].

This review records how graphene materials enhance critical fuel cell components such as electrodes, electrolytes, and bipolar plates. It also records new progress, such as 3D graphene electrodes and graphene-supported single-atom catalysts, that are enhancing the efficiency of fuel cells. This work also discusses in detail the important parameters that influence how well graphene materials operate in fuel cells. It records how they function, how their operation can be maximized through ingenuity of design, and how they can enhance the efficiency and life of fuel cells [16, 17].

2 Properties of Graphene Oxide for Fuel Cell Applications

Graphene oxide (GO) is a compound that has increased in popularity in recent years due to the prospect of being used in fuel cells. GO is a derivative of graphene with excellent electrical conductivity but with oxygen-functional groups, thereby being highly water dispersible and biocompatible with different components of fuel cells [18].

Graphene oxide (GO) has typical characteristics like large surface area, functional Zable oxygen functionalities, and reasonable mechanical stability, thereby qualifying it as a potential material to enhance proton-exchange membranes (PEMs) and enhance electrode performance in fuel cells. One of the most worthy benefits of graphene oxide (GO) for fuel cells is the capacity to offer proton transport, an extremely important feature of energy production. GO's hydrophilic functional groups are the ones responsible for the formation of micro-scale water channels and hence facile proton transport when used in polymer electrolyte membranes (PEMs). This is most useful in polymer electrolyte membrane fuel cells (PEMFCs) because it improves their performance and overall performance [19, 20].

Moreover, GO is very well recognized as a good catalytic support. It is a low-cost alternative to conventional carbon-based supports for Pt catalysts, improving the catalyst dispersion and reducing Pt loading but maintaining high catalytic activity [21]. That is only one aspect of the fuel cell cost-effectiveness and durability. Another highly significant aspect of GO in fuel cells is its effect on durability and stability. Due to its mechanical robustness and resistance to oxidation, GO-reinforced membranes are chemically more stable and less susceptible to mechanical stress, improving the working life of the fuel cell [22].

Furthermore, oxygen functionalities in GO could be engineered with improved compatibility of other fuel cell materials to obtain the best overall performance. Graphene oxide also possesses multiple positive attributes while utilized in fuel cells, e.g., improving proton conductivity, improving catalyst support, and increasing durability. R&D seeks the optimal GO functionalization and merging with composites to achieve top fuel cell efficiency and commercializability.

3 Fabrication of Graphene Oxide-Based Hybrid Composite Films

The chemically tunable, high surface area, and excellent electrical conductivity properties of graphene oxide (GO)-based hybrid composite sheets have received considerable interest in the use of fuel cell applications. For their enhanced performance in electrochemical energy conversion devices and utmost structure, various fabrication techniques have been reported. The most common techniques utilized are in situ polymerization, electrochemical deposition, solution processing, and layer-by-layer assembly. Each method is versatile to a particular fuel cell application due to the distinctive advantages of material integration, structural manipulation, and functionalization for each approach.

3.1 Solution Processing

Solution processing is the most convenient and scalable method to prepare GO-based hybrid composite film. GO is first dispersed in an appropriate solvent, like water, ethanol, or N,N-dimethylformamide (DMF), in order to have good dispersion and to prevent aggregation. Functional materials like conductive polymers (like polyaniline or polypyrrole), metal nanoparticles, or metal oxides are then added to the solution. The resulting solution is then cast onto a substrate by techniques like spin-coating, drop-casting, or vacuum filtration, thus achieving a thin and homogeneous film [23]. It is simple to control the thickness and composition of the film in this method and thus is highly suitable to be used in fuel cells. Moreover, the addition of catalytic nanomaterials into the GO matrix promotes proton conductivity and electron transfer enhancement, ultimately leading to effective fuel cells [24]. Nevertheless, the achievement of long-term stability and mechanical strength is the biggest challenge, which requires further modification through post-treatment or cross-linking.

3.2 Electrochemical Deposition

One of the successful ways of hybridizing GO with metal components or metal oxides to improve catalytic performance and electron transport functionality is by using the electrochemical deposition method. GO is deposited simultaneously in this case by a regulated electrochemical reaction onto a conductive substrate, such as carbon paper or platinum electrodes. An externally applied voltage provides opportunity for metal cation decrease (i.e., Pt, Ni, or Co), being present at the same time inside the GO matrix, leading to increased electrochemical hybrid composite film [25].

This method is particularly beneficial for the production of highly active electrode materials used in fuel cells. The catalytically active sites are co-deposited metal particles, and the graphene oxide matrix is a more conductive electron transfer network. In addition, electrochemical deposition allows one to create the thickness, structure, and film composition in strict control through careful adjustment of significant parameters including voltage, pH levels, and precursor concentration [26]. In spite of all these advantages, there remains a striking disadvantage of this method that includes the need for sophisticated electrochemical equipment and optimized processing conditions for homogeneous deposition.

3.3 Layer-by-Layer Assembly

The structure and microstructure of GO-based hybrid composite films can be effectively controlled by the highly sensitive layer-by-layer (LbL) assembly process. In the process, GO and other functional components, e.g., metal nanoparticles, protein, or oppositely charged polyelectrolytes, are sequentially deposited onto a substrate [27]. Adhesion of the layers is provided through electrostatic attraction, hydrogen bonding, or covalent bonding, and the final product is the creation of an ordered multilayer film. The Layer-by-Layer (LbL) method offers a number of benefits to fuel cells. It allows control over thicknesses of thin films at a micro-level, thus optimizing proton transport and improving the durability and stability of proton-exchange membranes (PEMs) [28]. In addition, sequential deposition of graphene oxide (GO) with proton-conductive polymers such as Nafion or sulfonated polyether ether ketone

(SPEEK) improves the ionic conductivity of the membrane while reducing fuel crossover. Nonetheless, despite its ability to control material composition carefully, LbL assembly tends to be time-consuming and consists of several cycles of deposition to achieve the desired film properties.

3.4 *In Situ Polymerization*

Another ordinary manufacturing process, which involves embedding GO directly in polymer matrices when polymerizing them, is in situ polymerization. This produces highly stable hybrid composite films by enabling substantial interfacial contacts between the GO and polymer, unlike through simple blending or physical mixing [29]. Beforehand, in order to increase GO's compatibility with the precursor polymer, the reactive groups are usually introduced in the first step. Subsequent to this step, the process of polymerization is carried out in the presence of GO such that it can be uniformly distributed in the network of the polymer.

This method has been widely used in the synthesis of proton-conducting membranes and fuel cell electrode materials. The resulting graphene oxide–polymer composites exhibit enhanced mechanical strength, thermal stability, and proton conductivity due to the synergistic interactions between the polymer matrix and the graphene oxide nanosheets [30]. In addition, in situ polymerization reduces phase separation and agglomeration very efficiently, which are typical problems of conventional blending. However, careful optimization of the reaction conditions including temperature, concentration of the monomer, and time of polymerization is essential to obtain the desired film properties.

4 Applications of Graphene Oxide-Based Hybrid Films in Fuel Cells

4.1 *Proton Exchange Membrane Fuel Cells (PEMFCs)*

Graphene oxide (GO)-based hybrid films have drawn considerable interest in recent years on the basis of their potential to enhance the performance of proton-exchange membrane (PEM) fuel cells. Several methods have been proposed to incorporate GO into the systems in an attempt to circumvent some of the drawbacks of the conventional membrane materials. One of the most promising uses of GO in PEM fuel cells is the use of GO in Nafion membranes, the most established polymer electrolyte membranes with enhanced proton conductivity and chemical stability [31]. The use of graphene oxide (GO) incorporation into Nafion membranes has been reported to lead to substantial improvement in the performance and the lifetime of polymer electrolyte membrane (PEM) fuel cells. Among the most notable advantages of application is the ability to provide proton mobility, which is essential to fuel cell operation. The distinct GO structure, high concentration of oxygen-containing functional groups, is responsible for the creation of additional proton transfer channels, thereby enhancing the system efficiency.

The enhanced channels enable enhanced proton transport through the membrane, ultimately leading to enhanced power output and enhanced overall fuel cell efficiency. In addition to enhancing proton conductivity, the incorporation of graphene oxide (GO) has been reported to suppress fuel crossover extensively [3]. Fuel crossover is a critical issue in direct methanol fuel cells (DMFCs), as it can lead to efficiency loss as well as loss of performance. Diffusion of methanol or hydrogen through the membrane can lead to their presence on the cathode side, where they can react with other chemicals in unwanted manners, thereby leading to reduced overall fuel consumption and unwanted byproducts. The nanostructure of GO as a layered barrier inhibits fuel molecule permeability, essentially inhibiting crossover and fuel cell lifespan extension [32].

In addition, GO incorporation significantly enhances the mechanical stability of Nafion membranes. Conventional Nafion membranes are prone to swelling and mechanical degradation when utilized for a long period, especially under high-temperature and high-humidity conditions. The GO incorporation enhances the structural firmness of the membrane with superior mechanical stability and resistance to degradation. Such a higher durability factor is crucial in the long-term operation of PEM fuel cells and hence GO-based hybrid films can be a viable candidate for commercial and industrial application [33].

Besides these main advantages, scientists have explored other possible advantages of GO–Nafion hybrid membranes. For example, GO may enhance the retention of water in the membrane, which is essential in maintaining high proton conductivity, particularly under high-temperature conditions. GO regulates the hydration level of the membrane, ensuring the best working conditions and inhibiting excessive dehydration that can lead to performance decline. Other than that, GO incorporation into Nafion membranes has also provided new opportunities for fuel cell technology optimization. There has been works that involved functionalizing GO with other functional groups or mixing it with other nanomaterials in an attempt to further optimize its performance. For instance, GO–metal oxide doping or sulfonic acid group functionalization has demonstrated even superior improvements in proton conductivity and thermal stability [34].

GO-based hybrid films are a future potential breakthrough in PEM fuel cell technology. GO addition to Nafion membranes provides a multidimensional strategy towards performance enhancement by increasing proton conductivity, minimizing fuel crossover, and enhancing mechanical stability. With the advancement of this area, continuous innovation in GO-based materials will continue to enhance the efficiency, lifespan, and cost-effectiveness of fuel cell systems. As the world continues to demand clean and sustainable energy, GO–Nafion hybrid membranes can be one of the key contributors to fuel cell technology development and enhancement for the future.

4.2 Electrocatalysts for Oxygen Reduction Reaction (ORR)

Graphene materials are universally applied to stabilize metal nanoparticles aimed at enhancing electron mobility to the electrode surface. N-doped graphene is one of the marvels in the process as it can distribute and stabilizing Pt–Co alloy nanoparticles. N-doped graphene enables improved fuel cells as it alters how the nanoparticles interact, grow, and interface with the support material as well [35–37].

A good example among these is a proton-exchange membrane fuel cell (PEMFC) with a Pt–Co/N-doped graphene cathode. The device exhibited a very high peak power density of 805 mW cm^{-2} at 60 °C, four times higher than that of a normal PEMFC with a conventional Pt/C cathode. This shows the ability of graphene-based materials to better the performance and efficiency of fuel cells by a significant margin. In a move to make fuel cells affordable and commercially viable, scientists are seeking catalysts that are PGM-free. One of these is a compound that is a combination of Co_3O_4 and reduced graphene oxide (rGO) and has been said to perform better, having an oxygen reduction reaction (ORR) onset potential of around 0.83 V compared to the reversible hydrogen electrode (RHE). Its long-term stability makes it more significant as well. A study by Liang et al. reported that Co_3O_4/rGO exhibited ORR performance for 25,000 s, much better than conventional platinum-based catalysts in terms of durability [38]. Hence, it may be a good candidate to employ in fuel cells in the future. Metal oxide-graphene interaction facilitates improved electron transfer and reaction kinetics and hence better performance. Additionally, high surface area three-dimensional (3D) graphene with high electrical conductivity and highly interconnected porous architecture offers favorable mass transfer and stable support to metal oxide nanoparticles. In the first breakthrough research, Hu et al. used a supercritical fluid process to fabricate a Pt–Fe/HSG electrode for attempting to probe the functionality of three-dimensional honeycomb-structured graphene in supporting the oxygen reduction reaction (ORR) [39]. The study proved that $Pt_{40}Fe_{60}$/HSG had 1.70 A $mgPt^{-1}$ mass activity, which was more than 14 times higher than conventional performances of Pt/C catalysts. Such evidence confirms the potential of promising performance by emerging graphene-based catalysts to achieve fuel cell performance far superior to the current ones.

Later on, researchers designed a platinum group metal-free (PGM-free) catalyst in attempting to further improve the oxygen reduction reaction (ORR). This was achieved through the use of iron (Fe), nitrogen (N), and sulfur (S)-doped Fe_3O_4 nanoparticles dispersed in a three-dimensional honeycomb-like graphene network (Fe_3O_4/FeNSG-3).They used melamine–formaldehyde resin as a nitrogen source and as a soft template to control material alignment in synthesizing this catalyst. This resulted in a high-performance catalyst, which had a remarkable onset potential of 0.951 V versus RHE in alkaline solution and even superior performance than commercial Pt/C catalysts. Its optimal performance was through its dense active sites, precisely ordered 3D porous framework, and enhanced surface area of 530.5 m^2 g^{-1}. In addition to its enhanced performance, Fe_3O_4/FeNSG-3 was very stable and maintained 71.5% of its initial current even after extended use for 30,000 s. In comparison,

Pt/C remained stable to only 54.4% for the same amount of time. These results show that heteroatom-doped graphene is a suitable support for stabilizing a single metal atom and is therefore a suitable potential candidate for future fuel cell technology [40, 41]. SACs supported on graphene have been emerging as the best potentials to replace traditional candidates for oxygen reduction reaction (ORR) in acidic and alkaline solution with enhanced four-electron pathway selectivity and stability.

Their excellent performance has been attributed to strong metal atom–graphene interaction, high active site density, and charge redistribution in the graphene network. Ru atoms dispersed on N-doped graphene in 0.1 mol/L $HClO_4$ solution displayed superior oxygen reduction reaction (ORR) activity, stability, and higher resistance to carbon monoxide and methanol poisoning than traditional Pt/C catalysts. This shows the potential of single-atom catalysts (SACs) to be effective but long-term alternatives for application in fuel cells (Table 1) [42]. These catalysts are long-term because stable chemical bonds between the metal atoms and the supporting graphene are established. In a paper by Shu et al. [43], single Co atoms were anchored on sulfuric acid-treated N-doped graphene. This modification provided a highly stable catalyst that retained 94% of the initial current density after application for 7,200 s in methanol solution and outperformed traditional Pt/C catalysts. Similarly, Mou et al. have concluded that $Cu–N_2$ and $Cu–N_4$ groups in nitrogen-doped graphene play a critical role in the high activity for oxygen reduction reaction (ORR) [44].

According to these strategies, researchers have successfully immobilized a variety of atomically dispersed metals—such as cobalt, iron, copper, manganese, scandium, rhodium, and ruthenium—on nitrogen-doped graphene using a simple annealing process. This illustrates the universality of graphene-based single-atom catalysts (SACs) in fuel cell technology. Metal salts and graphene oxide (GO) are sonicated in water during synthesis, and then water is removed under a rotary evaporator. The composite is further exposed to thermal pyrolysis in ammonia-rich conditions for the synthesis of SACs. Surprisingly, Fe–N–G catalysts have attracted tremendous attention because of their stability, effectiveness in ORR, and low cost. Atomic level placement of iron at pyridinic nitrogen sites results in the creation of many active moieties, such as $Fe–N_4$ and $Fe–N_2 + 2$, responsible for offering superior catalytic activity. Scientists are attempting to figure out the best sites in Fe–N–G catalysts for oxygen reduction reactions (ORR). They've identified that $Fe–N_2 + 2$ scaffolds, which are prepared via calcination of iron salts, graphitic carbon nitride, and reduced graphene oxide (rGO), are essential in maximizing ORR activity [45–58].

A breakthrough among them was a single-iron-atom catalyst, which was prepared by a ball-milling method and exhibited better stability and endurance towards sulfur oxides (SO_x), nitrogen oxides (NO_x), and poisoning by methanol. This is because iron atoms were confined in the framework of the graphene nanosheet. Advanced techniques like Mössbauer spectroscopy and X-ray absorption near-edge spectroscopy (XANES) have helped scientists identify the mechanism through which these catalysts work. High-resolution imaging (STEM) has confirmed that $Fe–N_4$ structures at the edges of the graphene play an important role in oxygen uptake and breaking its bonds, improving ORR activity [59].

Table 1 ORR performance of single-atom catalysts (SACs) supported by graphene

Graphene-based SACs	Eonset, V versus RHE	$E_{1/2}$, V versus RHE	Electron transfer number	Durability	Electrolyte	Refs.
Fe@NG-750	1.03	0.90	~4.0	1.6% loss in 25,000 s	0.1 mol/L KOH	[46]
Fe@NG-750	0.83	0.72	NA	< 9% loss in 14 h	0.5 mol/L H_2SO_4	[46]
Fe–N_x ISAs/GHS	1.05	0.87	~4.0	27.2% loss in 12,000 s	0.1 mol/L KOH	[56]
Fe_1–N–NG/rGO	0.96	0.84	~4.0	<5 mV loss in $E_{1/2}$ after 15 000 cycles	0.1 mol/L $HClO_4$	[52]
FeSA-G	0.950	0.804	3.8–3.9	20 mV loss in $E_{1/2}$ after 5000 cycles	0.1 mol/L $HClO_4$	[55]
SA-Fe/NG	NA	0.88	3.83–3.97	9 mV loss in $E_{1/2}$ after 5000 cycles	0.1 mol/L KOH	[53]
SA-Fe/NG	0.9	0.8	~3.94	~8 mV loss in $E_{1/2}$ after 5000 cycles	0.5 mol/L H_2SO_4	[53]
Fe–N–C/rGO	0.94	0.81	3.8–4.0	7.2% loss in 13,000 s	0.1 mol/L KOH	[54]
Co@NG-750	0.97	0.87	~4.0	6.5% loss in 25,000 s	0.1 mol/L KOH	[45]
Co@NG-750	0.79	0.69	NA	<1% loss in 50,000 s	0.5 mol/L H_2SO_4	[45]
Co–N/GS	0.89	NA	3.4	6% loss in 7200 s	0.1 mol/L KOH	[43]
0.7%Cu@NG-750	~0.94	0.84	4.0	4.0% loss in 25,000 s	0.1 mol/L KOH	[47]
0.7%Cu@NG-750	NA	0.551	NA	0.7% loss in 13 h	0.1 mol/L $HClO_4$	[47]
Cu_1/N-Graphene	0.869	0.779	~3.8	NA	0.1 mol/L KOH	[44]
Mn@NG	0.95	0.82	~3.86	~12.7% loss in 14 h	0.1 mol/L KOH	[48]
Sc@NG-750	0.99	0.89	~3.9	<2% loss in 50,000 s	0.1 mol/L KOH	[49]
Sc@NG-750	0.82	0.72	~4.0	5.8% loss in 50,000 s	0.5 mol/L H_2SO_4	[49]
Rh@NG	NA	0.848	3.82–4.0	12.5% loss in 25,000 s	0.1 mol/L KOH	[50]

(continued)

Table 1 (continued)

Graphene-based SACs	*E*onset, V versus RHE	$E_{1/2}$, V versus RHE	Electron transfer number	Durability	Electrolyte	Refs.
Rh@NG	0.85	0.74	~4.0	3.9% loss in 25,000 s	0.1 mol/L $HClO_4$	[50]
0.4-Ru@NG-750	0.945	0.826	3.8	<5.7% loss in 25,000 s	0.1 mol/L KOH	[51]
0.4-Ru@NG-750	0.893	0.723	~4.0	~22% loss in 50,000 s	0.1 mol/L $HClO_4$	[51]
Ru–N/G-750	0.89	0.75	~3.9	10% loss in 10,000 s	0.1 mol/L $HClO_4$	[42]

To determine how morphology affects the performance of graphene-supported SACs, scientists used DFT calculations to construct predictive models. Xu et al. built a special measure (φ) that considers factors such as the way atoms are bonded, their electronegativity, and how many valence electrons they have. Based on their work, this measure is closely related to ORR onset potentials and adsorption energies. Based on this, they have discovered materials like Au-SV-C_3, Tc-pyrrole-N_4, Fe-pyridine-N_4, and Fe-pyrrole-N_4 to be promising for ORR applications (Fig. 2) [60].

Another study by Xiao et al. demonstrated that Ir-based SACs enable ORR at a higher rate than those synthesized with Mn, Fe, or Co [61]. This is because iridium forms stronger bonds with reaction intermediates. Researchers also found that nickel (Ni) atoms, placed in a pentagon–octagon–pentagon graphene structure, had the lowest ORR overpotential among six tested metals. This suggests nickel could be a strong candidate for improving catalyst performance. These discoveries are helping

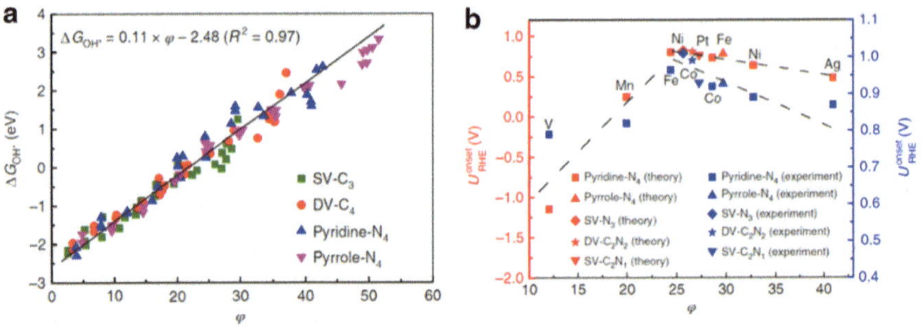

Fig. 2 **a** Adsorption free energy of OH* (ΔG OH*) versus the descriptor φ. **b** Theoretical and corresponding experimental *E*onset for ORR versus the descriptor φ for single metal atoms supported on graphene. Four coordination environments were considered, including single vacancy with three atoms (SV-C3), double vacancy with four carbon atoms (DV-C4), four pyridine N atoms (pyridine-N4), and four pyrrole N atoms (pyrrole-N4) [60]. Reproduce with Open Access @ (2018) Springer Nature

scientists develop better graphene-based ORR catalysts, paving the way for more efficient and affordable fuel cell technology [62].

4.3 Hydrogen Storage Materials

Graphene oxide (GO)-hybrid films are showing immense potential as hydrogen storage media. Hydrogen storage is a big thing in fuel cells because the manner in which we release and store hydrogen affects how efficiently these energy systems work. Because of its enormous surface area, tunable chemical nature, and porousness, GO is a wonderful material to boost hydrogen release and storage. One of the primary advantages of GO-based hybrid films is that they possess vast surface area, which offers ample space for hydrogen molecules to stick. GO's oxygen-functional groups also cause hydrogen to stick better, so it's more effective for storage. Also, because the films are porous, hydrogen can freely enter and exit, which is important for enabling fuel cells to run smoothly [63].

Scientists have been mixing GO with other materials to increase its hydrogen storage capacity. Metal–organic frameworks (MOFs), which have highly ordered pores, provide extra binding sites for hydrogen when mixed with GO. Palladium and platinum metal nanoparticles act as catalysts, enabling hydrogen to bind at lower pressures, thereby making storage systems more efficient and easier to mass-produce [64].

GO's chemistry itself can be also tuned so that it could be made more reactive with hydrogen. Nitrogen or boron could be doped and that has been reported to enhance the storage of hydrogen as they react with hydrogen molecules to form denser bonds. The chemical tuning also renders the GO-based material stable and recoverable for utilizations. The second significant property in hydrogen storage is how the material handles heat. GO-based hybrid films are highly thermally stable, i.e., they can release stored hydrogen with high efficiency without any degradation. This is essential for fuel cells because controlled hydrogen release will yield a stable power supply [65].

Evidently, issues must also be resolved before hybrid GO films may be used everywhere. One of them is that they don't hold as much hydrogen as under normal conditions would be needed. Researchers are looking into ways of solutions like layering GO sheets, adding new ingredients, and combining GO with other nanomaterials so storage capacity could be enhanced. Researchers are also working on lower and more feasible means of producing the films to enable their usage on a commercial basis. GO-based hybrid films are much potential for hydrogen storage. With their large surface area, tunable chemistry, and rapid hydrogen absorption and desorption, they are good contenders for use in fuel cells. With the ongoing research, researchers are continuing to make steady progress, bringing us closer to the day when hydrogen energy will be more viable and prevalent [66–70].

4.4 Protective Coatings and Stability Enhancement

Graphene oxide (GO)-composite films are increasingly taking center stage to increase the longevity of fuel cells and enhance their performance. With increasing demands for clean energy, fuel cells are becoming a vital solution. However, the challenge is massive, with crucial fuel cell components weakening over time as a result of corrosion and exposure to adverse environments. For this purpose, scientists are seeking the help of GO-based films as protective coatings to enhance performance and durability [71].

GO-based composite films possess the best possible combination of chemical stability and strength for which they are a suitable choice to employ in fuel cells. GO-based composite films possess the ability to act as an armor shield that protects vital components like the catalyst layer and bipolar plates from material corrosion damage. It is of utmost importance in proton-exchange membrane fuel cells (PEMFCs) and solid oxide fuel cells (SOFCs). Fuel cells are exposed to aggressive chemicals for an extended period of time, which corrode material and decrease service life [72, 73].

Chemical stability is one of the biggest strengths of GO. Hydrophobic nature and oxygen functionalities of GO repel corrosion and avoid penetration of water, eliminating water corrosion. Researchers have found that GO layers on fuel cell materials made of metal were capable of blocking corrosion and enhancing service life of fuel cells. GO-based films also toughen fuel cell components. They are strong and flexible and hence do not allow materials to be weakened by heat and mechanical stresses. Fuel cells are subject to alternating charge and discharge cycles, and this could lead to surface wear and cracking. The application of GO protective layers averts this problem and maintains the system functioning as intended. Scientists are also covalently binding GO with other nanomaterials such as metal oxides and polymers to enhance the protective ability of GO. GO–metal oxide composites enhance corrosion resistance against acid, and such films are increasingly suitable for application in fuel cells. GO-polymers integrate flexibility and longevity, keeping the fuel cells stable over time[74–77].

Albeit these benefits, there are always issues. One of them is the deposition of uniform, flaw-free coatings onto complex fuel cell components. Improved manufacturing techniques such as spray-coating, chemical vapor deposition, and electrochemical deposition are being developed by scientists to improve the quality of coating. Another issue is how to produce GO economically so that it can be implemented in large-scale fuel cell systems. GO composite films are a breakthrough for fuel cells. They protect against wear and corrosion, allowing fuel cells to last longer and function optimally. Improving technology in the future is likely to come up with even more efficient GO-based coatings based on new blends of materials and improved manufacturing practices, paving the way for longer-lasting, durable fuel cell technology [78].

5 Challenges and Future Prospects

Fuel cells are becoming more of the favorite alternative source of energy because they are efficient and eco-friendly. Of all the materials that have been experimented with in fuel cells, graphene oxide (GO)-based hybrid films are among the most promising materials. They possess good electrical conductivity, high surface area, as well as strength. There are challenges, however, that must be addressed so that they can be utilized on a large scale [31].

One of the major issues is stability. GO films deteriorate with time upon exposure to water, heat, and chemical reactions, lowering their performance. Moreover, oxygen groups within GO at times hinder proton conduction, thus decreasing the efficiency of fuel cells. The issue is how to leave GO films free but stable and strong but not free to conduct electricity [3].

The second is fabrication. It is challenging to deposit smooth, defect-free GO films under ideal control of material and process. Conductivity, robustness, and chemical stability in the high-volume production environment are hard to obtain, and the integration of GO with other materials like metal nanoparticles, polymers, or metal oxides entails high-end engineering for compatibility assurance [79].

Even with such challenges, GO-based films hold a promising future for fuel cell technology. Scientists are experimenting with new methods to stabilize and make them efficient, for instance, by introducing specific chemical treatments or combining GO with nanostructured catalysts. They are also developing improved production methods such as roll-to-roll processing and 3D printing that could streamline mass production and become cost-effective. If they pan out, GO-based films could overshadow the fuel cells of the future [80–83].

6 Conclusion

Graphene oxide hybrid films are assisting fuel cells to operate and sustain for longer periods. They have the right properties in a balance that ensures maximum efficiency and sustainability. Improved electrical conductivity is one of their greatest assets. That allows for the transfer of charge to be faster, thus the functioning of fuel cells more efficient in energy conversion. Fuel cells that function more efficiently in real usage are the outcomes. The materials are resilient and tough in themselves as well. The fuel cells must also deal with changes in temperature, water, and mechanical stress. Graphene oxide toughens up the materials so that they won't break down in the long run. That means fewer repair work and fewer expenses.

Another important advantage is enhanced catalytic performance. Fuel cells need chemical reactions to provide power, and these must be rapid. Graphene oxide films allow these to occur more easily by incorporating tiny particles of metals or specific polymers. This not only increases the production of power but reduces the utilization of expensive metals such as platinum, rendering fuel cells cheaper and greener. There are still problems. Scientists must develop ways to mass-produce these materials and store them stably over the long term. They are studying new methods of producing and integrating these materials so they will be even more suitable for fuel cell applications.

As the globe continues its transition to cleaner power, fuel cells are going to be a major player. Graphene oxide hybrid films are the major breakthrough towards their improvement. More research later can make these materials more strong, efficient, and environmentally friendly fuel cells.

References

1. Sharaf, O.Z., Orhan, M.F.: An overview of fuel cell technology: fundamentals and applications. Renew. Sustain. Energy Rev. **32**, 810–853 (2014). https://doi.org/10.1016/j.rser.2014.01.012
2. Novoselov, K.S., Fal'ko, V.I., Colombo, L., Gellert, P.R., Schwab, M.G., Kim, K.: A roadmap for graphene. Nature **490**(7419), 192–200 (2012). https://doi.org/10.1038/nature11458
3. Dreyer, D.R., Park, S., Bielawski, C.W., Ruoff, R.S.: The chemistry of graphene oxide. Chem. Soc. Rev. **39**(1), 228–240 (2010). https://doi.org/10.1039/B917103G
4. Su, H., Hu, Y.H.: Recent advances in graphene-based materials for fuel cell applications. Energy Sci. Eng. **9**(7), 958–983 (2021). https://doi.org/10.1002/ese3.833
5. Li, S., Cheng, C., Thomas, A.: Carbon-based microbial-fuel-cell electrodes: from conductive supports to active catalysts. Adv. Mater. **29**(8), 1602547 (2017). https://doi.org/10.1002/adma.201602547
6. Zhou, X., Qiao, J., Yang, L.; Zhang, J.: A review of graphene-based nanostructural materials for both catalyst supports and metal-free catalysts in PEM fuel cell oxygen reduction reactions. Adv. Energy Mater. **4**(8), 1301523 (2014). https://doi.org/10.1002/aenm.201301523
7. Yang, L., Shui, J., Du, L., Shao, Y., Liu, J., Dai, L., Hu, Z.: Carbon-based metal-free ORR electrocatalysts for fuel cells: past, present, and future. Adv. Mater. **31**(13), 1804799 (2019). https://doi.org/10.1002/adma.201804799
8. Shao, Y., Jiang, Z., Zhang, Q., Guan, J.: Progress in nonmetal-doped graphene electrocatalysts for the oxygen reduction reaction. ChemSusChem **12**(10), 2133–2146 (2019). https://doi.org/10.1002/cssc.201900060
9. Zhu, Y., Murali, S., Cai, W., Li, X., Suk, J.W., Potts, J.R., Ruoff, R.S.: Graphene and graphene oxide: synthesis, properties, and applications. Adv. Mater. **22**(35), 3906–3924 (2010). https://doi.org/10.1002/adma.201001068
10. Ambrosi, A., Chua, C.K., Bonanni, A., Pumera, M.: Electrochemistry of graphene and related materials. Chem. Rev. **114**(14), 7150–7188 (2014). https://doi.org/10.1021/cr500023c
11. Navalon, S., Dhakshinamoorthy, A., Alvaro, M., Garcia, H.: Carbocatalysis by graphene-based materials. Chem. Rev. **114**(12), 6179–6212 (2014). https://doi.org/10.1021/cr4007347
12. Ambrosi, A., Chua, C.K., Latiff, N.M., Pumera, M.: Graphene and its electrochemistry—an update. Chem. Soc. Rev. **45**(9), 2458–2493 (2016). https://doi.org/10.1039/C6CS00136J
13. Chen, D., Feng, H., Li, J.: Graphene oxide: preparation, functionalization, and electrochemical applications. Chem. Rev. **112**(11), 6027–6053 (2012). https://doi.org/10.1021/cr300115g
14. Kong, X.K., Chen, C.L., Chen, Q.W.: Doped graphene for metal-free catalysis. Chem. Soc. Rev. **43**(8), 2841–2857 (2014). https://doi.org/10.1039/C3CS60401B
15. Huang, H., Yan, M., Yang, C., et al.: Graphene nanoarchitectonics: recent advances in graphene-based electrocatalysts for hydrogen evolution reaction. Adv. Mater. **31**(48), e1903415 (2019). https://doi.org/10.1002/adma.201903415
16. Shaari, N., Kamarudin, S.K.: Graphene in electrocatalyst and proton conducting membrane in fuel cell applications: an overview. Renew. Sustain. Energy Rev. **69**, 862–870 (2017). https://doi.org/10.1016/j.rser.2016.07.044

17. Iqbal, M.Z., Rehman, A.-U., Siddique, S.: Prospects and challenges of graphene-based fuel cells. J. Energy Chem. **39**, 217–234 (2019). https://doi.org/10.1016/j.jechem.2019.02.009
18. Chen, J., Li, H., Zhao, X.: Enhanced durability of graphene oxide-based proton exchange membranes for fuel cells. J. Energy Mater. **45**(3), 567–580 (2020). https://doi.org/10.1007/s10800-020-01372-3
19. Lee, S., Kim, Y.: Proton conductivity improvements in graphene oxide composite membranes. Int. J. Hydrog. Energy **44**(15), 8002–8010 (2019). https://doi.org/10.1016/j.ijhydene.2019.01.123
20. Smith, A., Brown, P., Green, C.: Graphene oxide as a multifunctional material for fuel cells. Adv. Mater. Res. **56**(2), 210–225 (2020). https://doi.org/10.4028/www.scientific.net/AMR.56.210
21. Wang, D., Zhou, L., Xu, W.: Catalytic performance of graphene oxide-supported platinum for fuel cells. Fuel Cell Technol. J. **12**(1), 45–58 (2022). https://doi.org/10.1002/fuce.202200045
22. Zhang, R., Chen, X., Wu, J.: High-performance fuel cells with graphene oxide-modified membranes. Nano Energy **78**, 105249 (2021). https://doi.org/10.1016/j.nanoen.2020.105249
23. Liu, Y., Wang, J., Zhang, H.: Solution-processed graphene oxide-based composite films for energy applications. J. Mater. Sci. **55**(10), 4152–4164 (2020). https://doi.org/10.1007/s10853-019-04271-3
24. Kim, S.H., Lee, J.H., Park, Y.T.: Enhancing the proton conductivity of graphene oxide composite membranes for fuel cells. Electrochim. Acta **298**, 195–203 (2019). https://doi.org/10.1016/j.electacta.2018.12.132
25. Zhao, X., Sun, W., Liu, P.: Electrochemical deposition of graphene oxide-metal composites for catalytic applications. ACS Appl. Mater. Interfaces **13**(5), 6789–6798 (2021). https://doi.org/10.1021/acsami.0c21745
26. Chen, Z., Li, M., Zhou, J.: Electrodeposition of Pt/GO hybrid films with enhanced electrocatalytic performance for fuel cells. Appl. Surf. Sci. **509**, 144917 (2020). https://doi.org/10.1016/j.apsusc.2020.144917
27. Wang, H., Lin, Y., Xu, D.: Layer-by-layer assembled graphene oxide hybrid films for proton exchange membranes. J. Power Sources **503**, 230001 (2021). https://doi.org/10.1016/j.jpowsour.2021.230001
28. Tang, W., Hu, X., Liu, B.: Improving mechanical and proton conduction properties of LbL-assembled graphene oxide membranes. Int. J. Hydrog. Energy **44**(22), 11445–11454 (2019). https://doi.org/10.1016/j.ijhydene.2019.03.041
29. Singh, R., Kumar, S., Gupta, V.: In-situ polymerization of graphene oxide-based hybrid membranes for electrochemical applications. Chem. Eng. J. **429**, 131819 (2022). https://doi.org/10.1016/j.cej.2021.131819
30. Mao, H.-N., Wang, X.-G.: Use of in-situ polymerization in the preparation of graphene/polymer nanocomposites. New Carbon Mater. **35**(4), 336–343 (2020). https://doi.org/10.1016/S1872-5805(20)60018-1
31. Novoselov, K.S., Geim, A.K., Morozov, S.V., Jiang, D., Zhang, Y., Dubonos, S.V., Grigorieva, I.V., Firsov, A.A.: A roadmap for graphene. Nature **490**(7419), 192–200 (2012). https://doi.org/10.1038/nature11458
32. Xu, X., Chen, Y., Zhang, J.: Graphene-based materials in proton exchange membrane fuel cells. Prog. Mater. Sci. **90**, 1–45 (2017). https://doi.org/10.1016/j.pmatsci.2017.07.002
33. Kamarudin, M.A., Kamarudin, S.K., Masdar, M., Daud, W.R.W.: Direct methanol fuel cell: an overview. Int. J. Hydrog. Energy **38**(22), 9438–9453 (2013). https://doi.org/10.1016/j.ijhydene.2013.05.138
34. Kim, J.Y., Jang, J.H., Jung, M.H., Kim, H.: Functionalized graphene oxide/nafion composite membranes for high-performance polymer electrolyte membrane fuel cells. J. Power. Sources **274**, 823–834 (2015). https://doi.org/10.1016/j.jpowsour.2014.10.148
35. Vinayan, B.P., Nagar, R., Rajalakshmi, N., Rama Prabhu, S.: Novel platinum-cobalt alloy nanoparticles dispersed on nitrogen-doped graphene as a cathode electrocatalyst for PEMFC applications. Adv. Funct. Mater. **22**(16), 3519–3526 (2012). https://doi.org/10.1002/adfm.201102544

36. Gao, Z., Li, M., Wang, J., Wang, Z., Ma, L., Wang, B., Wu, B., Zhang, Z., Wang, X., Wu, Z., Zhang, J., Wang, J.: Pt nanocrystals grown on three-dimensional architectures made from graphene and MoS_2 nanosheets: highly efficient multifunctional electrocatalysts toward hydrogen evolution and methanol oxidation reactions. Carbon **139**, 369–377 (2018). https://doi.org/10.1016/j.carbon.2018.06.058
37. Yan, M., Jiang, Q., Zhang, T., Wang, Y., Wang, H., Wang, H., Zhang, X., Wang, W., Zhang, Y., Wang, J.: Three-dimensional low-defect carbon nanotube/nitrogen-doped graphene hybrid aerogel-supported Pt nanoparticles as efficient electrocatalysts toward the methanol oxidation reaction. J. Mater. Chem. A **6**(37), 18165–18172 (2018). https://doi.org/10.1039/C8TA05124K
38. Liang, Y., Li, Y., Wang, H., Zhou, J., Wang, J., Regier, T., Dai, H.: Co_3O_4 nanocrystals on graphene as a synergistic catalyst for oxygen reduction reaction. Nat. Mater. **10**(10), 780–786 (2011). https://doi.org/10.1038/nmat3087
39. Zhou, Y., Yen, C.H., Hu, Y.H., Wang, C., Cheng, X., Wai, C.M., Yang, J., Lin, Y.: Making ultrafine and highly-dispersive multimetallic nanoparticles in three-dimensional graphene with supercritical fluid as excellent electrocatalyst for oxygen reduction reaction. J. Mater. Chem. A **4**(47), 18628–18638 (2016). https://doi.org/10.1039/C6TA08508C
40. Li, Y., Zhou, Y., Zhu, C., Hu, Y.H., Gao, S., Liu, Q., Cheng, X., Zhang, L., Yang, J., Lin, Y.: Porous graphene doped with Fe/N/S and incorporating Fe_3O_4 nanoparticles for efficient oxygen reduction. Catal. Sci. Technol. **8**(20), 5325–5333 (2018). https://doi.org/10.1039/C8CY01328D
41. Zhang, Q., Guan, J.: Single-atom catalysts for electrocatalytic applications. Adv. Funct. Mater. **30**(31), 2000768 (2020). https://doi.org/10.1002/adfm.202000768
42. Zhang, C., Sha, J., Fei, H., Liu, M., Yazdi, S., Zhang, J., Zhong, Q., Zou, X., Zhao, N., Yu, H., Jiang, Z., Ringe, E., Yakobson, B.I., Dong, J., Chen, D., Tour, J.M., Duan, X., Hu, W., Luo, J., Ajayan, P.M.: Single-atomic ruthenium catalytic sites on nitrogen-doped graphene for oxygen reduction reaction in acidic medium. ACS Nano **11**(7), 6930–6941 (2017). https://doi.org/10.1021/acsnano.7b02148
43. Shu, Y., Miyake, K., Quflez-Bermejo, J., Zhang, X., Wang, X., Wang, H., Zhang, Y., Wang, J., Yang, J., Lin, Y.: Rational design of single atomic Co in CoN_x moieties on graphene matrix as an ultra-highly efficient active site for oxygen reduction reaction. ChemNanoMat **6**(2), 218–222 (2020). https://doi.org/10.1002/cnma.201900597
44. Mou, X., Wang, D., Liu, X., Liu, W., Cao, L., Yao, T.: XAFS study on single-atomic-site Cu_1/N-graphene catalyst for oxygen reduction reaction. Radiat. Phys. Chem. **175**, 108230 (2020). https://doi.org/10.1016/j.radphyschem.2020.108230
45. Wen, X., Bai, L., Li, M., Guan, J.: Atomically dispersed cobalt- and nitrogen-codoped graphene toward bifunctional catalysis of oxygen reduction and hydrogen evolution reactions. ACS Sustain. Chem. Eng. **7**(10), 9249–9256 (2019). https://doi.org/10.1021/acssuschemeng.9b00620
46. Bai, L., Duan, Z., Wen, X., Guan, J.: Bifunctional atomic iron-based catalyst for oxygen electrode reactions. J. Catal. **378**, 353–362 (2019). https://doi.org/10.1016/j.jcat.2019.09.001
47. Bai, L., Hou, C., Wen, X., Guan, J.: Catalysis of oxygen reduction reaction on atomically dispersed copper- and nitrogen-codoped graphene. ACS Appl. Energy Mater. **2**(7), 4755–4762 (2019). https://doi.org/10.1021/acsaem.9b00765
48. Bai, L., Duan, Z., Wen, X., Si, R., Guan, J.: Atomically dispersed manganese-based catalysts for efficient catalysis of oxygen reduction reaction. Appl. Catal. B **257**, 117930 (2019). https://doi.org/10.1016/j.apcatb.2019.117930
49. Wen, X., Duan, Z., Bai, L., Guan, J.: Atomic scandium and nitrogen-codoped graphene for oxygen reduction reaction. J. Power Sources **431**, 265–273 (2019). https://doi.org/10.1016/j.jpowsour.2019.05.084
50. Rahul, Bhattacharya, B., Singh, P.K., Singh, R., Khan, Z.H.: Perovskite sensitized solar cell using solid polymer electrolyte. Int. J. Hydrog. Energy **41**(15), 6503–6510 (2016). https://doi.org/10.1016/j.ijhydene.2015.12.093
51. Johari, R., Shambhavi, Kumar, U., Sonker, R.K., Kumar, P., Siddhartha, Singh, R., et al.: Perovskite-based gas sensors. In: Smart Nanostructure Materials and Sensor Technology, pp. 245–259. Springer Nature Singapore, Singapore (2022)

52. Sonker, R.K., Shastri, R., Johari, R.: Superficial synthesis of CdS quantum dots for an efficient perovskite-sensitized solar cell. Energy Fuels **35**(9), 8430–8435 (2021). https://doi.org/10.1021/acs.energyfuels.1c00514
53. Sonker, R.K., Sabhajeet, S.R.: ZnO nanoneedle structure-based dye-sensitized solar cell utilizing solid polymer electrolyte. Mater. Lett. **223**, 133–136 (2018). https://doi.org/10.1016/j.matlet.2018.04.104
54. Ansari, M.Z., Wahid, M.A., Johari, R., Qureshi, M.A., Siddiqi, W.A.: Novel honey-mediated green synthesis of graphene@Ag nanocomposite and its two-dimensional application in photovoltaic and anti-microbial activity. Mater. Res. Express **6**(11), 115071 (2019). https://doi.org/10.1088/2053-1591/ab3c2e
55. Sonker, R., Sabhajeet, S., Yadav, B., Johari, R.: Liquefied petroleum gas detection using SnO_2, PANI-SnO_2, and Ag-SnO_2 composite film fabricated by chemical route. Int. J. Electroact. Mater. **5**, 6–12 (2017)
56. Sonker, R.K., Shukla, P.K., Singh, P.K., Khan, Z.H.: Experimental and characterization techniques. In: Composite Materials, pp. 99–123. CRC Press (2021)
57. Sonker, R.K., Shabajeet, S.R., Johari, R., Yadav, B.: Design and growth of metal oxide film as liquefied petroleum gas sensors. In: Gas Sensors. IntechOpen (2019). https://doi.org/10.5772/intechopen.85153
58. Singh, P.K., Singh, R., Singh, V., Tomar, S.K., Bhattacharya, B., Khan, Z.H.: Effect of crystal and powder of CH_3NH_3I on the $CH_3NH_3PbI_3$ based perovskite sensitized solar cell. Mater. Res. Bull. **89**, 292–296 (2017). https://doi.org/10.1016/j.materresbull.2017.01.022
59. Ahmed, S., Parvaz, M., Johari, R., Rafat, M.J.M.E.: Studies on activated carbon derived from neem (*Azadirachta indica*) bio-waste, and its application as supercapacitor electrode. Mater. Res. Express **5**(4), 045601 (2018). https://doi.org/10.1088/2053-1591/aaac66
60. Xu, H., Cheng, D., Cao, D., Zeng, X.C.: A universal principle for a rational design of single-atom electrocatalysts. Nat. Catal. **1**(5), 339–348 (2018). https://doi.org/10.1038/s41929-018-0062-1
61. Xiao, Y., Zhang, W.: High-throughput calculation investigations on the electrocatalytic activity of codoped single metal-nitrogen embedded in graphene for ORR mechanism. Electrocatalysis **11**(4), 393–404 (2020). https://doi.org/10.1007/s12678-020-00579-z
62. Mao, X., Kour, G., Yan, C., Zhu, Z., Du, A.: Single transition metal atom-doped graphene supported on a nickel substrate: enhanced oxygen reduction reactions modulated by electron coupling. J. Phys. Chem. C **123**(6), 3703–3710 (2019). https://doi.org/10.1021/acs.jpcc.8b11213
63. Smith, J., et al.: Advances in graphene oxide for hydrogen storage. Energy Mater. J. **15**(3), 245–260 (2020)
64. Brown, T., Lee, K.: Surface functionalization of GO for enhanced hydrogen storage. Nanotechnol. Lett. **12**(2), 78–85 (2019)
65. Wang, Y., et al.: Porous graphene oxide films for efficient hydrogen adsorption. Int. J. Hydrog. Energy **46**(4), 565–580 (2021). https://doi.org/10.1016/j.ijhydene.2020.10.178
66. Kim, H., Zhao, X.: MOF-GO hybrid materials for hydrogen storage. Mater. Chem. Rev. **10**(1), 102–119 (2022)
67. Patel, R., et al.: Metal nanoparticles and graphene composites in hydrogen storage applications. J. Appl. Nanomater. **18**(6), 213–227 (2020)
68. Ahmed, S., Kumar, D.: Heteroatom doping of graphene oxide for energy applications. Adv. Mater. Res. **25**(7), 332–345 (2021)
69. Lin, C., et al.: Thermal stability of GO-based hydrogen storage films. Fuel Cell Sci. Technol. **32**(5), 432–447 (2023)
70. Martinez, F., Gomez, L.: Next-generation hydrogen storage: challenges and opportunities. Renew. Energy Adv. **14**(3), 156–173 (2022)
71. Johnson, M., et al.: Graphene oxide as a corrosion barrier in fuel cells. J. Energy Mater. **18**(4), 233–247 (2021)
72. Lee, T., Kim, H.: Protective coatings for PEMFC components: a review. Nanotechnol. Adv. **14**(2), 89–105 (2020)

73. Wang, Y., et al.: Hydrophobicity and corrosion resistance of GO-based films. Int. J. Hydrog. Energy **47**(5), 678–690 (2022). https://doi.org/10.1016/j.ijhydene.2021.10.190
74. Patel, R., et al.: Metallic component protection in fuel cells using graphene oxide. J. Appl. Nanosci. **16**(3), 312–329 (2019). https://doi.org/10.1007/s13204-018-0825-7
75. Ahmed, S., Kumar, D.: Enhancing mechanical stability of fuel cells with GO coatings. Adv. Mater. Res. **25**(7), 410–423 (2021)
76. Lin, C., et al.: Hybrid GO-metal oxide films for acidic environments. Mater. Chem. Eng. **32**(5), 150–163 (2023)
77. Martinez, F., Gomez, L.: Polymer-graphene oxide composites for energy applications. Renew. Energy Sci. Technol. **14**(3), 210–225 (2022)
78. Brown, T., et al.: Deposition techniques for GO-based protective layers in fuel cells. Surf. Coat. Technol. **30**(8), 405–418 (2021)
79. Xu, X., Chen, Y., Zhang, J.: Graphene-based materials in proton exchange membrane fuel cells. Prog. Mater. Sci. **90**, 1–45 (2017). https://doi.org/10.1016/j.pmatsci.2017.07.003
80. Kamarudin, M.A., Kamarudin, S.K., Masdar, M., Daud, W.R.W.: Direct methanol fuel cell: an overview. Int. J. Hydrog. Energy **38**(22), 9438–9453 (2013). https://doi.org/10.1016/j.ijhydene.2013.05.117
81. Kim, J.Y., Jang, J.H., Jung, M.H., Kim, H.: Functionalized graphene oxide/nafion composite membranes for high-performance polymer electrolyte membrane fuel cells. J. Power Sources **274**, 823–834 (2015). https://doi.org/10.1016/j.jpowsour.2014.10.094
82. Arof, A.M., Shamsuddin, N.K., Ramesh, R.: Advancements in graphene-based nanocomposites for fuel cell applications. Electrochim. Acta **329**, 135–150 (2019). https://doi.org/10.1016/j.electacta.2019.01.102
83. Bose, S., Kuila, T., Mishra, A.K., Kim, N.H., Lee, J.H.: Carbon-based nanostructured materials and their composites as supercapacitor electrodes. J. Mater. Chem. **22**(3), 767–784 (2012). https://doi.org/10.1039/C1JM13271K

Environmental Impact of Nanostructured Materials for Energy Storage Devices

Sanjay Kumar, Gaurav Awasthi, and Pawan Kumar

Abstract The twenty-first century has introduced major environmental challenges, rising energy demands, and a greater focus on sustainable architectural practices, all of which are crucial for the long-term preservation of resources and the improvement of overall quality of life. Smart materials are at the forefront of these advancements, offering innovative solutions for energy storage and environmental sustainability. In this chapter, we explore the environmental impact of smart materials used in energy storage devices, particularly batteries and supercapacitors. Key materials discussed include metal–organic frameworks (MOFs) for ion transport and adsorption, graphene-based materials for high-performance electrodes, phase change materials (PCMs) for thermal energy storage, and shape-memory alloys for adaptive energy systems. Furthermore, we examine critical environmental aspects such as the physicochemical nature of these materials, potential toxicity, carcinogenic organic solvents, recyclability, lifespan, and waste management. Additionally, challenges related to nuclear contamination, lack of established reliability, market fluctuations, and inconsistencies in environmental regulations are addressed. By evaluating both the benefits and potential risks of smart materials, this chapter aims to provide a comprehensive understanding of their role in advancing sustainable energy solutions.

Keywords Environment · Smart materials · Energy devices · Threats

S. Kumar · G. Awasthi · P. Kumar (✉)
Materials Application Research Laboratory, Department of Nano Sciences and Materials, Central University of Jammu, Rahya Suchani, UT of Jammu & Kashmir, India
e-mail: pawannano10@gmail.com; pawan.nsm@cujammu.ac.in

P. Kumar
Department of Chemistry and Chemical Sciences, Central University of Jammu, Rahya Suchani, UT of Jammu & Kashmir, India

R. K. Sonker et al. (eds.), *Low-Dimensional Multifunctional Hybrid Nanostructured Materials*, Engineering Materials, https://doi.org/10.1007/978-981-96-7952-2_11

1 Introduction

The simple definition of energy is that it is a dynamic indication of a country's degree of development [1]. By the end of the twenty-first century, the global population is anticipated to double [2]. As economic expansion has accelerated in emerging nations over the past few decades, so has energy demand. The global energy consumption is anticipated to rise by 54%, increasing from 404 quadrillion British thermal units (Btu) in 2001 to an estimated 623 quadrillion Btu by 2025 [3]. In the view of increasing energy demand the storage of produced energy is essential that's why the global energy storage capacity from 2010 to 2022 increased as shown in Fig. 1.

The identification and execution of economically viable and environmentally friendly energy preservation and conversion facilities is essential, as the importance and attraction of energy storage, particularly at the grid scale, grows in tandem with the increasing supply of renewable energy. Energy storage systems broadly can be classified in the five types; (i) Chemical energy storage system, (ii) Thermal energy storage system, (ii) Mechanical energy storage system, (iv) Hybrid energy storage system, and (v) Electrochemical energy storage system which are depicted in the Fig. 2 [4].

It is believed that the development of smart materials marked a turning point that has attracted a lot of attention. Devices made of smart materials are more precise and important than those made of conventional materials [5]. These increased and upgraded attributes have a significant impact on the development of smart materials devices with a variety of features like high energy density, high efficiency, catalytic property, and sensing [6].

In the near future, we anticipate a considerable rise in the use of smart materials. Hence, energy storage and conversion devices appear to have fascinating uses as

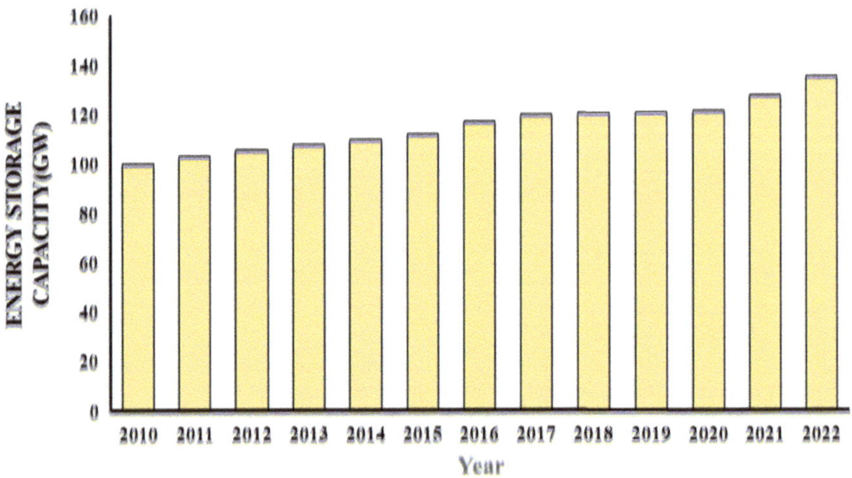

Fig. 1 Global demand of energy storage devices during last decade

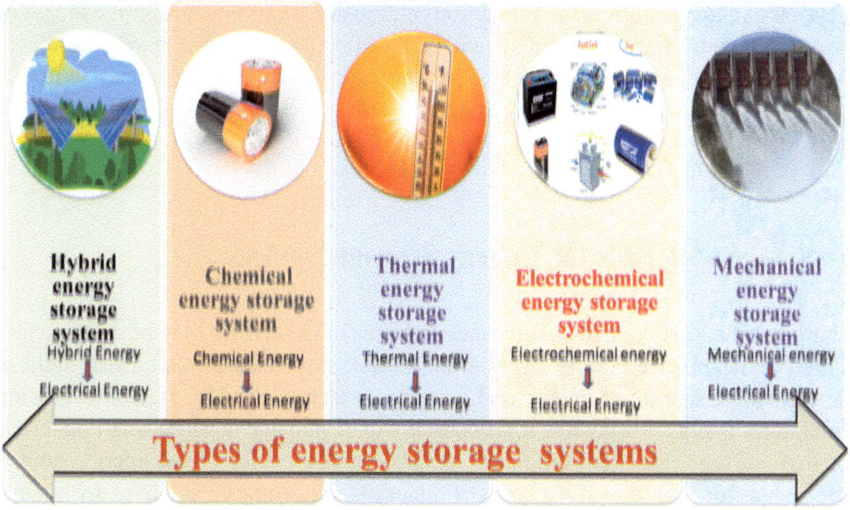

Fig. 2 Types of energy storage systems

novel power source categories. Supercapacitors, Lithium-ion batteries (LiBs), solar cells, and other high-performance techniques for storing electricity have experienced considerable breakthroughs in recent years [7, 8]. Before mass manufacturing can begin, a number of studies are required. Nevertheless, environmental considerations may be incorporated into the planning process at an early stage. Decisions made in the early stages of design are thought to have an impact on more than 80% of a product's environmental impact [9]. It becomes sense to assume that an earlier assessment of environmental issues has a comparable effect on developing new technology. Applying the constitution alone is no longer sufficient to address environmental concerns in today's society.

The advancement of effective storage systems is crucial for the effective execution of the energy revolution. The steady rise in the world's energy needs has a major impact on the socioeconomic progress and wellbeing of people. India, the world's first most populated nation, is experiencing fast urbanization, economic growth, and development; as a result, conventional fossil fuel resources and energy storage techniques cannot meet all of the country's energy needs. In order to develop alternative energy conversion, a variety of renewable energy sources, including hydropower, geothermal, solar power, tidal power, wind power, and smart materials for energy storage, have been investigated. Batteries, fuel cells, supercapacitors, and other energy storage technologies using intelligent materials are examples of alternative energy storage devices.

This chapter aims to talk about new smart materials used in energy storage devices and how they affect the environment. It examines the effects on people and the environment of various materials and processing methods. In the present chapter, we first study the smart materials used for energy storage devices like supercapacitors,

LiBs, nanomaterial, ferroelectric materials etc. In next section, we describe the types of environmental impact of smart materials. After that, we have given the recent progress and techniques to reduce environmental impact of smart materials to for energy storage devices. At last, we conclude with some remarks and future research directions.

2 Smart Materials for Energy Storage Devices

Smart materials are created as substances whose characteristics may be dramatically altered under precise control by external stimuli, for instance, an electric field in the context of storage of energy and conversion technology [5]. Due to increasing demand for energy worldwide, it is necessary to design and develop smart materials for high storage, excellent efficiency, and better life cycles of the energy storage devices. In this regard, various types of materials are synthesized like phase change materials, supercapacitors, LiBs, ferroelectric materials and nanomaterials.

2.1 Phase Change Materials (PCMs)

Phase Change Materials (PCMs) are the materials utilized for Latent Heat Energy Storage (LHES) [10]. PCMs are a class of materials with an inherent capacity for absorption. During phase transition cycles, heat is produced and released, causing charging and discharging [11]. PCMs constitute one of the finest efficient alternatives for enhancing the use of thermal energy obtained from sources that are renewable. Current literature indicates that a significant focus of research efforts aimed at enhancing thermal energy storage is centered on the advancement of phase change materials [12]. The thermal conductivity, thermal energy storage density, and melting point of inorganic, organic, and eutectic PCMs are the main factors in choosing different thermal energy storage materials appropriate for a wider working temperature range [13].

PCMs for LHES can be categorized (Fig. 3) based on the phase transition temperature and phase change mechanism. In the phase change phenomenon LHES systems are classified into sub-categories according to solid–gas, liquid–gas, solid–liquid, and solid–solid phase changes and vice versa [14]. The significant volume requires, volumetric shifts associated with phase transitions, and increased pressure conditions found in systems such as the water–steam system make them commercially impracticable for large-scale thermal energy storage (TES). TES systems reliant on liquid–gas or solid–gas phase transitions are unfeasible. In solid–solid phase change materials, thermal energy storage occurs due to the heat associated with the material's transition between crystalline forms. The thermal energy associated with phase transitions in solid PCMs is less than that observed in solid–liquid PCMs [15]. Polyalcohol [16], polyurethanes [17], and cross-linked polyethylene [14], along with other polymers, have been extensively researched for thermal energy storage (TES). A significant number of phase change materials (PCMs) used in TES are solid–liquid phase transition materials, categorized based on their composition as organic, inorganic, or eutectic. As illustrated in Fig. 4, PCMs are further classified based on their phase transition temperature into low-, intermediate-, and high-temperature categories.

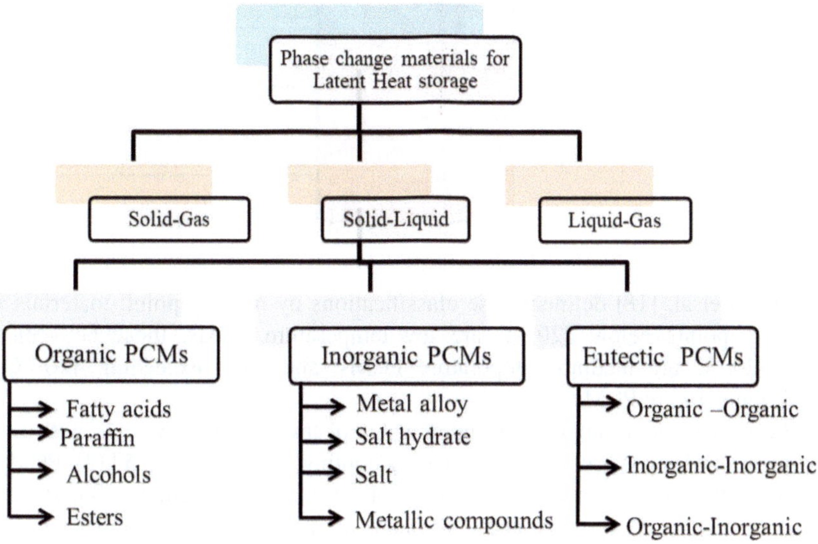

Fig. 3 Classification of phase change materials based on phase change mechanism for latent heat energy storage

Fig. 4 Classification of phase change materials based on phase transition temperature

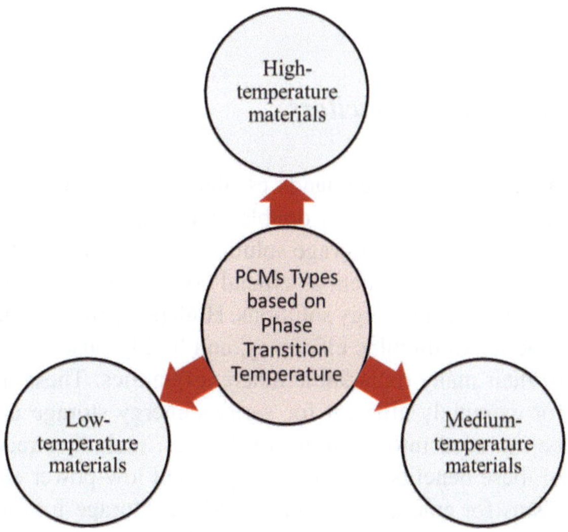

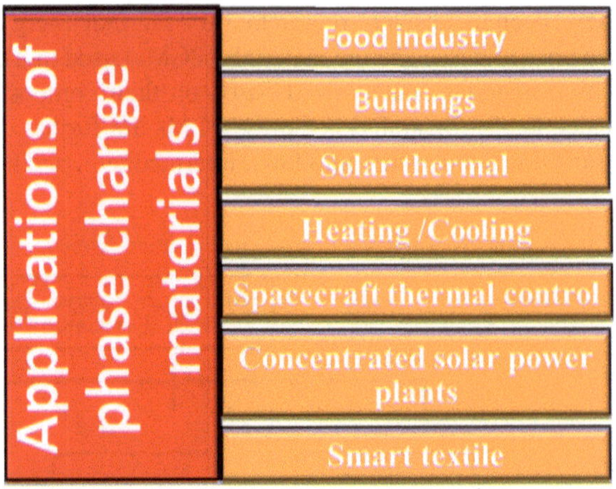

Fig. 5 Show various applications of phase change materials

Hoshi et al. [18] defined these classifications by melting point: materials with melting points below 220 °C are low-temperature PCMs, those between 220 and 420 °C are medium-temperature PCMs, and those exceeding 420 °C are high-temperature PCMs.

PCMs serve as a component or integral part of cascaded TES systems, with diverse applications for passive and active heating/cooling, as seen in Fig. 5 [13]. PCMs are hired on the selection criteria's discussion of the system requirements. Smart thermal grid, compact TES systems, and CSP plants are just a few of the numerous uses of PCMs that are now at the research and development stage.

TES systems can effectively capture excess energy and address the disparity between demand and supply by delivering energy during periods of peak consumption when energy generation surpasses usage [19].

2.2 *Supercapacitors*

As global energy demands escalate, supercapacitors—commonly known as ultracapacitors or electrical double-layer capacitors—are becoming an essential technology for energy storage solutions. Their capacity for efficient energy storage and release positions them as crucial components in addressing the increasing demand for sustainable energy solutions. High power density, fast charge and discharge rates, superior Coulombic efficiency, and highly raised service existences are only a few of their many remarkable inherent qualities. These qualities render supercapacitor for extremely efficient for various energy storage uses [20]. Supercapacitors will be included into a variety of devices, including regenerative braking, as a result of these benefits. Systems independent low-power electronics, high-power mechanisms for emergencies, and grid-level storage for electricity [21]. The emergence of a double layer of electricity at the electrodes that are positively and negatively charged underlies the charge storage process of supercapacitors, which makes use of substantial surface areas (SAs) frequently offered by carbon substances [22]. Compared to normal capacitors, supercapacitors have a larger capacitance because they use electrode materials with a large surface area and thin dielectrics [23].

Supercapacitors are classified in three categories according to the electrode materials, i.e., electrochemical double-layer supercapacitors (ECDLS), pseudosupercapacitors, and hybrid supercapacitors [24]. The ECDLS, further sub-classified as carbon nanotubes, activated carbon, and carbon aerogels. Pseudocapacitors as metal oxides and conducting polymers, whereas hybrid supercapacitors are divided in asymmetric hybrid, composite hybrid and battery-type hybrid supercapacitors [25].

For electric vehicles and digital communication equipment, electrochemical double-layer supercapacitors (ECDLS) have been regarded as a possible high-power energy source. When compared to contemporary secondary batteries, the ECDLS has the advantage of a longer cycle life and better rate capability. ECDLS utilizes the double layer that develops at the interface between the electrode and electrolyte. Here, ions of opposite charges are arranged on the electrolyte side, while electric charges accumulate on the surfaces of the electrodes. In order to allow for rapid ionic mobility and electrolyte wetting, Electrode materials for ECDLS must exhibit an extensive surface area to facilitate charge accumulation. Currently, the ECDLS electrode materials are either molecular-sieving carbons or activated carbons [25].

Between a battery and an electrochemical double-layer supercapacitor is a pseudocapacitor, or supercapacitor. Additionally, it has two electrodes that are divided by an electrolyte. The electrode experiences a chemical reaction, while the electrical charge is maintained electrostatically, ensuring no direct contact occurs across an electrode and the charged ions. One source of pseudocapacitance is an ion that has been adsorbed and desolvated, which results in a transfer of electron charges from the electrode to the electrolyte. There is one electron per charge unit [26]. Hybrid electrode arrangements, which combine two separate electrodes composed of various materials, have a lot of promise. Composite electrodes are electrodes that have one type of material incorporated into another.

2.3 Lithium-Ion Batteries (LiBs)

One of the most significant battery innovations has been the development of LiBs. In general, lithium-intercalation chemicals are the foundation of LiB technology [27]. Lithium ions move through the electrolyte in the LiBs, which is situated between the anode and cathode [28]. Lithium ions are easily diffused into delighted cathode and discharged from the anode throughout the discharge process, which is connected to the reduction and oxidation of two electrodes, respectively [29].

Due to the inexpensive price and widespread availability of carbon, graphite-based materials are used as anodes in LiBs. Because of its stability to support the lithium insertion, graphite is frequently used in commercial LiBs. Because LiBs have a low coefficient of thermal expansion, they are stable and can continue functioning as intended even after many lengthy discharge and recharge cycles. Recent advancements have focused on enhancing graphite-based anode materials, leading to novel anode materials such as silicon, alloys, and metal oxides. A complex litigated molecule makes up cathodes, especially various lithium metal oxides as $LiMn_2O_4$, $LiCoO_2$, and $LiFePO_4$ [30].

Liquid electrolytes and solid-state/semisolid electrolytes are the two categories into which the electrolytes in LiBs are separated. Lithium salts, such as $LiBF_4$, $LiN(CF_3SO_2)_2$, $LiPF_6$, and LiBOB, are solubilized in organic carbonates, including carbonates derived from propylene, ethylene, dimethyl, ethyl methyl, and their various variants, for producing the liquid electrolytes [31]. The high-molecular-weight polymer matrices (such as poly (ethylene oxide), polyvinylidene fluoride, and polyvinylidene fluoride-hexafluoropropylene) and lithium salts as the conducting salts make up the solid-state semisolid electrolytes [32].

2.4 *Ferroelectric Materials*

Relaxor ferroelectrics exhibit narrow hysteresis loops and minimal remnant polarization. These characteristics enable the attainment of elevated saturated polarization and outstanding energy conversion efficiency. Due to these benefits, they are gaining traction in energy storage applications, providing elevated discharge energy densities and rapid discharge rates [33]. Ferroelectric polymers also a great deal of potential use in dielectric energy storage applications due to their high dielectric constant and electrical displacement. The most recognized and commonly utilized ferroelectric materials for energy storage include multilayer-structured ferroelectric polymer nanocomposites, nanofillers such as lead zirconate titanate, and polymeric ferroelectric materials like nylon, polylactic glycolic acid, polylactic acid, and poly (vinylidene fluoride) along with its copolymers [34].

2.5 *Nanomaterials*

The decrease of electrode material dimensions and the availability of effective methods to produce diverse surface coatings and functionalization layers on electrode materials are important applications of nanotechnology in energy storage devices [35].

Several important benefits abound from one-dimensional nanomaterials including nanowires, nanofibers, nanorods, nanobelts, and nanotubes. In applications of energy storage, their greater surface areas additional surface-active sites and enhanced permeability can substantially elevate power density, energy density, and cycling performance [36]. Because of their broad potential for use in supercapacitors and rechargeable batteries, nanomaterials have attracted a lot of attention.

2.5.1 Nanomaterials for Anode

Silicon: One excellent illustration of how nanotechnology might address material challenges in battery research is the Si anode. With a substantial theoretical capacity of 4200 mAh g^{-1} when considering the mass of silicon, which exceeds that of graphite by over tenfold at 372 mAh g^{-1} silicon stands out as the leading anode material for the advancement of next-generation batteries. The significant volume expansion exceeding 300% during lithiation, coupled with the high capacity derived from its alloying mechanism—where each silicon atom can accommodate up to 4.4 lithium ions—results in compromised cyclability [37].

Li Metal Anode: Research on Li metals has recently had a rebirth thanks to advances in nanotechnology. The substantial theoretical capacity of lithium metal, measured at 3860 mAh g^{-1} due to its unique "hostless" structure, combined with its minimal electrochemical potential of 3.04 V relative to the standard hydrogen electrode, positions it as the anode of choice [38]. The nanostructuring of Li enhances its surface area, potentially resulting in increased side reactions during the initial cycles. A viable approach to passivating the lithium surface involves applying a solution-based coating of an engineered solid–electrolyte interphase [39].

2.5.2 Nanomaterials for Cathode

Sulfur is an excellent candidate for cathode material, attributed to its significantly higher specific energy of 2600 Wh kg^{-1} and its elevated specific heat capacity. With a capacity of 1675 mAh g^{-1}, this material exhibits performance that is five times superior to that of current cathodes [40]. Carbon materials are frequently utilized as hosts for confining sulfur, owing to their excellent electrical conductivity and significant pore capacity [41]. Alongside porous carbon, 2D-nanostructured graphene demonstrates significant promise owing to its extensive surface area and diverse surface chemistry. The graphene used by the Dai group was coated with carbon-black. To encapsulate sulfur particles covered with poly (ethylene glycol) [42] Coaxial carbon nanotube (CNT) structures covered with sulfur and vertically aligned CNT-sulfur composites have both been employed to enhance electrochemical performance [43].

Nanomaterials, including $Li_{3x}La_{2/3-x}TiO_3$ nanowires, are employed in composite solid electrolytes. These, in conjunction with protective nanopolymers, efficiently inhibit direct contact between fire retardants and the electrolyte, hence augmenting safety and performance in energy storage applications [44, 45]. A novel class of materials based on Prussian blue counterparts utilized as a nanomaterials for grid-scale uses. The typical formula for this class of compounds is AxPR-$(CN)_6 \cdot {}_nH_2O$, where A is an alkali cation, P is a transition-metal cation, and R-$(CN)_6$ is a hexacyanometalate anion. Through the modification of the composition, it is possible for both P and R-$(CN)_6$ sites to exhibit electrochemical activity [46].

3 Environmental Impact

The usage of more or fewer significant volumes of acids and solvents for synthesis and cleaning, as well as gases for depositing coatings of metals and materials, depends on the type of devices used for energy storage, it is generally acknowledged. The bulk of these substances pose serious risks to both human health and the environment. Researchers and scientists are paying attention to some of the environmental and health issues caused by chemical risks related to the toxicity, flammability, and carcinogenic qualities of materials utilized for energy storage technologies are listed in Table 1. Here, we'll talk about several common materials used in energy gadgets and how they affect the environment and people's health.

3.1 *Resources Extraction*

The price of lithium will rise due to the growing scarcity, and it will be possible to exploit diluted supplies with negative environmental effects [47]. Chile and Argentina are the major lithium producing hubs in South America. The extraction and utilization of lithium in this region could significantly alter the availability of freshwater and lead to contamination, resulting in detrimental impacts on local biodiversity and the health of society. Ore mining and Evaporative ponds associated with lithium extraction have an influence on the environment, biodiversity, and human health. The extraction of lithium is projected to have considerable implications for water quality, potentially leading to detrimental impacts on native biodiversity and facilitating outbreaks of cyanobacteria. Large shallow ponds walled with polyvinyl chloride (PVC) are used to create lithium carbonate, which is then evaporated and washed with sodium carbonate in salt lakes [48]. Environmental effects of evaporative lithium extraction are poorly

Table 1 List of various types of environmental hazardous caused by materials used in energy storage devices during their manufacturing, application, and disposal time

S. No.	Impact on environment and human beings
1	Acidification
2	Toxicity of cell
3	Harm to genetic materials due to oxidation stress or due to triggering allergies
4	Photochemical air pollution
5	Eutrophication
6	Global air pollution
7	Global worming
8	Ozone depletion
9	Abiotic depletion
10	Formation of particulate matter during manufacturing

understood, relative to ore mining, but must be thoroughly assessed. Research has shown that water contamination, alterations in landforms, and the introduction of sediments substantially detriment aquatic biodiversity in the Neotropics [49].

3.2 Toxicity

The PCMs based on their types have different level of toxicity, health hazards and environmental impact. The vegetables oil-based PCMs, microencapsulated lauric acid, a sustainable and biocompatible are safe, carbon-free, environmentally friendly, renewable but highly flammable [50]. Organic phase change materials derived from commercial grade paraffin wax release harmful toxins upon combustion, exhibiting non-biodegradable properties and flammability, as well as being carcinogenic in nature. There are seven significant toxins are identified in paraffin, including benzene, naphthalene, toluene, and methyl ethyl ketone. Vinyl chloride and formaldehyde are examples of volatile chemicals found in the organic PCMs made from commercial paraffin wax. Due to the presence of the carcinogenic benzene and toluene components, prolonged exposure to their vapors could be dangerous [51]. They pose an environmental risk because they are non-renewable and non-biodegradable, making their disposal a problem. Salt hydrates based PCMs have varying grade of toxicity. The salt-dust of Glauber irritates eyes and causes asthma [51], $Mg(NO_3)_2$ is harmless but non-biodegradable; $MgCl_2{\cdot}6H_2O$ is non-toxic but causes irritation to the skin and eyes; $CaCl_2{\cdot}6H_2O$ is harmful if eaten and irritates the eyes [52]. Additionally, the toxicity of fatty acids varies, and the toxicity of eutectic is contingent upon whether it is composed of organic or organic (fatty acid/paraffinic) components.

Through a variety of exposure pathways, such as ingestion, injection, skin or eye contact, and inhalation, certain metals and non-metals used in battery manufacture can provide serious health hazards. For instance, cadmium (Cd) can be absorbed through consumption and inhalation, lead (Pb) through ingestion, inhalation, and skin contact, and mercury (Hg) through ingestion, inhalation, and skin contact. These elements are known carcinogens, along with chromium (Cr) and arsenic (As). In general, heavy metals pose major risks to the environment and public health. Metals, non-metals, hard rubbers (ebonite), electrolytes, and plastics can all contribute to solid waste, emissions of greenhouse gas, hazardous gasses, suspended particles, and at different phases of battery recycling. According to an examination of the damage categories in the EI99 H/A framework, the production of LiBs has a greater impact on resource quality (39%) and human health (44%), with ecosystem quality (17%) being less affected. Inorganic pollutants including PM_{10}, SO_2, and NO_x that damage the respiratory system are the main causes of harm to human health. The use of minerals and fossil fuels comes next, which worsens the effects on the environment overall [53]. An assessment of saltwater habitats in the US found that nothing is known about cyanobacteria's toxicity in hypersaline environments [54]. Microcystin production was detectable in 85% of all species. A poison called microcystin, which is produced by different cyanobacteria, can be devastating to both people and wildlife [55].

3.3 Carcinogenic Organic Solvents

High-performance devices frequently include hazardous solvents that are either carcinogenic, poisonous to the reproductive system, or bad for the environment. Usually, these solvents are classified as non-halogenated aromatic solvents. During various phases of nanomaterial synthesis, hazardous solvents such as Dimethylformamide (DMF), N-methyl-2-pyrrolidone (NMP), and γ-Butyrolactone (GBL), among others, are employed. These solvents play a crucial role in processes such as solvent annealing, interfacial layer deposition, precursor dissolution, and additive incorporation, all aimed at improving film quality, crystallization, morphology, and ultimately the performance and stability of nanomaterials [56]. The minimal exposure thresholds of these solvents, coupled with their ability to mix with water, heighten the risk of absorption via oral intake and dermal contact, making them particularly critical in industrial applications. Given that DMSO enhances the permeability of cell membranes, this risk is made considerably more obvious.

Organic solvents utilized in devices necessitate meticulous treatment due to their intrinsic risks. Effective management commences with the classification of solvents according to their flammability, toxicity, and reactivity properties. This classification facilitates accurate risk evaluation and the execution of targeted safety policies, encompassing suitable storage, handling, and spill response strategies. Organizations must adhere to regulatory standards, such as NFPA and UL norms, while establishing comprehensive safety measures that encompass appropriate labeling, containment systems, ventilation, and personnel training to avert accidents and maintain environmental compliance [57, 58].

3.4 Recycling and Lifespan

The life cycle of the renewable energy storage devices based on smart materials results in environmental burden [59]. Energy storage systems exemplify electronic components with a limited lifespan, as detailed in Table 2. The exponential growth in the mass of used batteries and supercapacitors, PCMs and ferroelectric materials and nanomaterials which create negative environmental impact, is a challenge highlighted by the growing application of electrical energy storage devices. There is a problem of rate performance, cycle life, interface compatibility, and voltage attenuation with electrolyte which is attributed to the elevated electrolyte-to-sulfur ratio and the shuttle effect [60]. Due to the abundance of toxic metals (such as Cd, Pb, and Ni), device components (such as polyvinylidene fluoride (PVDF) binders and $LiPF_6$ electrolytes), F^- containing electrolytes, corrosive fluids (such as H_3PO_4 and

Table 2 An assortment of energy storage techniques along with their cycle efficiency and lifespan

S. No.	Energy storage method	Lifetime (years)	Cycle efficiency (%)
1	Lithium-ion batteries	10–15	65–95
2	Compressed air energy storage	20–30	60–80
3	Sodium–sulfur batteries	10–15	70–90
4	Lead–acid batteries	3–15	65–80
5	Vanadium batteries	5–20	70–80

H_2SO_4), and fire hazards from organic electrolytes, disposing of batteries is a significant challenge [61]. When compared to general garbage incineration and landfill storage, battery disassembly and recycling are expensive and inefficient methods that are favored due to the components' negative environmental impacts and numerous health issues that can be caused by short-term or long-term exposure. These concerns are not specific to batteries and could also arise with modern supercapacitors materials [62]. The life cycle of lithium-ion and vanadium flow batteries give a greater concern regarding to environmental hazards.

Recycling of energy storage devices utilizes four primary methodologies: physical recycling for mechanical separation, direct recycling to maintain material integrity, pyrometallurgical processing for high-temperature treatment, and hydrometallurgical techniques for aqueous extraction. Each method presents unique benefits: physical recycling ensures minimal energy usage, direct recycling preserves material integrity, pyrometallurgical processing efficiently manages mixed materials, and hydrometallurgical techniques facilitate selective metal extraction. Collectively, these approaches facilitate ecological transition by diminishing resource exploitation and emissions while preserving essential materials. This holistic strategy propels the expanding waste battery recycling sector, fostering sustainable technological advancement and circular economy concepts [63].

3.5 *Chemical Hazard*

The electrolyte employed in lithium-ion batteries is a flammable and volatile liquid. It can ignite if a lithium dendrite extends to the cathode, leading to a short circuit, which poses significant risks to both human safety and the environment [64]. Over the coming years, it is anticipated that the amount of e-waste we already produce over 50 million metric tons annually will expand tremendously [65]. In addition to the carcinogenic toxicity effects on humans, there are various other toxicity effects present, including those impacting freshwater, marine, and terrestrial ecosystems, as well as non-carcinogenic toxicity in humans.

3.6 Resources Consumption

Due to utilization in areas including industry, transportation, construction, and electric generating as well as an improvement in quality of life in many nations, the demand for energy has greatly expanded in recent years [66]. Therefore, the consumption of resources increased exponentially to fulfill their needs for energy. But the availability of resources is limited. That's why consumption of resources is very fast.

3.7 Acidification

The environmental effects of acidification from large-scale and small-scale lithium-ion battery manufacture are more severe. The integration of the reference scenario with the primary metals yields emissions related to acidification in the small-3.7 and Giga-3.7 models, approximately measuring 0.9 and 0.6 kg SO_2-eq/kWh, respectively. The repercussions of acidity are primarily attributable to upstream material production. The two main components that most strongly influence the effects of acidification are the active cathode material and copper foil, which is utilized as a current collector. The supply chain for nickel sulfate plays a key role in the effects of the manufacture of active cathode materials [67]. Large-scale LiBs production has been reported to produce 0.5 kg SO_2-eq/kWh by several research [68]. Other studies that focus on large-scale manufacturing only yields inventory outcomes for pollutants that contribute to acidification, such as SO_2 and NO_x. Accordingly, the consequences of acidification range between 0.8 kg SO_2-eq/kWh [69] and 1.3 kg SO_2-eq/kWh [70]. Along with the effects of acidification, other regional effects such as freshwater and marine eutrophication simultaneously change. Eutrophication emissions are influenced by upstream material processing and manufacturing energy usage.

3.8 Formation of Particulate Matter During Manufacturing

The mixture of particles and gases that makes up atmospheric PM is extremely complicated and varied. While secondary particles are created in the atmosphere as a result of gaseous emissions, primary particles are released directly from sources. Despite small size, airborne particles are not harmless. But as per estimates, 811,000 individuals worldwide pass away before their time due to exposure to high levels of particulate matter (PM) in cities, and 1.8 million are equally harmed by smoke from the use of solid fuels indoors. They are comparable to "overweight and obesity" as factors of early death when combined. Some environmental effects, such as the production of particulate matter, ground-level ozone, ionizing radiation and stratospheric ozone depletion fluctuate in tandem with emissions of global warming.

Lithium-ion (Li-ion) batteries, which are increasingly common in a variety of applications, could be vulnerable to thermal runaway under certain abusive conditions. A thermal runaway event that results in fire causes a massive discharge of gaseous and particle pollutants that pose a serious threat to public safety and human health. When lithium-ion phosphate and nickel manganese cobalt oxide are subjected to thermal heat and over charging emit the particulate matter mass (PM 2.50) [71]. Thermal runaway gives very high particle emissions. So, batteries produce particulate matter which is a serious issue regarding the health of people.

4 Significant Actions to Mitigate the Impact of Smart Materials on the Environment

The International Renewable Energy Agency (IRENA) must make an estimate of the number of raw materials that can be retrieved through recycling and reusing in addition to the recycling of energy equipment. By reusing materials and substrates, worldwide projects, businesses, and others must work toward achieving a 100% renewable and sustainable economy. Emphasis ought to be placed on the creation of innovative products and the enhancement of technology, alongside the formulation of effective strategies for the recycling and repurposing of the waste currently produced. (DMF, a hazardous solvent, and Pb, an element that is frequently used in the process of device fabrication, certain materials that remain in the final product exemplify "substances of very high concern" (SVHC), as outlined by the European Chemical Agency's Registration and Evaluation framework. The primary goal of the authorization process is the replacement of SVHC and solvents.

4.1 Waste Reduction and Recycling

Due to environmental consequences and resource exploitation costs can be decreased, recycling of lithium will be advantageous to investors, local residents, and natural resources. Recycling can't stop resource depletion; lithium technology needs to be supported with alternative energy theories. Because the demand for lithium-powered devices is constantly increasing and must therefore be sustainable, proper recycling is only required in this situation [47]. The flexible electronics industry must specifically address the following issues: (i) reducing the power consumption of electronic devices; (ii) reducing electronic waste (e-waste); and (iii) minimizing or eradicating harmful chemicals and poisonous compounds utilized in electronics production [72].

In the United States, recycling electric vehicle batteries involves the reclassification of recovered strategic metals, such as cobalt and nickel, as domestic resources. The reclassification under the Inflation Reduction Act allows battery manufacturers to be eligible for federal incentives, hence enhancing domestic battery production in the United States. Recycling diminishes dependence on foreign supply chains, so fostering a more sustainable and resilient domestic energy storage sector. This case illustrates how policy-driven recycling initiatives can improve resource security and foster circular economy practices in energy storage [73]. In the another study, the implementation of Cathode-Healing™ in battery recalls offers a scalable industrial framework for the recycling of lithium-ion batteries, encompassing those from consumer devices and electric vehicles (EVs). This procedure entails a multiphase recovery technique, commencing with electrolyte extraction by carbon dioxide, succeeded by industrial shredding, electrode retrieval, and froth flotation. The principal innovation—Cathode-Healing™ rehabilitates deteriorated cathode materials, enabling their reutilization in new battery cells in conjunction with recovered anodes. The initial complete cells constructed using directly recycled cathodes and anodes have exhibited encouraging performance, confirming the efficacy of this recycling method. Furthermore, knowledge acquired by recycling standard battery chemistries is already being utilized in formulations pertinent to electric vehicles, facilitating the advancement of more sustainable, closed-loop battery manufacturing. This concept markedly diminishes reliance on raw materials, improves battery sustainability, and fosters the circular economy in energy storage [74].

4.2 Design of Sustainable and Extended Lifespan Materials

The current energy transition scenario calls for the development of effective, secure, and cost-effective batteries. Redox-flow batteries are ideal options for economical static storage, particularly for prolonged discharges and longer storage durations, owing to their distinctive ability to decouple power from energy [75]. Research on redox-flow technology is advancing due to the incorporation of sources of clean energy and the consequent demand for energy storage.

4.3 Circular Economy Approach

In light of a circular economy, the search for more plentiful, non-toxic, and biodegradable materials should be supplemented by environmentally friendly production techniques that permit recycling and reusing [72]. The circular economy, anchored by a transition to electricity from renewable sources, materials that are sustainable, and ecologically sound technology, prioritizes sustainable energy storage. The redox-flow battery is one of the most promising substitutes for grid-scale energy storage [76]. According to the USEPA Reuse and repurposing provide sustainable alternatives to the premature disposal of lithium-ion batteries, prolonging their lifespan prior to recycling. Degraded electric vehicle batteries can be used for lower-range vehicles or utilized for renewable energy storage. These solutions eliminate waste, conserve resources, and encourage a circular economy in energy storage [77].

4.4 Environment Regulation and Standard

Lead–acid batteries exhibit constrained energy density and are inadequate for frequent cycling, necessitating the advancement and implementation of alternative storage technologies. Energy storage has emerged as an essential component in the management of renewable energy, facilitating the discharge of energy into the grid during peak demand intervals, when it is most advantageous [78]. Also, the resources of lithium is limited and extraction of the lithium form their site leads to the problems to the native of that place and increasing use of lithium leads to the extraction of lithium from the sites of low availability so to maintain the environment regulation and standard we have to use and design sustainable and eco-friendly materials for energy devices.

5 Conclusion

Over the past few decades, the pace of economic growth in emerging countries has quickened, and with it, so has energy demand. To fulfill the need of energy, demand the advance research achieved the great success by formulating and designing the smart materials for high storage and good efficiency and better life cycle of the energy storage devices. Smart materials enable more precise and significant devices than ordinary materials do. Phase change materials play a key role in thermal energy storage. Lithium-ion batteries, supercapacitors are used for the storage of electrochemical energy and have good storage efficiency and life cycle. Ferroelectric materials and nanomaterials-based electrode and electrolyte enhanced the storage efficiency and life of these devices.

This chapter has provided a broad overview of the environmental impacts associated with the use of smart materials in energy storage devices. While smart materials offer promising solutions for addressing the growing demand for sustainable energy, it is crucial to carefully consider their potential environmental consequences throughout their entire lifecycle. Several key environmental concerns have been highlighted:

- **Resource Depletion**: As discussed in Sect. 2, the extraction of raw materials for many smart materials, such as lithium, cobalt, nickel, and rare earth elements, can lead to habitat destruction, water pollution, and social conflicts. This is particularly relevant for LiBs and certain types of supercapacitors.
- **Toxicity**: Some smart materials, as noted in Sect. 2, contain toxic elements (e.g., heavy metals in LiB cathodes) or require the use of hazardous chemicals during manufacturing (e.g., organic solvents in supercapacitors and LiB electrolytes), posing risks to human health and the environment.
- **Manufacturing Impacts**: The production of smart materials can contribute to acidification and eutrophication through emissions and wastewater discharge. The

use of strong acids and bases in the production of activated carbon for supercapacitors, as well as the energy-intensive processes involved in manufacturing PCMs and LiB components, can have significant environmental impacts.
- **Waste Management**: The end-of-life disposal of smart materials presents significant challenges due to their complex composition and the presence of hazardous substances. The low recycling rates for LiBs, as well as the lack of effective recycling technologies for many other smart materials, lead to the accumulation of battery waste in landfills, posing a long-term environmental risk.

Addressing these challenges requires a multifaceted approach that prioritizes sustainability at every stage of the smart material lifecycle. Several actionable steps can be taken:

- **Sustainable Design**: Integrating environmental considerations into the early stages of smart material design is crucial. This includes selecting materials with lower environmental footprints (e.g., bio-based PCMs, biomass-derived carbon for supercapacitors), optimizing manufacturing processes to reduce waste and energy consumption, and designing for recyclability and reuse.
- **Responsible Sourcing**: Ensuring that raw materials are sourced responsibly, with minimal environmental and social impacts, is essential. This can involve supporting ethical mining practices, promoting the use of recycled materials, and diversifying supply chains to reduce reliance on conflict-affected regions.
- **Improved Recycling Technologies**: Investing in the development of more efficient and cost-effective recycling technologies for smart materials is critical. This will reduce the reliance on virgin materials, minimize the amount of hazardous waste sent to landfills, and recover valuable resources from end-of-life products.
- **Extended Producer Responsibility (EPR)**: Implementing EPR schemes that hold manufacturers responsible for the end-of-life management of their products can incentivize sustainable design and recycling practices. This can involve setting collection targets, providing financial incentives for recycling, and promoting consumer awareness of recycling programs.

References

1. Berktay, A., Demirbaş, A., Koçak, S., Nas, B.: Electrical energy prices and losses respect to Turkish social-economic situations. Energy Explor. Exploit. **22**(3), 195–206 (2004). https://doi.org/10.1260/0144-5987.22.3.195
2. Demirbaş, A.: Energy conservation and storage systems. Energy Explor. Exploit. **20**(5), 391–399 (2002). https://doi.org/10.1260/014459802760170411
3. Conti, J., Holtberg, P., Diefenderfer, J., LaRose, A., Turnure, J.T., Westfall, L.: International Energy Outlook 2016 with Projections to 2040; DOE/EIA-0484(2016); U.S. Department of Energy, Energy Information Administration, Washington, DC (2016)
4. Dehghani-Sanij, A.R., Tharumalingam, E., Dusseault, M.B., Fraser, R.: Study of energy storage systems and environmental challenges of batteries. Renew. Sustain. Energy Rev. **104**, 192–208 (2019). https://doi.org/10.1016/j.rser.2019.01.023

5. Julien, C.M., Mauger, A., Vijh, A., Zaghib, K.: Smart materials for energy storage in Li-ion batteries. AIMS Mater. Sci. **3**(1), 137–148 (2016). https://doi.org/10.3934/matersci.2016.1.137
6. Shinde, S.J., Waikar, M.R., Sonker, R.K., Sonkawade, R.G.: Optical sensors based on polymeric materials. In: Advanced Functional Materials for Optical and Hazardous Sensing: Synthesis and Applications, pp. 221–251. Springer Nature Singapore, Singapore (2023). https://doi.org/10.1007/978-981-19-9356-6_9
7. Gurav, S.R., Waikar, M.R., Rasal, A.S., Sonker, R.K., Sonkawade, R.G.: Current development and challenges in textile-based flexible supercapacitors. In: Smart and Flexible Energy Devices, pp. 539–550. CRC Press (2022). https://doi.org/10.1201/9781003205286-22
8. Johari, R., Sonker, R.K., Victor, O., Khan, Z.H., Aggarwal, D., Gupta, S., Kumar, S.: Optoelectronic study of polymer electrolyte incorporated perovskite sensitized solar cell. Macromol. Symp. **407**(1), 2200126 (2023). https://doi.org/10.1002/masy.202200126
9. Díaz-Ramírez, M.C., Calderón, A.J., Vázquez-Rowe, I., Quispe, I., Torres, D.: Environmental assessment of electrochemical energy storage device manufacturing to identify drivers for attaining goals of sustainable materials 4.0. Sustainability **12**(1), 342 (2020). https://doi.org/10.3390/su12010342
10. Barnes, F.S., Levine, J.: Large Energy Storage Systems. Taylor & Francis Group, New York (2011). https://doi.org/10.1201/b10894
11. Giro-Paloma, J., Martínez, M., Cabeza, L.F., Fernández, A.I.: Types, methods, techniques, and applications for microencapsulated phase change materials (MPCM): a review. Renew. Sustain. Energy Rev. **53**, 1059–1075 (2016). https://doi.org/10.1016/j.rser.2015.09.040
12. Zhao, B.-C., Yang, Y.-P., Wang, C., Wang, J.-Q., Chen, X.-L.: Cyclic thermal characterization of a molten-salt packed-bed thermal energy storage for concentrating solar power. Appl. Energy **195**, 761–773 (2017). https://doi.org/10.1016/j.apenergy.2017.03.060
13. Sharma, A., Tyagi, V.V., Chen, C.R., Buddhi, D.: Review on thermal energy storage with phase change materials and applications. Renew. Sustain. Energy Rev. **13**(2), 318–345 (2009). https://doi.org/10.1016/j.rser.2007.10.005
14. Aldoss, T.K., Rahman, M.M.: Comparison between the single-PCM and multi-PCM thermal energy storage design. Energy Convers. Manag. **83**, 79–87 (2014). https://doi.org/10.1016/j.enconman.2014.03.049
15. Chandra, D., Chellappa, R., Chien, W.-M.: Thermodynamic assessment of binary solid-state thermal storage materials. J. Phys. Chem. Solids **66**(2–4), 235–240 (2005). https://doi.org/10.1016/j.jpcs.2004.07.019
16. Wu, C.-B., Zhao, C.-Y., Wang, W.-Q., Zhang, G.-H., Wang, X.-L., Wang, J.-L.: Preparation of microencapsulated medium temperature phase change material of tris(hydroxymethyl) methyl aminomethane@SiO_2 with excellent cycling performance. Appl. Energy **154**, 361–368 (2015). https://doi.org/10.1016/j.apenergy.2015.05.019
17. Cao, Q., Liu, P.: Hyperbranched polyurethane as novel solid-solid phase change material for thermal energy storage. Eur. Polym. J. **42**(11), 2931–2939 (2006). https://doi.org/10.1016/j.eurpolymj.2006.07.020
18. Hoshi, A., Mills, D.R., Bittar, A., Saitoh, T.: Screening of high melting point phase change materials (PCM) in solar thermal concentrating technology based on CLFR. Sol. Energy **79**(3), 332–339 (2005). https://doi.org/10.1016/j.solener.2004.02.012
19. Yang, Z., Garimella, S.V.: Cyclic operation of molten-salt thermal energy storage in thermoclines for solar power plants. Appl. Energy **103**, 256–265 (2013). https://doi.org/10.1016/j.apenergy.2012.09.043
20. Simon, P., Gogotsi, Y.: Charge storage mechanism in nanoporous carbons and its consequence for electrical double layer capacitors. Philos. Trans. R. Soc. A **368**(1923), 3457–3467 (2010). https://doi.org/10.1098/rsta.2010.0109
21. Lu, M.: Supercapacitors: Materials, Systems, and Applications. Wiley-VCH, Weinheim, Germany (2013). https://doi.org/10.1002/9783527646661
22. Conway, B.E.: Electrochemical Supercapacitors: Scientific Fundamentals and Technological Applications. Springer, Boston, MA (2013). https://doi.org/10.1007/978-1-4757-3058-6

23. Burke, A.: Ultracapacitors: why, how, and where is the technology. J. Power. Sources **91**(1), 37–50 (2000). https://doi.org/10.1016/S0378-7753(00)00485-7
24. Song, B., et al.: Systematic study on structural and electronic properties of diamine/triamine functionalized graphene networks for supercapacitor application. Nano Energy **31**, 183–193 (2017). https://doi.org/10.1016/j.nanoen.2016.11.026
25. Afif, A., et al.: Advanced materials and technologies for hybrid supercapacitors for energy storage—a review. J. Energy Storage **25**, 100852 (2019). https://doi.org/10.1016/j.est.2019.100852
26. Garthwaite, J.: How ultracapacitors work (and why they fall short). Earth2Tech, GigaOM Network (2011). [Online]. https://gigaom.com/2011/01/24/how-ultracapacitors-work-and-why-they-fall-short/
27. Vincent, C.A.: Lithium batteries: a 50-year perspective, 1959–2009. Solid State Ion. **134**(1–2), 159–167 (2000). https://doi.org/10.1016/S0167-2738(00)00347-5
28. Dunn, B., Kamath, H., Tarascon, J.-M.: Electrical energy storage for the grid: a battery of choices. Science **334**(6058), 928–935 (2011). https://doi.org/10.1126/science.1212741
29. Hannan, M.A., et al.: A review of lithium-ion battery state of charge estimation and management system in electric vehicle applications: challenges and recommendations. Renew. Sustain. Energy Rev. **78**, 834–854 (2017). https://doi.org/10.1016/j.rser.2017.05.001
30. Chen, T., Jin, Y., Lv, H., et al.: Applications of lithium-ion batteries in grid-scale energy storage systems. Trans. Tianjin Univ. **26**, 208–217 (2020). https://doi.org/10.1007/s12209-020-00242-7
31. Sasaki, Y.: Organic electrolytes of secondary lithium batteries. Electrochemistry **76**(1), 2–15 (2008). https://doi.org/10.5796/electrochemistry.76.2
32. Meyer, W.H.: Polymer electrolytes for lithium-ion batteries. Adv. Mater. **10**(6), 439–448 (1998). https://doi.org/10.1002/(SICI)1521-4095(199805)10:6%3c439::AID-ADMA439%3e3.0.CO;2-K
33. Zhao, P., et al.: High-performance relaxor ferroelectric materials for energy storage applications. Adv. Energy Mater. **9**(17), 1803048 (2019). https://doi.org/10.1002/aenm.201803048
34. Jiang, Y., et al.: Ferroelectric polymers and their nanocomposites for dielectric energy storage applications. APL Mater. **9**(2), 021107 (2021). https://doi.org/10.1063/5.0039455
35. Sonker, R.K., Shastri, R., Johari, R.: Superficial synthesis of CdS quantum dots for an efficient perovskite-sensitized solar cell. Energy Fuels **35**(9), 8430–8435 (2021). https://doi.org/10.1021/acs.energyfuels.1c00520
36. Kumar, U., Sikarwar, S., Sonker, R.K., Yadav, B.C.: Carbon nanotube: synthesis and application in solar cell. J. Inorg. Organomet. Polym. Mater. **26**, 1231–1242 (2016). https://doi.org/10.1007/s10904-016-0366-3
37. Su, X., et al.: Silicon-based nanomaterials for lithium-ion batteries: a review. Adv. Energy Mater. **4**(1), 1300882 (2014). https://doi.org/10.1002/aenm.201300882
38. Tarascon, J.-M., Armand, M.: Issues and challenges facing rechargeable lithium batteries. Nature **414**(6861), 359–367 (2001). https://doi.org/10.1038/35104644
39. Zhong, Y., et al.: Architecture design principles for stable electrodeposition behavior—towards better alkali metal (Li/Na/K) anodes. Energy Storage Mater. **45**, 48–73 (2022). https://doi.org/10.1016/j.ensm.2021.10.012
40. Bruce, P.G., et al.: Li–O_2 and Li–S batteries with high energy storage. Nat. Mater. **11**(1), 19–29 (2012). https://doi.org/10.1038/nmat3191
41. Evers, S., Nazar, L.F.: New approaches for high energy density lithium–sulfur battery cathodes. Acc. Chem. Res. **46**(5), 1135–1143 (2013). https://doi.org/10.1021/ar400036g
42. Wang, H., et al.: Graphene-wrapped sulfur particles as a rechargeable lithium–sulfur battery cathode material with high capacity and cycling stability. Nano Lett. **11**(7), 2644–2647 (2011). https://doi.org/10.1021/nl200658a
43. Zhou, G., et al.: A flexible nanostructured sulphur–carbon nanotube cathode with high rate performance for Li–S batteries. Energy Environ. Sci. **5**(10), 8901–8906 (2012). https://doi.org/10.1039/C2EE22766C

44. Liu, W., et al.: Ionic conductivity enhancement of polymer electrolytes with ceramic nanowire fillers. Nano Lett. **15**(4), 2740–2745 (2015). https://doi.org/10.1021/acs.nanolett.5b00303
45. Liu, K., et al.: Electrospun core–shell microfiber separator with thermal-triggered flame-retardant properties for lithium-ion batteries. Sci. Adv. **3**(1), e1601978 (2017). https://doi.org/10.1126/sciadv.1601978
46. Pasta, M., et al.: Full open-framework batteries for stationary energy storage. Nat. Commun. **5**, 3007 (2014). https://doi.org/10.1038/ncomms4007
47. Wanger, T.C.: The lithium future—resources, recycling, and the environment. Conserv. Lett. **4**(3), 202–206 (2011). https://doi.org/10.1111/j.1755-263X.2011.00166.x
48. Garrett, D.E.: Handbook of Lithium and Natural Calcium Chloride. Elsevier, Amsterdam (2004)
49. Donohue, I., Garcia Molinos, J.: Impacts of increased sediment loads on the ecology of lakes. Biol. Rev. **84**(4), 517–531 (2009). https://doi.org/10.1111/j.1469-185X.2009.00085.x
50. Sami, S., Sadrameli, S.M., Etesami, N.: Thermal properties optimization of microencapsulated a renewable and non-toxic phase change material with a polystyrene shell for thermal energy storage systems. Appl. Therm. Eng. **130**, 1416–1424 (2018). https://doi.org/10.1016/j.applthermaleng.2017.11.126
51. Chandel, S.S., Agarwal, T.: Review of current state of research on energy storage, toxicity, health hazards and commercialization of phase changing materials. Renew. Sustain. Energy Rev. **67**, 581–596 (2017). https://doi.org/10.1016/j.rser.2016.09.062
52. Paris, J., Falardeau, M., Villeneuve, C.: Thermal storage by latent heat: a viable option for energy conservation in buildings. Energy Sources **15**(1), 85–93 (1993). https://doi.org/10.1080/00908319308909038
53. Notter, D.A., et al.: Contribution of Li-ion batteries to the environmental impact of electric vehicles. Environ. Sci. Technol. **44**(17), 6550–6556 (2010). https://doi.org/10.1021/es903729a
54. Hudnell, H.K., et al.: Cyanobacterial harmful algal blooms: state of the science and research needs. In: Cyanobacterial Harmful Algal Blooms: State of the Science and Research Needs, pp. 259–273. Springer, New York (2008). https://doi.org/10.1007/978-0-387-75865-7_16
55. Hamilton, T.: Lithium battery recycling gets a boost. Technol. Rev. (2009). [Online]. https://www.technologyreview.com/2009/02/23/219512/lithium-battery-recycling-gets-a-boost/
56. Rahul, S., Shruti; Singh, P.K., Sharma, J.P., Kakroo, S., Sonker, R., Khan, Z.H.: Encompassing environment synthesis, characterization and photovoltaic utilization of cadmium sulphide quantum dots. Mater. Today Proc. **34**, 767–770 (2021). https://doi.org/10.1016/j.matpr.2020.03.320
57. De Liso, B.A., Pio, G., Salzano, E.: Fire behaviour of liquid solvents for energy storage applications. Process. Saf. Environ. Prot. **188**, 726–734 (2024). https://doi.org/10.1016/j.psep.2024.02.025
58. Conzen, J., Lakshmipathy, S., Kapahi, A., Kraft, S., DiDomizio, M.: Lithium ion battery energy storage systems (BESS) hazards. J. Loss Prev. Process Ind. **81**, 104932 (2023). https://doi.org/10.1016/j.jlp.2022.104932
59. da Silva Lima, L., et al.: Life cycle assessment of lithium-ion batteries and vanadium redox flow batteries-based renewable energy storage systems. Sustain. Energy Technol. Assess. **46**, 101286 (2021). https://doi.org/10.1016/j.seta.2021.101286
60. Maleki Kheimeh Sari, H., Li, X.: Controllable cathode–electrolyte interface of $Li[Ni_{0.8}Co_{0.1}Mn_{0.1}]O_2$ for lithium ion batteries: a review. Adv. Energy Mater. **9**(39), 1901597 (2019). https://doi.org/10.1002/aenm.201901597
61. Archuleta, M.M.: Toxicity of materials used in the manufacture of lithium batteries. J. Power Sources **54**(1), 138–142 (1995). https://doi.org/10.1016/0378-7753(94)02094-A
62. Bernardes, A.M., Espinosa, D.C.R., Tenório, J.A.S.: Recycling of batteries: a review of current processes and technologies. J. Power Sources **130**(1–2), 291–298 (2004). https://doi.org/10.1016/j.jpowsour.2003.12.026
63. Kang, Z., Huang, Z., Peng, Q., Shi, Z., Xiao, H., Yin, R., Fu, G., Zhao, J.: Recycling technologies, policies, prospects, and challenges for spent batteries. iScience **26**(10), 107324 (2023). https://doi.org/10.1016/j.isci.2023.107324

64. Ma, J., et al.: The 2021 battery technology roadmap. J. Phys. D Appl. Phys. **54**(18), 183001 (2021). https://doi.org/10.1088/1361-6463/abeb5d
65. Salnikova, A., Raslambekova, N.: Analysis of the legal and regulatory framework of the electrical and electronic equipment waste management. Econ. Bull. IPU RAS **2**(4), 22–28 (2021)
66. Royo, P., et al.: Hybrid diagnosis to characterise the energy and environmental enhancement of photovoltaic modules using smart materials. Energy **101**, 174–189 (2016). https://doi.org/10.1016/j.energy.2016.02.004
67. Mistry, M., Gediga, J., Boonzaier, S.: Life cycle assessment of nickel products. Int. J. Life Cycle Assess. **21**, 1559–1572 (2016). https://doi.org/10.1007/s11367-016-1104-1
68. Sun, X., et al.: Life cycle assessment of lithium nickel cobalt manganese oxide (NCM) batteries for electric passenger vehicles. J. Clean. Prod. **273**, 123006 (2020). https://doi.org/10.1016/j.jclepro.2020.123006
69. Dai, Q., et al.: Life cycle analysis of lithium-ion batteries for automotive applications. Batteries **5**(2), 48 (2019). https://doi.org/10.3390/batteries5020048
70. Kim, H.C., Wallington, T.J., Arsenault, R., Bae, C., Ahn, S., Lee, J.: Cradle-to-gate emissions from a commercial electric vehicle Li-ion battery: a comparative analysis. Environ. Sci. Technol. **50**(14), 7715–7722 (2016). https://doi.org/10.1021/acs.est.6b00830
71. Premnath, V., et al.: Detailed characterization of particle emissions from battery fires. Aerosol Sci. Technol. **56**(4), 337–354 (2022). https://doi.org/10.1080/02786826.2021.1994042
72. Rahul, S., Singh, P.K., Parvaz, M., Ahmed, S., Sonker, R.K., Bhattacharya, B., Khan, Z.H.: Less toxic tin incorporated perovskite solar cell using polymer electrolyte processed in the air. Optik **169**, 166–171 (2018). https://doi.org/10.1016/j.ijleo.2018.05.064
73. EV Battery Recycling and the Role of Battery Energy Storage Systems. https://www.exro.com/industry-insights/ev-battery-recycling-and-the-role-of-battery-energy-storagesystems. Accessed 13 Feb 2025
74. Sloop, S., Crandon, L., Allen, M., Koetje, K., Reed, L., Gaines, L., Sirisaksoontorn, W., Lerner, M.: A direct recycling case study from a lithium-ion battery recall. Sustain. Mater. Technol. **25**, e00152 (2020). https://doi.org/10.1016/j.susmat.2020.e00152
75. Sánchez-Díez, E., Ventosa, E., Guarnieri, M., Trovò, A., Flox, C., Marcilla, R., Soavi, F., Mazur, P., Aranzabe, E., Ferret, R.: Redox flow batteries: status and perspective towards sustainable stationary energy storage. J. Power Sources **481**, 228804 (2021). https://doi.org/10.1016/j.jpowsour.2020.228804
76. Thielke, M., Sobrido, A.J.: Multifunctional sustainable materials for energy storage. In: Sustainable Energy Storage in the Scope of Circular Economy: Advanced Materials and Device Design, pp. 203–224. Wiley (2023). https://doi.org/10.1002/9781119817741.ch9
77. Lithium-Ion Battery Recycling. https://www.epa.gov/hw/lithium-ion-battery-recycling. Accessed 13 Feb 2025
78. Ibrahim, H., Ilinca, A., Perron, J.: Energy storage systems—characteristics and comparisons. Renew. Sustain. Energy Rev. **12**(5), 1221–1250 (2008). https://doi.org/10.1016/j.rser.2007.01.023

Evolution of Energy Storage System and Its Life-Saving Applications

Abhisikta Bhaduri and Shakti Singh

Abstract The evolution of energy storage systems has transformed the procedure of energy capturing, storing, and utilization across miscellaneous sectors, from commercial electronics to crucial life-saving equipment. Primarily fixated on simple energy retention, energy storage technologies have advanced considerably with innovations in mechanical, thermal, chemical, electrochemical, and hybrid systems. The improvements has managed to reach higher energy densities, improved efficiency, faster response times, and longer lifespans. Technologies like Li-ion and Li–S batteries have facilitated compact and effective solutions for medical devices, emergency equipment, and wearable health monitors, while large-scale solutions like grid storage systems supply trustworthy power for hospitals, industries and critical infrastructures. Furthermore, hybrid storage systems, combining distinct energy storage systems, deliver robust power for electric vehicles and uninterruptible power supply (UPS) systems that are crucial in disaster-stricken regions and for serious medical care. As renewable energy sources become predominant, innovative energy storage systems play a vital role in storing intermittent energy, safeguarding an unwavering power resource for remote areas, and healthcare facilities. This chapter conveys the evolution of these systems, and their next-generation applications.

1 Introduction

The evolution of energy storage systems (ESS) has played a pivotal role in advancement of technology, starting from powering conventional mechanical systems towards allowing latest sustainable energy usage and electric conveyance. With the course of time, energy storages have transitioned from basic mechanical ones to advanced thermal, chemical, and electrochemical systems (Fig. 1), all motivated by

A. Bhaduri (✉)
Department of Physics, North Bengal St. Xavier's College, Rajganj, West Bengal, India
e-mail: abhisikta.bhaduri03@gmail.com

S. Singh
Department of Chemical Engineering, University of Seoul, Dongdaemun-gu, Seoul, Republic of Korea

R. K. Sonker et al. (eds.), *Low-Dimensional Multifunctional Hybrid Nanostructured Materials*, Engineering Materials, https://doi.org/10.1007/978-981-96-7952-2_12

the requirement for more versatile, reliable, efficient, easily accessible, and adaptable energy storage.

This evolvement has been enraged by the rising demand for the renewable energy sources and the necessity of balancing the intermittent replenishable energy generation from sustainable sources. Progresses in battery technology, especially the escalation of Li-ion batteries, have modernized portable electric vehicles, electronic devices, and large-scale energy storages. Budding technologies like flow batteries, solid-state batteries, and quantum batteries give the assurance to thrust the boundaries of the possibilities in ESS, making it a crucial element for the global transition en route for sustainable energy [1].

1.1 The Inauguration of Basic Energy Storage Systems

The most primitive forms of energy storage systems, were based on thermal energy. People used to store thermal energy for cooking and heating purposes, by methods like water heating, wood and coal burning. But, it did not last until the lead-acid batteries were invented by Gaston Planté in the year 1859, when the world saw the first rechargeable electrochemical energy storage system [2]. Since then, the lead–acid battery became contributory in prompt automobile applications, and embarked a journey towards a major landmark in energy storage, by bringing portability in electrical power supply.

1.2 Evolution of Electrochemical Storage Systems

By twentieth century, further innovations led to the development of new electrochemical storages. Nickel–cadmium (NiCd) batteries, launched in the 1900s, revealed greater energy density, with ecological downsides due to usage of toxic constituents. The major revolution in the history of electrochemical storages took place in the 1970s, when lithium-based technology was introduced, forecasting the revolutionary introduction of lithium-ion batteries (LIBs) in the 1990s, which provided prolonged cycle life and high energy density. These batteries became the foundation for modern portable electronics and later expanded into electric vehicles (EVs) and large-scale energy storage solutions [3].

1.3 The Rise of Modern Storage Technologies

Last few couple of years have witnessed rapid progressions in storage systems. Lithium–sulfur (Li–S), and lithium iron phosphate (LFP) batteries were developed

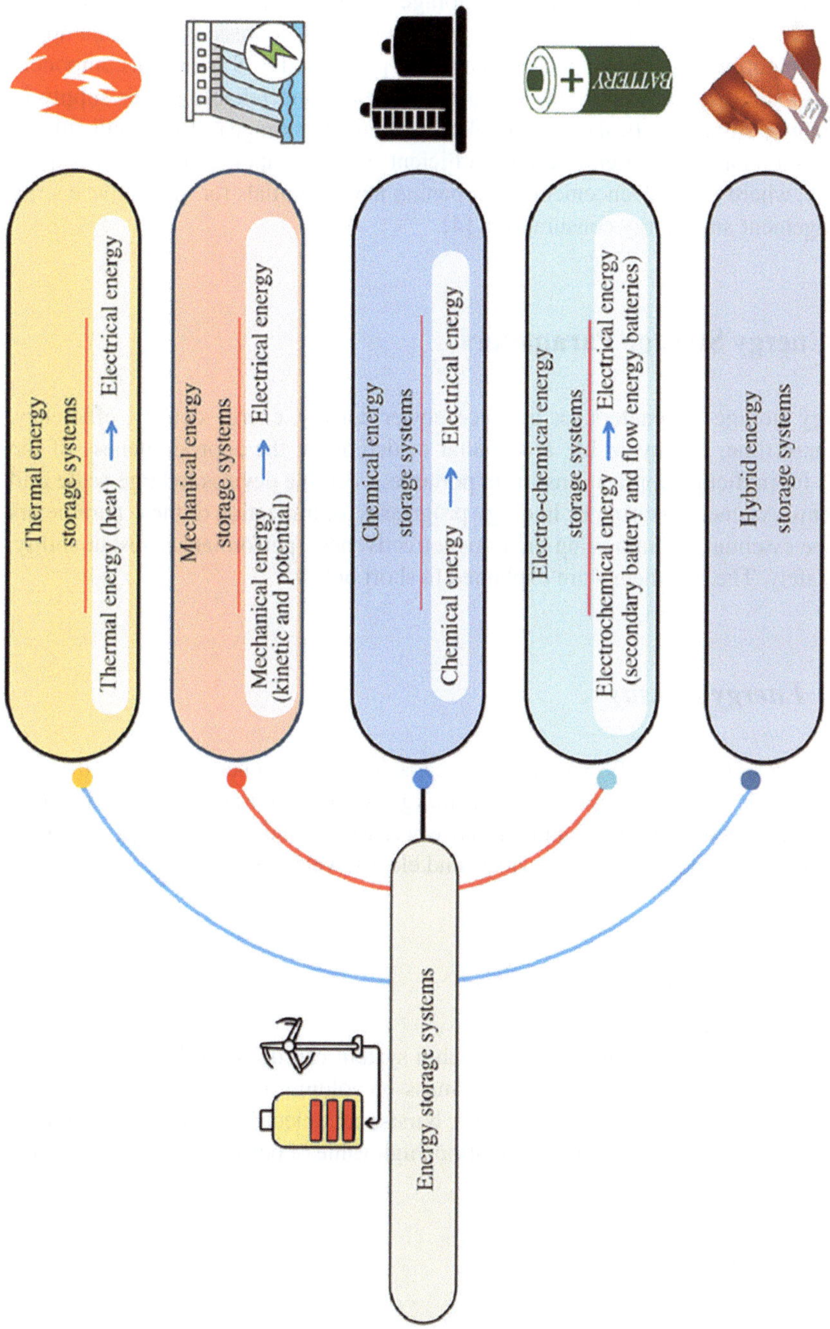

Fig. 1 Classification of major energy storage systems [1]. Reproduced with permission Copyright @ (2022) Elsevier

to increase energy capacity, safety, and sustainability. In conjunction with electrochemical storage systems, mechanical storage technologies, thermal energy storage, electrical storage systems, and hybrid systems emerged, addressing particular application requirements. These advancements have positioned ESS as critical component for grid stabilization, renewable energy integration, and reliable power supply in emergency situations. In short, the evolution of energy storage reflects civilization's unceasing chase for better, and more efficient ways to capture, store, and employ energy, where each advancements are opening new potentials for innovative energy management and energy consumption [4].

2 Energy Storage Parameters

Energy storage parameters for instance power density, energy density, efficiency, response time, and cycle life are crucial to determine the appropriateness of the ESSs for particular applications, from portable, wearable devices to large-scale grid systems. As energy storing technology progresses, optimization of these parameters will be essential for improving their cost-effectiveness, performance, sustainability, and safety. The parameters are explained in short below.

2.1 Energy Density

Energy density of any ESS narrates its capacity to store energy per unit mass or volume. It is usually measured in watt-hr/kg (Wh/kg), or watt-hr/ltr (Wh/L). This parameter is essential for the applications where weight and space are restricted, e.g., portable electronics, wearable devices, and electric vehicles.

2.2 Power Density

Power density indicates the rate at which a system can deliver/release power. It is measured in power delivered per unit mass or volume, i.e., W/kg or (W/L). For the applications demanding rapid power bursts, e.g., electric vehicles acceleration, power tools, and grid frequency regulation, high value of power density is necessary.

2.3 Efficiency

Efficiency of an ESS is defined as the ratio of output energy to input energy, usually conveyed as percentage. For instance, 80% efficiency denotes 80% of the energy stored in that ESS can be fetched. In practice, high efficiency is always preferred as it effectively cuts energy losses during energy storing and release, which is very crucial for economic and environmental-friendly applications.

2.4 Cycle Life

The numeral value of continuous, complete charging–discharging cycles an ESS can withstand before its capability drops below a certain fraction (often 80%) of its original capacity. Extended cycle life lessens the requirement of frequent replacements of ESSs that is vital for the sustainability, and cost-effectivity of ESSs.

2.5 Self-Discharge Rate

The rate at which an ESS loses its charge when not in use, or even not connected to the external circuit. Obviously, low self-discharge rate is preferable for practical applications. As example, Li-ion batteries have a self-discharge rate of about 1–2% per month, whereas nickel-based batteries may self-discharge at rates as high as 10–20% per month.

2.6 Charging/Discharging Rates

The rate at which any ESS charges/discharges with respect to time is defined as the charging/discharging rate.

2.7 Response Time

The time taken by an ESS to respond to the supply signal, usually determined in milliseconds to minutes. Fast response time is always necessary for almost all applications, especially in frequency regulation, where the ESS must retort immediately in order to balance supply and demand.

3 Types and Current Developments of Energy Storage Systems

3.1 *Mechanical Energy Storage (MES) Systems*

Here, the electrical energy is stored by transforming physical energies in the form of kinetic, potential, or gravitational energy. MES systems are generally durable, efficient, cost-effective, and reliable, suitable for grid-level energy storing applications and other large-scale applications. MES systems generally have great cycle life, long lifespans, and most of the MES systems do not use any hazardous gases/chemicals. The early developments in MES systems include pumped hydro storage, flywheel energy storage, and compressed air energy storage (Fig. 2), while the modern innovations include gravity-based energy storage, rail energy storage systems, liquid air energy storage, and elastic energy storage systems.

3.1.1 Early Developments in MES Systems—Pumped Hydroelectric Storage (PHS)

PHS system, the most primitive form of MES system, came into act in the nineteenth century and remains the most widely utilized large-scale ESS. These systems store energy by transferring water between two reservoirs at different elevations. During low electricity demand, water is pumped from the lower reservoir to the upper one. During high demand, water is allowed to pour down from the upper reservoir to the lower one, which drives the turbines, and generates electricity. PHS have witnessed

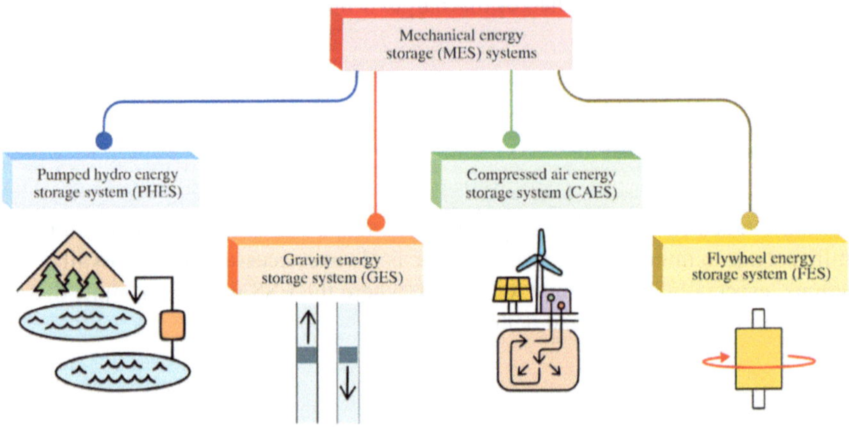

Fig. 2 Classification of mechanical energy storage systems [1]. Reproduced with permission Copyright @ (2022) Elsevier

extensive acceptance during mid-twentieth century with the growing demand for reliable energy storage. PHS maintenance is inexpensive, they own ~40–60 years long lifetime, and great energy capacity. Over years, improvements in reservoir design, turbine technology, and controlling units have upgraded its efficiency and scalability. Today, PHS accounts for more than 90% of global energy storage capacity, offering efficient, stable and durable grid support. However, they need substantial initial investment, and explicit geographical settings (water convenience and elevation difference). Despite these shortcomings, PHS systems are extensively used in real-life applications [5].

3.1.2 Compressed Air Energy Storage (CAES)

By squeezing air inside a high-pressure vessel, or, an underground cavity, energy is stored in a CAES system. During high demand, the compacted air is freed, extended, and heated up for driving turbines that produces electricity. This technology emerged around the 1970s, addressing the necessity of efficient large-scale energy storage that accompanies PHS system. The first commercial CAES plant was constructed in the year 1978, in Huntorf, Germany, followed by a plant in McIntosh, Alabama, in 1991. Traditionally, CAES systems used to elevate the air temperature by combustion of natural gas. To reduce carbon emissions, and enhance efficiency, adiabatic CAES was introduced to stock up the heat generated while compressing the air, and recycles that during expansion. These systems are still under improvement, and their efficiency is moderate (~50–70%). Like PHS systems, CAES systems are also restricted by the geographical constraints [6].

3.1.3 Flywheel Energy Storage (FES)

Flywheels store energy in terms of rotational kinetic energy, are used since the 1990s in the applications demanding constant, stable output energy. During charging, the flywheel is spun at high speeds by the electricity spins. When discharging, the flywheel decelerates, due to which its kinetic energy converts back into electricity via a generator system. The flywheel systems possess extraordinary cycle life (>20,000 cycles), fast response time, high power density, and marginal maintenance needs. Primarily, flywheels were restricted by high frictional losses (causing high self-discharge rate), and inadequate energy densities. Progresses in material usage, like the use of carbon fiber-based composites, and improvements in the magnetic bearings have ominously decreased the frictions, and improved energy density, as well as rotational speed. These days, flywheels offer prompt charging/discharging cycles, which make them perfect for grid stability, and frequency regulation applications [7].

3.1.4 Recent Innovations and Hybrid Systems

The advancements of MES has continued in recent few years with the fabrication of hybrid systems that associate MES systems with other technologies. For instance, PHS is often integrated renewable energy sources [8], and CAES systems are combined with thermal energy storages [9]. The target of these hybrid systems is abridged emissions, and optimized efficiency, to make them apt for renewable energy integration. Improvements in materials research have also empowered scalable CAES systems, and high-speed flywheels, which offer multipurpose energy storing preferences.

3.1.5 Next-Generation MES Systems

Gravity-Based Energy Storage

The working mechanism of this system is similar to PHS, but minus water. This system uses a heavyweight item (e.g., solid weights, or, concrete blocks) elevated at a higher altitude during times of spare electricity. When in need of energy, the item is dropped, which in turn drives a generator to generate electricity. It is a comparatively new technology with inadequate large-scale placement involvement [10].

Rail Energy Storage Systems

These systems store energy by utilizing the additional electricity in shifting trains carrying hefty loads to a place situated at higher altitudes. When energy is desired, the trains move back down, transforming gravitational potential energy in electrical energy via recovering braking [11].

Liquid Air Energy Storage (LAES)

LAES uses electricity to cool down atmospheric air at extremely low temperatures, till it turns into liquid. The liquid air is stored inside insulated chambers. During energy discharge, that stored liquefied air is heated up, which makes it expand speedily, driving a turbine, which then generates electricity. Unlike CAES, there is no geographical limitations in these systems, and the energy can be stored for days. However, they suffer high energy losses during the vaporization and liquefaction processes [12].

Elastic Energy Storage Systems

These systems wield the compression of the elastic materials like, springs to store the corresponding strain energy. Whenever external work is done to compress any elastic material, the work done is stored as elastic potential energy, or strain energy. The energy releases when the material is permitted to expand. This system is employed mainly in toys, and certain tools where short-term energy storing is needed. As a consequence of the mechanism, these systems are not eligible for large-scale/long-term energy storage applications [13].

In short, the evolution of MES systems replicates an unceasing adaptation of technical advancements with the rising necessity of sustainable, competent energy storage solutions. Starting from the initial PHS technologies to emerging hybrid and innovative modern systems, MES plays an essential role in guaranteeing reliable and stable energy storage covering a wide range of present applications, predominantly for the integration of renewable energy harvesters with the grids.

3.2 *Thermal Energy Storage (TES)*

TES systems have experienced substantial evolution, becoming chief solutions for the management of energy in residential, renewable, and industrial divisions. These systems store energy in the form of heat, which later would be used in cooling, heating, power generation, and industrial practices (Fig. 3). TES systems are advantageous in stabilizing grids, handling energy requirements, and enhancing energy efficiency, exclusively when coupled with renewable energy sources. The progression of TES systems can be traced through numerous technical improvements and applications in both high-temperature (solar-thermal, industrial heat storage) and low-temperature (space heating, cooling) fields.

3.2.1 Early Developments in TES Systems—Hot Water Tanks, and Ice Storage

The most initial form of TES systems includes usage of hot water tanks, and ice storage. Since centuries, ice storage is used as a natural technique for cooling and conserving food. During nineteenth century, ice used to be commercially garnered and stockpiled to be utilized as a useful TES system in primitive industrial applications. Likewise, hot water tanks have been employed in domestic heating, particularly in colder provinces, where excess heat used to be stored and exploited as per requirement. These systems established the commencement of TES by facilitating basic heating/cooling applications [15].

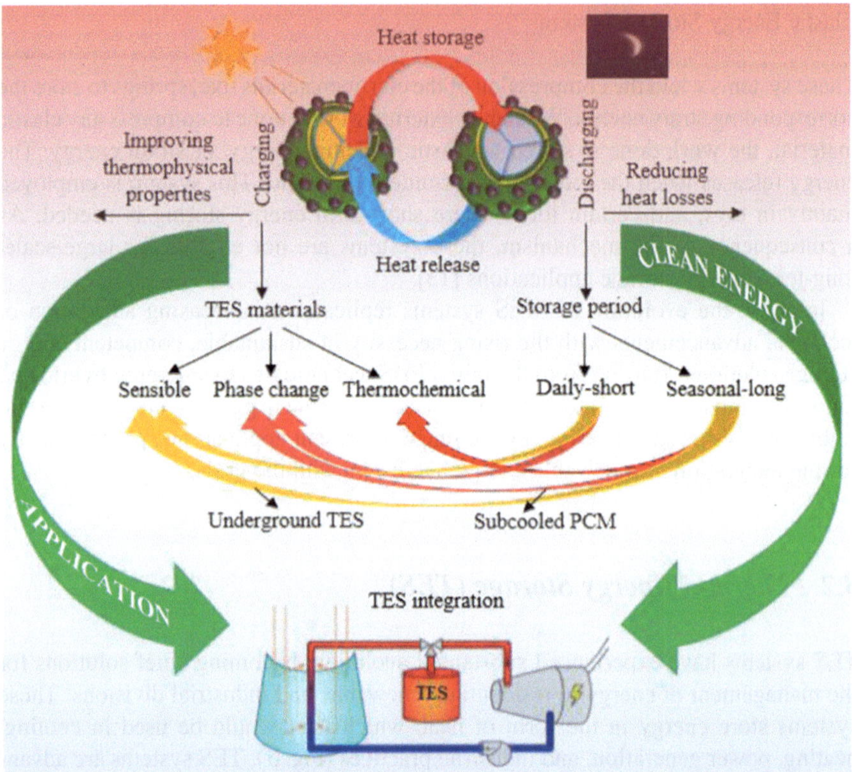

Fig. 3 Graphical abstract of the present study from material characterization to real-world TES applications. Reprinted from [14]. Reproduced with permission Copyright @ (2020) Wiley

3.2.2 Mid-Twentieth Century—Latent Heat Storage (LHS)

During mid-twentieth century, LHS started receiving interests. These units employ phase changing materials (PCMs) to absorb or liberate thermal energy while transitioning from one phase to another (e.g., solid → liquid or, liquid → gas) at specific temperature. PCMs store energy during the liquefaction, or gasification, and release energy during solidification. Early researches on LHS employed salt hydrates, and paraffin wax that work well between the temperatures 20 and 100 °C, and hence were used in building heating/cooling systems, and HVAC units. During 1960–1970, advancements in PCMs-based TES were observed, applied mainly in refrigeration applications, cold storages, buildings and industry heating and cooling units, and in the practices demanding stable temperatures. Most common PCMs employed in these systems are eutectic salts and hydrated salts that are suitable for high-temperature applications such as concentrated solar power (CSP). These devices deliver impressively high energy density, and are pliable towards numerous temperature arrays, although most of the PCMs are expensive [16].

3.2.3 1970–1990—Solar-Thermal Energy Storage

The energy crisis during the 1970s incited attention towards renewable energy harvest and storage that led to the development of solar-thermal storage systems. CSP plants collect sunlight, and produce heat (high-temperature). These storage units store thermal energy within molten salts, usually potassium or sodium nitrates, as these compounds possess high heat capacity and can retain heat for extended periods. These compounds absorb thermal energy from the CSP, and then release it to produce steam for power generation when sunlight is unavailable. The first commercial solar-thermal plant-based TES system was constructed around 1980 in Mojave Desert, California, establishing the feasibility of molten salt in long-term, high-temperature storage applications. This improvement arranged the base for large-scale TES systems, allowing CSP plants in generating consistent, and dispatchable energy [17].

3.2.4 The 2000s—Sensible Heat Storage (SHS) and Thermochemical Energy Storage (TCES)

The early 2000s experienced innovative expansions regarding SHS (heat is stored without any change in phase of the material) and thermochemical energy storage (heat is stored by means of reversible chemical reactions). SHS systems turned widely popular in industrial processes, district heating units, solar-thermal power plants, and building heaters/coolants, contributing in inexpensive and scalable TES solutions. These systems store thermal energy by elevating the temperature of a storage medium, usually a solid (e.g., concrete, rocks, or molten salt), or, liquid (e.g., water). In domestic heating/cooling, water is used as storage medium in SHS owing to high specific heat capacity. On the other hand, concrete and rocks are used in large-scale TES for industrial applications and thermal power generation. Molten salt is widely used in the concentrated solar power (CSP) plants. These systems have simple design, and are budget-friendly, and appropriate for high-temperature storage applications in power plants. However, large space is needed for substantial storage capacity [15].

TCES systems, still in pilot phases, attained attention for great energy density, and long-term heat storing ability without any loss. TCES are mainly preferred in storing seasonal energy, as they can store heat in summer months and release in winters. Relying on the reversible chemical reactions, TCES absorbs and stores thermal energy generated during the breakage of chemical compounds into constituents, and releases the heat energy when they recombine. Generally, metal oxides are used in TCES for high-temperature industrial applications, while, ammonia and salt hydrates are used for low to moderate temperature applications [18].

3.2.5 Recent Developments—Advanced Materials and Cryogenic Energy Storage (CES)

The last few years have witnessed rapid encroachments in TES materials and hybrid units. New PCMs formulated with improved conductivity, stability, and recyclability have developed, making them more feasible for industrial applications. To address the slow heat release problems, and other issues, nanomaterials have been integrated.

Cryogenic engines were first invented by the end of the 1800s. Recently, these are gaining attentions in the field of large-scale ESS. CES systems store energy via cooling gas or, air at cryogenic temperatures, where it liquefies. While discharging, the liquid is made to vaporize and expand, which thereby yields a turbine to generate electricity. These units exhibit small round-trip efficiency, as non-negotiable energy gets lost during liquefaction and vaporization [19].

3.2.6 TES, Recent Applications, and Future Directions

With rising demand for renewable energy sources, TES has become critical for additional energy storage and safeguarding grid stability. Molten salt storages are nowadays extensively used in CSP plants, whereas large-scale water and PCM-based TES systems have been incorporated in district cooling/heating units. Some contemporary TES systems are aimed to cooperate with battery and grid storage systems, allowing a multidirectional methodology for energy management [15].

3.2.7 Forthcoming Researches—Thermochemical and Seasonal Storage

Current research emphases on innovative thermochemical storages for season-based applications, exclusively in substantial temperature variations. Materials that store energy via reversible chemical reactions, are being inspected for long-term, high-density storage solutions. Seasonal TES have the ability to harvest solar energy during summer seasons, to utilize in winters, particularly advantageous in areas with extreme weather conditions. Furthermore, TES systems are being fabricated to incorporate with AI-based smart grids and other energy management systems, with improved energy storage and release depending on requirements [20].

In short, the advancement of thermal energy storage systems replicates evolution from simple ice and water tanks to sophisticated equipment skilled with high-density energy storage for prolonged periods. From molten salt in CSP plants to advanced materials and hybrid storage strategies, TES systems now hold a crucial place in sustainable energy solutions, increasing the dependency of renewable energy, while contributing in next-generation durable storage systems across residential, industrial, and utility divisions.

3.3 Chemical/Electrochemical Energy Storage Systems

Chemical Energy Storage (CES) Systems utilize the energy residing in chemical bonds, and provide possibility of capturing, retaining, and releasing energy via chemical interactions, while Electrochemical Energy Storages (EES) transform and store electrical energy from electrochemical interactions. These systems are the center of modern energy applications sprouting from simple battery cells to progressive energy storage technologies, supporting grid storage, portable electronics, renewable energy integration, and electric vehicles, as overviewed below.

3.3.1 Primary EES—Lead–Acid Batteries (1859)

In batteries, chemical energy is stored in form of electrical energy by means of electrochemical reactions. This procedure happens in an enclosed system containing two electrodes (anode and cathode), separated by electrolyte solution. Batteries are commonly of two types: primary (non-rechargeable) and secondary (rechargeable). The first practical EES invented by Gaston Planté in 1859 was lead–acid-based rechargeable battery, which became extensively useful for steady and consistent power-demanding applications, like UPS [2]. Despite of being heavy, and exhibiting low energy density, these persist being used till date owing to their affordability, and reliability.

3.3.2 Battery Cells—Zn–C, and Alkaline Batteries (the Late Nineteenth–Twentieth Centuries)

Non-rechargeable Zn–C, and alkaline batteries were introduced during late nineteenth century, and obtained commercial admiration by twentieth century. Zn–C cells, developed in the 1890s, delivered a compressed, inexpensive power source for radios, flashlights, and portable electronics, whereas alkaline batteries, developed in the 1950s, provided high efficiency, longer shelf life, and improved energy density.

3.3.3 Ni-Based Rechargeable Batteries—NiCd and NiMH (Early–Mid-Twentieth Century)

NiCd-based rechargeable batteries, first prepared in early twentieth century, provided high output power, and improved cycling stability, but suffered with low energy density, and ecological influence of toxic cadmium. Addressing this, NiMH (nickel–metal hydride) batteries were introduced in the 1980s that offered environmental safety, with better energy density.

3.3.4 Lithium-Ion Batteries (the 1970s–the 1990s)

Lithium-ion batteries (LIBs) have become the governing ESS technology owing to their high efficiency, energy density, and extended cycle life. LIBs function by the movement of Li-ions among anode–cathode via an electrolyte, allowing energy storage/release. The battery generally contains a graphite anode, a lithium metal oxide cathode, a lithium salt-based electrolyte, and a separator to prevent short circuits while permitting ion drift (Fig. 4). When charging, an external power source drives Li-ions from cathode to anode via the electrolyte, where they accumulate within graphite, while electrons flow over the external circuit. When discharging, the Li-ions flow back towards cathode, producing electron drift in the circuit that powers associated devices. This reversible ion transfer process happens for repeated charging/discharging cycles, making LIBs long-lasting, efficient, and apt for numerous applications, including electric vehicles, consumer electronics, and renewable energy storage [21]. Since their commercialization by Sony in 1991, endless advancements have upgraded their safety, performance, and applications. Primary versions encountered challenges like overheating, limited lifespan, and degradation, however material engineering have ominously improved their capabilities.

3.3.5 Development of Solid-State and Li-Polymer Batteries (the 2000s–Present)

Solid-state batteries (SSBs) are the advanced energy solution that substitutes the liquid/gel electrolyte in conventional LIBs with solid electrolytes. This design increases safety, energy density, and lifespan by reducing risks of thermal runaway, leakage, and dendrite formation.

Early research in SSBs began in the 1970s, which experienced challenges like high manufacturing costs, and low ionic conductivity. Over years, material advancements in polymeric, and ceramic electrolytes enhanced performance, but market adoption remained limited due to poor scalability. During the 2010s, breakthroughs in oxide, sulfide, and polymer-based solid electrolytes considerably enhanced conductivity, leading to alliances between research institutions and big companies like Samsung, Toyota, and QuantumScape. By the 2020s, SSBs achieved faster charging times, higher energy densities, and improved safety, making them favorable substitutions for consumer electronics, electric vehicles, and aerospace applications.

Lithium–polymer batteries (LPBs) are a type of LIBs that incorporate a gel/solid-like electrolyte instead of a liquid one, making them thinner, lighter, and flexible. By reducing leakage risks, and allowing customized battery designs these devices have secured ideal position in portable electronics, electric vehicles, and drones. Over the years, advancements in electrode materials, and polymer electrolytes have amended their charge cycles, energy density, and stability. By the 2010s, LPBs were extensively implemented in smartphones, unmanned aerial vehicles, and wearables, owing to their lightweight properties, and compact size. Recent innovations have incorporated solid-state electrolyte technology focusing on fast-charging capabilities,

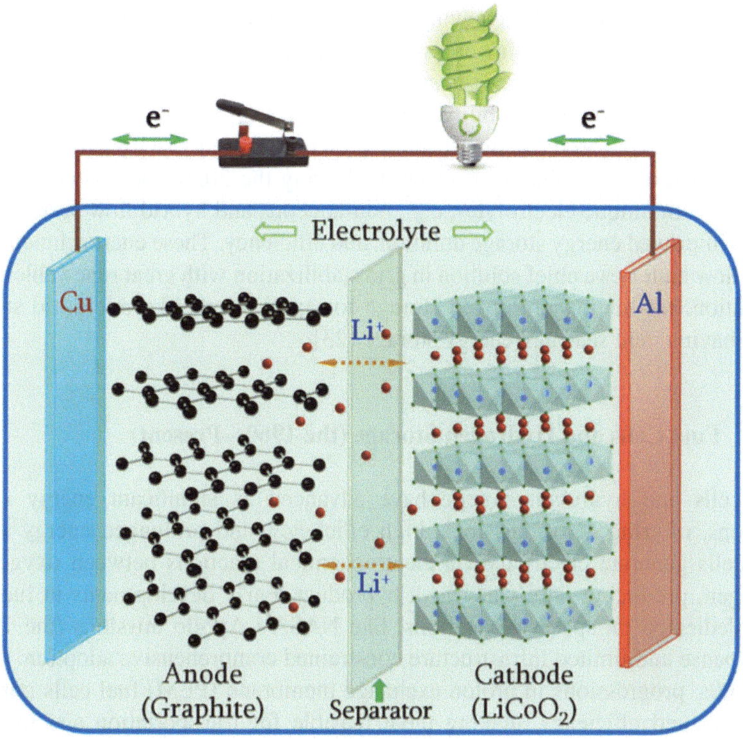

Fig. 4 Graphical representation of a battery. Reprinted from [22]. Reproduced with permission Copyright @ (2016) Elsevier

increased lifespan, and improved safety. The development of silicon, and lithium–sulfur anodes has further boosted their energy capacity, making them more apt for high-performance applications. These innovations have made LPBs chief energy solution for next-generation electric mobility, consumer electronics, and aerospace applications [21].

3.3.6 Advanced Flow Batteries (the 1970s–Present)

Advanced flow batteries are electrochemical ESS that store energy within liquid electrolytes kept in external tanks, consenting scalable and prolonged storage. Unlike conventional ones, flow batteries separate energy capacity from power output, which makes them ideal for large-scale grid storage. Early flow battery technologies, like iron–chromium systems emerged in the 1970s, were expensive, and poorly efficient, restricting their prevalent acceptance. During the 1980s and the 1990s, vanadium redox flow batteries (VRFBs) appeared, contributing enhanced efficiency, extended lifespans, and better scalability owing to their capability of using the same electrolyte for both electrodes, decreasing cross-contamination.

Modern developments are dedicated to enhance energy density, reduce expenses, and increase operational flexibility. Innovations like organic flow batteries, iron–air systems, and zinc–bromine flow batteries, own upgraded sustainability, and cycle life via low-cost, and abundant materials. During the 2020s, next-generation flow batteries with unique electrolytes, e.g., non-aqueous and hybrid flow systems, have further improved energy storage duration, and efficiency. These encroachments have made flow batteries a chief solution in grid stabilization with great renewable energy infiltration, facilitating prolonged storage for applications like microgrid support, peak shaving, and seasonal energy storage [23].

3.3.7 Fuel Cells and Hydrogen Storage (the 1960s–Present)

Fuel cells and hydrogen storage have advanced as significant energy storage solutions, offering zero emissions, high efficiency, and prolonged energy supply. Fuel cells generate electricity via electrochemical reactions between oxygen and hydrogen, producing only water as a byproduct. Early developments in fuel cells were dedicated on space applications, like NASA's Apollo missions (the 1960s), but expense and limited infrastructure constrained comprehensive adoption. During the 2000s, progressions in proton exchange membrane (PEM) fuel cells improved durability and efficiency, making them feasible for transportation and immobile power generation. The 2010s witnessed progress in hydrogen production, storage, and distribution, with advances in liquid organic hydrogen carriers (LOHC), metal hydrides, and high-pressure tanks improving safety and storage capacity. By the 2020s, hydrogen fuel cells gained trust in heavy transport, electric vehicles, and grid-scale energy storage, reinforced by cost reduction and global investments in hydrogen infrastructure. Recent ongoing research focuses on solid-state hydrogen storage, green hydrogen production through electrolysis, and cutting-edge catalysts to ease dependence on rare metals. With nonstop advances, fuel cells and hydrogen storage are becoming essential in clean energy alterations, aiding renewable incorporation, carbon-free mobility, and industrial applications [24, 25].

3.3.8 Next-Generation EES Systems

Next-generation EES are evolving beyond traditional lithium-ion technology by addressing sustainability, cost, and performance challenges. **Sodium-ion batteries** are the inexpensive alternatives, use abundant materials while provide equivalent energy density, making them apt in grid storage. **Magnesium-ion batteries** assure improved safety, and higher energy density owing to their dendrite-free nature, though face challenges in electrolyte section. **Zinc–air batteries** clout oxygen from the air as a reactant, provide cost-effectivity, and high energy density, making them suitable for prolonged storage [26].

Organic batteries have materialized as sustainable storage solutions by swapping conventional metal-based electrodes with organic ones. Early research during the 2000s fixated on employing organic redox-active compounds, however challenges like limited cycle life, and poor conductivity constrained their applicability. By the 2010s, improvements in stabilized organic electrodes, quinone-based materials, and conducting polymers, have significantly improved their performance, embarking organic batteries more feasible in consumer electronics, and grid storage. Recently, implementations of recyclable and biodegradable materials have boosted lifespan, efficiency, and sustainability. With ongoing advances in stability and energy density, organic batteries are emerging as a key contestant in green energy storage technology, decreasing dependence on rare metals, and supporting renewable energy integration.

Hybrid batteries combine multiple electrochemical storage technologies to achieve optimized performance, enhanced energy density, and longevity. Early hybrid battery research during the 1990s focused on combining lithium-ion cells with supercapacitors to improve charging/discharging rates. By the 2000s, advancements in lithium–air, and lithium–sulfur hybrids amplified energy-storing capacity, aiming applications like aerospace and electric vehicles. The 2010s witnessed the rise of hybrid flow batteries, incorporating redox chemistry with solid electrodes for elongated cycle life and enhanced scalability. Recent innovations include lithium–supercapacitor hybrids for rapid-charging devices and metal–air hybrids for protracted storage proficiencies. With ongoing innovations in materials, and design, hybrid batteries are evolving as a chief solution for next-generation energy storage, supporting grid stability, electric mobility, and renewable energy applications.

Hybrid EES

Hybrid electrochemical systems combine batteries with supercapacitors/other energy storages, have arisen to stabilize high energy density and power output. These are advantageous in applications demanding steady energy storage and rapid energy release, e.g., electric vehicle acceleration and regenerative braking.

4 Emergence of ESS in Next-Generation Applications

The emergence of EES in next-generation applications is reforming multiple sectors, augmenting resilience, efficiency, and sustainability. In microgrids, advanced technologies like LIB and flow batteries facilitate decentralized energy management, guaranteeing stable power resource and grid independence, particularly in disaster-prone, and remote areas. Wearable electronics benefit from lightweight, flexible batteries, like solid-state, and lithium–polymer ones, providing longer lifespan and safer energy storage for medical devices, smartwatches, and fitness trackers. Renewable energy incorporation depends on large-scale storage solutions, including hydrogen, vanadium flow batteries, and next-generation lithium-ion systems, to stabilize intermittent wind and solar power, ensuring energy security, and grid stability. As improvements continue, ESS plays a crucial role in leading innovations across industries, supporting cleaner energy solutions and smarter, more resilient infrastructure.

4.1 *Utilization of Renewable Energy Sources*

The expense of electricity generation from the limited resources like fossil fuels have inspired the exploration of the renewable energy technologies as the better and sustainable alternatives of the conventional non-renewable ones. Presently, various renewable energy sources (e.g., solar radiation, wind blow, and water flow) are being utilized worldwide. The deployment of these technologies in the residential areas are rising swiftly. However, the environmental conditions lead to fluctuations and instabilities in the harvested energy. This inconsistency has led to the development of integrated systems that can harvest the surplus available energy, store, and release it as per requirement. Deployment of such devices decentralizes the energy managing systems. Hence, besides enhancing the energy efficiency and reliability, the incorporation of the renewable energy harvesters with ESS is also renovating the universal energy scenery towards a carbon–neutral and resilient future. As progressions continue, ESS is likely to enact further critical role in renewable energy management, while addressing the environmental challenges ahead.

Being one of the most utilized renewable energy source available globally, solar energy is used in numerous applications like transportation, desalination, irrigation, and residential. The recent commercial solar power technologies include the development and improvement of concentrated solar power technologies (CSP), and bifocal photovoltaic technologies (BP). CSPs generate both heat and electricity from solar irradiation, while BPs utilize both of front and backsides of the solar panels to harvest solar energy in form of electricity. Till date, quite a few large-scale solar projects have been implemented in various countries like India, China, Egypt, UAE, and Morocco. Among these, the Bhadla Solar Park project, effected in India in the year 2018, used first ever ultra-mega power plant (UMPP) with a goal of a total capacity of ~20 GW, which was later proposed to be increased up to ~40 GW.

Besides solar energy, the kinetic energy of wind is also harvested globally as sustainable and efficient approach. The most essential element of any wind energy harvesting system is the wind turbine (WT). With time and requirements, multiple refinements and advancements in the turbine designing has been done mainly in its components like the blades, tower, nacelle, controller, and gears. In general, the wind flow propels the stationary blades, and the generator from the nacelle collects the energy from turbine via driving shaft. Post conversion of wind kinetic energy into electrical energy, the final energy is delivered to the grid by the transformer. Primarily two types of turbines are used: vertical axis wind turbine (VAWT), and horizontal axis wind turbine (HAWT). VAWTs are positioned adjacent to the ground, and hence are less exposed to the wind, making them less efficient, and unreliable. On the contrary, HAWTs generate more electricity, and their installation is way cheaper than the VAWTs, thereby are currently dominating the majority of the wind energy sectors.

Other than the usual renewable energy sources, the mechanical energy harvesters aka nanogenerators are steadily receiving global attractions. These devices harvest energies from any type of mechanical stimuli in the form of electrical signals, following triboelectric/piezoelectric characteristics of the materials concerned [27, 28]. Nanogenerators are highly efficient and can power small devices like wearable devices, sensors, and medical implants.

Despite being sustainable and having nominal adversarial effects on the environment, all renewable energy sources possess intermittent energy production ability, and face major grid integration challenges, hence are entitled to be integrated with ESS. ESS play an essential role in ensuring grid stability, efficiency, and reliability, by storing additional energy during peak generation and releasing it when required. For example, LIBs in California store solar energy during the daytime and supply that in the evening. Other ESS like CSP plants are often used to collect sunlight, and concentrate to heat-up molten salt for long-term heat retaining. ESS also supports in regulating grid frequency, averting instability triggered by fluctuations in renewable energy output, as realized in Tesla's Hornsdale Power Reserve in Australia. Furthermore, ESS permits load shifting, and peak shaving, decreasing demand hikes and optimizing energy usage, as observed in Norway's pumped hydro storage (PHS). The battery-backed solar microgrids in Hawaii demonstrate enhanced grid resilience with the provision of backup power during outages.

Universally governments are prompting ESS via tax credits, subsidies, and mandates, like Germany's KfW Program, as well as U.S. Inflation Reduction Act. Regulatory reorganizations, containing dynamic pricing and grid code updates, expedite storage integration into markets, as perceived in the FERC Order 841 in the U.S. Decentralized storage solutions, like Virtual Power Plants (VPPs) in Australia, combine domestic batteries to stabilize the grids. Emerging technologies like hydrogen storage, and vanadium flow batteries are being utilized to guarantee long-term energy storage, for instance the 100 MW Vanadium Flow Battery project of China. In short, incorporating storages with renewables is the key to attain a resilient, stable, and sustainable grids, demanding continuous policy provision, regulatory progressions, and infrastructure investments.

4.2 *Flexible Wearable Electronics*

The modern living style and societal requirements have incorporated portable, lightweight, wearable, multifunctional goods. On that note, the wearable electronic system has been experienced a paradigm shift due to introduction of flexible electronic systems. Flexible electronics systems, regarded as the combination of multiple flexible electrical components, have gone through significant encroachments due to the incorporation of nanotechnology. Current advances have introduced extremely miniaturized and smart constituents, like nanosensors, integrated circuits, microprocessors, signal transmitters, and circuit chips. Such evolution has empowered the production of multifunctional, highly sophisticated, flexible electronic systems, escalating their utilization in miscellaneous fields like wearable technology, biomedical devices, and smart textiles. ESS are vital in wearable electronics, facilitating a broad collection of applications ranging from smartwatches and fitness trackers to smart textiles and medical devices. Wearable EES must be lightweight, compact, durable, and flexible, to fit effortlessly with the consumer's physique, while supplying consistent power.

Portable/wearable electronics, e.g., roll-up displays, wireless sensors, flexible smart-devices, and electronic skins, are lightweight, and miniaturized. For achieving independent, completely self-sustainable, wireless, and efficient wearable electronic devices, it is necessary to incorporate lightweight, robust, and flexible power sources. There are many requirements of the next-generation ESS for this particular application. It should be safe enough for using under any circumstances, and inexpensive for commercial availability. The mechanical consistency of the flexible ESS embraces good deformation tolerance, and retention of electrical performance. The system needs to be robust and flexible to withstand any geometrical deformation without affecting its performance. Besides these, the ESS also needs to possess extensive cycle life, high energy density, compatible constituent materials, and exceptional rate competency. Old-style ESS, colossal coin cells and inelastic rechargeable batteries generally demoralize the mechanical convenience and compliance of wearable units. Enthused by this, the flexible ESS like flexible zinc–carbon batteries, alkaline batteries, ion-based rechargeable batteries (Li-ion, Na-ion, Li–S ion, etc.), polymeric batteries, and flexible supercapacitors are being extensively explored. Among these, the ion-based flexible rechargeable batteries have exhibited certain limitations like slow charging–discharging rates, small cycle life, and poor power densities. Addressing low-power density issue, electrochemical capacitors, aka supercapacitors were employed for better energy storing performance. Besides high power density, the supercapacitors also delivered higher lifetime, and rapid charging/discharging. However, the conventional batteries and supercapacitors are made up of fragile and rigid electrodes, making them unfit in flexible electronic applications. Moreover, the electrochemical constituents are mostly considered as safety hazards that is obviously not preferable in wearable unit fabrication. As a result, safer materials were emerged in flexible EES formation, following two approaches: (a) preparing flexible freestanding films of active materials; (b) depositing rigid active materials on flexible non-conducting/conducting substrates, where resilient interaction among the substrate and the active material is needed for maintaining the mechanical reliability during any sort of deformation [29].

Flexible wearable electronics incorporated with advanced ESS are revolutionizing multiple industries globally. Researchers at Stanford University (United States) have developed stretchable LIBs using nanomaterials, enhancing the efficiency of smart textiles and medical wearables. Tsinghua University (China) has pioneered flexible supercapacitors that harvest energy from body movements, powering next-generation fitness and health-monitoring devices. The European Union's Smart2Go project concentrates on self-sustaining wearable electronics, combining flexible batteries with energy harvesting technologies to reduce frequent recharging. RIKEN research institute in Japan has developed ultra-thin biofuel cells that produce power from sweat, significantly advancing the health sensors. South Korea's KAIST has introduced foldable solid-state batteries, empowering seamless integration into smart wearables and soft robotics. These innovations lead global developments in self-powered wearables, influencing the future of sports, healthcare, and electronics market while encouraging sustainability via user-friendly, energy-efficient strategies.

4.3 *Microgrids*

As mentioned earlier, the renewable energy harvesters, coupled with ESS are getting widely endorsed as sustainable power solution. The poor lifetime of ESS batteries encouraged the deployment of hybrid ESS, microgrids. Microgrid is a small-scale grid system made of interconnected loads, ESS and distributed energy generators. The generators are used to give-and-take power with the central grid via a particular point, known as the point of common coupling. Microgrids can function in two different modes: Isolated mode (independently), and Grid-connected mode (with the power grid). The necessary components of a microgrid are the power generation systems (e.g., renewable energy sources), loads, energy converters, and protection units. The greatest substantial stake of renewable energy sources in microgrids is built on the basis of wind turbine, hydro units, and solar PV generation. Using microgrids, the reliability of the local energy network, and the energy security can be significantly improved by incorporating abundant renewable sources. Microgrids, connected with ESS can function independently during any calamities in any provinces, and provide an uninterrupted power supply.

Microgrids are the chief facilitators of the impending smart grids that have the prospective to shift the current centralized electricity networks in completely distributed architectures.

Microgrids architectures can be classified as direct current microgrid (DCMG), alternating current microgrid (ACMG), and hybrid microgrid (HMG). As the name suggests, The HMG fuses the benefits of the DCMG and ACMG architectures, as both the DC and AC ones are joined by a single energy converter.

Microgrid integration with ESS enhances energy reliability, resilience, and sustainability by facilitating localized energy generation and management. By integrating renewable sources (e.g., solar, wind) with ESS, microgrids lessen reliance on centralized grids, and provide advanced energy security during extreme weather events, or power outages. In off-grid, and remote regions, ESS permits microgrids to deliver stable power without depending on fossil fuels, as witnessed in rural electrification assignments across Africa and island nations, while usage of microgrids in urban cities like Tokyo and New York support high demand management, and decreasing stress on the main grid while reducing electricity expenses. Microgrids paired with ESS is being extensively used. Industrial facilities and military bases use microgrids-ESS for incessant power supply, improving operational efficiency and security. Progressions in smart grid technology allow real-time energy optimization, demand response, and continuous grid integration, driving the transition towards cleaner, and decentralized ESS globally [30].

5 Outlook

The evolution of ESS has renovated our approach to capture, store, and utilize energy, creating the possibility in harnessing renewable energy sources at large scale. A comparative analysis of all ESS are charted in Table 1 for better understanding. Addressing the challenges of irregular renewable energy and rising demand of electricity, the ESS have advanced significantly, from traditional mechanical and lead–acid batteries, to the cutting-edge Li-ion batteries, flow batteries, along with other advanced technologies, creating extra robust, sustainable, and flexible grids, with reduced emissions, and enhanced energy security. As the progression continues, the role of ESS will strengthen, allowing grid stability, assisting the extensive implementation of electrical vehicles, and increasing resilience in contrast to the climate-related disturbances. Next-generation technologies like long-duration energy storage, artificial intelligence and smart grid integration, and solid-state batteries are assuring more accessible, sustainable and efficient energy storage. In short, the evolution of ESS mirrors our pledge towards a cleaner, sustainable, and more resilient, future, encouraging an inclusive transition to renewable energy, and carbon–neutral economy.

Table 1 Comparative analysis of different ESS

Category	Technology	Energy density	Efficiency (%)	Life span	Charge/discharge speed	Advantages	Limitations	Best applications
Mechanical	Pumped hydro storage	Medium	70–85%	50+ years	Slow	Large-scale storage, Low operating cost	Needs definite geography (hills/water), High upfront cost	Grid-scale energy storage, renewable energy balancing
Mechanical	Gravity storage (energy vault)	Low	80–90%	25–50 years	Slow	Uses recycled materials, No chemical degradation, Sustainable	Requires space (large infrastructure), Not suitable for mobile applications	Grid storage, renewable energy buffering
Thermal	Molten salt storage	Medium	60–75%	30+ years	Slow	Ideal for solar-thermal plants, Low-cost materials	Low efficiency, Heat losses over time	Solar power plants, industrial heat supply
Chemical	Hydrogen fuel cells	Medium	40–60%	10–20 years	Moderate	Zero emissions, High energy storage potential, Quick refueling	Low efficiency, Requires hydrogen infrastructure, Expensive hydrogen production	Hydrogen cars, buses, backup power, aerospace

(continued)

Table 1 (continued)

Category	Technology	Energy density	Efficiency (%)	Life span	Charge/discharge speed	Advantages	Limitations	Best applications
Electrochemical	Lithium-ion batteries	High	90–95%	10–15 years	Fast	High energy density, Mature technology, Widely used in EVs and electronics	Risk of overheating/fire, Degrades over time, Expensive materials	EVs, smartphones, home energy storage, grid backup
Electrical	Supercapacitors	Very low	95–98%	1 M+ cycles	Ultra-fast	Extremely fast charge/discharge, Long lifespan, High efficiency	Low energy storage capacity, High self-discharge	Regenerative braking (trains, buses), burst power applications

References

1. Mitali, J., Dhinakaran, S., Mohamad, A.A.: Energy storage systems: a review. Energy Storage Saving **1**, 166–216 (2022). https://doi.org/10.1016/j.enss.2022.07.002
2. Kurzweil, P.: Gaston Planté and his invention of the lead-acid battery—the genesis of the first practical rechargeable battery. J. Power. Sources **195**(14), 4424–4434 (2010). https://doi.org/10.1016/j.jpowsour.2009.12.126
3. Abbas, Q., Mirzaeian, M., Hunt, M.R.C., Hall, P., Raza, R.: Current state and future prospects for electrochemical energy storage and conversion systems. Energies **13**(21), 5847 (2020). https://doi.org/10.3390/en13215847
4. Tiwari, P., Kafle, A., Bijukchhe, P., Bhattarai, A.: A review on energy storage systems. J. Phys. Conf. Ser. **2629**(1), 012024 (2023). https://doi.org/10.1088/1742-6596/2629/1/012024
5. Nikolaos, P.C., Marios, F., Dimitris, K.: A review of pumped hydro storage systems. Energies **16**(11), 4516 (2023). https://doi.org/10.3390/en16114516
6. Olabi, A.G., Wilberforce, T., Ramadan, M., Abdelkareem, M.A., Alami, A.H.: Compressed air energy storage systems: components and operating parameters—a review. J. Energy Storage **34**, 102000 (2021). https://doi.org/10.1016/j.est.2020.102000
7. Li, X., Palazzolo, A.: A review of flywheel energy storage systems: state of the art and opportunities. J. Energy Storage **46**, 103576 (2022). https://doi.org/10.1016/j.est.2021.103576
8. Mahmoud, M., Ramadan, M., Olabi, A.-G., Pullen, K., Naher, S.: A review of mechanical energy storage systems combined with wind and solar applications. Energy Convers. Manag. **210**, 112670 (2020). https://doi.org/10.1016/j.enconman.2020.112670
9. Gautam, K.R., Andresen, G.B., Victoria, M.: Review and techno-economic analysis of emerging thermo-mechanical energy storage technologies. Energies **15**(17), 6328 (2022). https://doi.org/10.3390/en15176328
10. Tong, W., Lu, Z., Chen, W., Han, M., Zhao, G., Wang, X., Deng, Z.: Solid gravity energy storage: a review. J. Energy Storage **53**, 105226 (2022). https://doi.org/10.1016/j.est.2022.105226
11. Moraski, J.W., Popovich, N.D., Phadke, A.A.: Leveraging rail-based mobile energy storage to increase grid reliability in the face of climate uncertainty. Nat. Energy **8**, 736–746 (2023). https://doi.org/10.1038/s41560-023-01276-x
12. Rabi, A.M., Radulovic, J., Buick, J.M.: Comprehensive review of liquid air energy storage (LAES) technologies. Energies **16**(17), 6216 (2023). https://doi.org/10.3390/en16176216
13. Pal, A., Goswami, D., Martinez, R.V.: Elastic energy storage enables rapid and programmable actuation in soft machines. Adv. Funct. Mater. **30**(14), 1906603 (2020). https://doi.org/10.1002/adfm.201906603
14. Sadeghi, G.: Energy storage on demand: thermal energy storage development, materials, design, and integration challenges. Energy Storage Mater. **46**, 192–222 (2022). https://doi.org/10.1016/j.ensm.2022.01.017
15. Sarbu, I., Sebarchievici, C.: A comprehensive review of thermal energy storage. Sustainability **10**(1), 191 (2018). https://doi.org/10.3390/su10010191
16. Frazzica, A., Palomba, V., Freni, A.: Development and experimental characterization of an innovative tank-in-tank hybrid sensible-latent thermal energy storage system. Energies **16**(4), 1875 (2023). https://doi.org/10.3390/en16041875
17. Medrano, M., Gil, A., Martorell, I., Potau, X., Cabeza, L.F.: State of the art on high-temperature thermal energy storage for power generation. Part 2—case studies. Renew. Sustain. Energy Rev. **14**(1), 56–72 (2010). https://doi.org/10.1016/j.rser.2009.07.036
18. Airò Farulla, G., Cellura, M., Guarino, F., Ferraro, M.: A review of thermochemical energy storage systems for power grid support. Appl. Sci. **10**(9), 3142 (2020). https://doi.org/10.3390/app10093142
19. Zhang, Q., Heuberger, C.F., Grossmann, I.E., Sundaramoorthy, A., Pinto, J.M.: Optimal scheduling of air separation with cryogenic energy storage. In: Computer Aided Chemical Engineering, vol. 37, pp. 2267–2272. Elsevier (2015). https://doi.org/10.1016/B978-0-444-63576-1.50072-8

20. Abdullah, M., Koushaeian, N.A., Shah, N.A., Chung, J.D.: A review on thermochemical seasonal solar energy storage materials and modeling methods. Int. J. Air-Cond. Ref. **32**(1), 1 (2024). https://doi.org/10.1007/s44189-023-00044-6
21. Reddy, M.V., Mauger, A., Julien, C.M., Paolella, A., Zaghib, K.: Brief history of early lithium-battery development. Materials **13**, 1884 (2020). https://doi.org/10.3390/ma13081884
22. Liu, C., Neale, Z.G., Cao, G.: Understanding electrochemical potentials of cathode materials in rechargeable batteries. Mater. Today **19**, 109–123 (2016). https://doi.org/10.1016/j.mattod.2015.10.009
23. Flow batteries for grid-scale energy storage. MIT News | Massachusetts Institute of Technology (2023). https://news.mit.edu/2023/flow-batteries-grid-scale-energy-storage-0407. Accessed 24 Nov 2024
24. Qasem, N.A.A., Abdulrahman, G.A.Q.: A recent comprehensive review of fuel cells: history, types, and applications. Int. J. Energy Res. **2024**, 7271748 (2024). https://doi.org/10.1155/2024/7271748
25. Fang, W., Ding, C., Chen, L., Zhou, W., Wang, J., Huang, K., Zhu, R., Wu, J., Liu, B., Fang, Q., Wang, X., Wang, J.: Review of hydrogen storage technologies and the crucial role of environmentally friendly carriers. Energy Fuels **38**, 13539–13564 (2024). https://doi.org/10.1021/acs.energyfuels.4c01781
26. Itani, K., De Bernardinis, A.: Review on new-generation batteries technologies: trends and future directions. Energies **16**, 7530 (2023). https://doi.org/10.3390/en16227530
27. Singh, S., Bhan, C., Bhaduri, A., Gupta, M.K., Yoon, J., Maiti, C., Singh, J., Yadav, B.C.: Self-powered fluoride detection and removal system using waste material utilization and self-healing capabilities. Mater. Today Commun. **41**, 111061 (2024). https://doi.org/10.1016/j.mtcomm.2024.111061
28. Bhaduri, A., Singh, S., Gupta, M.K.: Self-powered fluoride ion detectors based on piezoelectric nanogenerators with filler materials comprising Mn-doped $BaTiO_3$ nanostructures and carbon nanotubes. ACS Appl. Nano Mater. **6**, 6637–6652 (2023). https://doi.org/10.1021/acsanm.3c00340
29. Jahandar, M., Kim, S., Lim, D.C.: Transforming wearable technology with advanced ultra-flexible energy harvesting and storage solutions. Nat. Commun. **15**, 8149 (2024). https://doi.org/10.1038/s41467-024-52534-9
30. Georgious, R., Refaat, R., Garcia, J., Daoud, A.A.: Review on energy storage systems in microgrids. Electronics **10**, 2134 (2021). https://doi.org/10.3390/electronics10172134

MXene-Based Functional Materials for Supercapacitors Energy Device

Priyanka Chaudhary, Jitesh Pani, Hitesh Borkar, and Meng-Fang Lin

Abstract The rising energy demand has led to the development of supercapacitors designed to hold more energy and have an improved capacity than conventional batteries, with separate charging and discharging properties. The layered, two-dimensional structures of MXenes grant them high electrical conductivity, rich redox activity, and surface chemistry that can be modified, making them promising candidates for next generation supercapacitor energy storage devices. Their unique attributes enable rapid charge transport, substantial pseudocapacitance, and extraordinary rate capability, thus making MXenes ideal for high power energy storage systems. However, MXenes are still facing challenges such as surface termination instability, structural deterioration, narrow voltage windows, and layer restacking which limit fully utilizing their potential. To address these challenges, considerable works focused on applying transition metal oxides, carbon nanostructures, and conductive polymers onto MXenes to improve their mechanical stability and electrochemical performance. Additionally, advances in synthesis methods, electrolyte compatibility, and device integration have expanded their applications in flexible and wearable energy storage systems. By recapping recent progress, design approaches, and the main hurdles for MXene-based materials used in supercapacitors, this review aims to illuminate the gaps toward achieving high-performing, scalable, and environmentally safe energy storage technologies.

1 Introduction

Supercapacitors (SCs) are a type of energy storage devices that provides excessive power density, almost instantaneous charge/discharge speeds, and long cycle life when compared to regular batteries [1]. Renewable energy sources like solar, wind,

P. Chaudhary · J. Pani · M.-F. Lin (✉)
Department of Materials Engineering, Ming Chi University of Technology, New Taipei City, Taiwan
e-mail: mflin@mail.mcut.edu.tw

J. Pani · H. Borkar
Department of Physics, National Institute of Technology, Warangal, Telangana, India

R. K. Sonker et al. (eds.), *Low-Dimensional Multifunctional Hybrid Nanostructured Materials*, Engineering Materials, https://doi.org/10.1007/978-981-96-7952-2_13

thermal, and hydroelectric power play an intermittent role in minimizing concern of rapidly exhausting fossil fuels reserves, increasing environmental pollution and adverse climate changes. The adoption of such extensive renewable source technologies in turn manifests sustainable, cleaner, environment-friendly and economical, cheaper energy storage technologies. Electrochemical energy storage offers a great appeal as an omnipotent player to develop novel energy storage device with high energy and high power density with a good cycle ability [2]. Many strategies have been developed such as exploiting sustainable energy storage systems like solar cells and rechargeable batteries in combination with SCs. SCs are ideal for energy storage device application because of their high power, long-life cycles, environmentally friendly processes in fabrication and exploitation with high efficiency to recover energy [3]. Recently, lamellar type 2D materials in which the sheets are built only a few monolayers are subjects of intensive in field of energy storage device. 2D materials classified in Chalcogenides, Graphene, MXenes etc. MXene were found in this environment in 2011, and their name comes from their chemical compositions, M stands for transition metals like titanium, chromium, or vanadium, X denotes Carbon as well as Nitrogen, and the suffix "ene" designed for a link to graphene. Significantly, because there is a large range of transition metals that can be employed, many MXenes which can include a broad range of distinct atoms may be synthesized [4]. This characteristic sets of MXenes apart in the field of 2D materials, produced from the MAX phase, a group of more than 70 members with a range of compositions, which is a term for three-dimensional solids. Through the etching process, other atoms, such as oxygen, hydrogen, or fluorine, can be added to the surface of MAX sheets, giving MXenes their chemical diversity. The amount of these atoms can be adjusted, and their presence can also affect their properties [5, 6]. The chemical prosperity, excellent electrical conductivity, and many useful and adaptable qualities of MXenes are the causes of this enthusiasm. Their characteristics are determined by the chemical components found in the MAX phase, but they are also influenced by the end groups that are left over after exfoliation. Studies have indicated that morphology, plate size, and other factors affect MXene's characteristics. There are expected to be many more MXenes developed in future, bringing the total to about thirty. These diverse chemical compositions provide a wide range of potential characteristics for various MXenes are a 2D family of layered transition metal carbides, nitrides, or carbonitrides that are selectively etched using a wet chemical process to separate from their parent MAX phases [7]. Because of their comparable properties, graphene and MXene share a name. Figure 1 depicts diverse elements from the periodic desk representing MAX phases [8]. On the other hand, the metallic M–A bond is slightly weaker than the M–X bond, so it is easier to chemically etch the A layer to form MXene. Various elements of the periodic table exhibit the MAX phase. The MAX phase is a member of the space group D6h4-P63/mmc. The crystal structure of MXene confirmed by a hexagonal layered structure with the M and Al layers interspersed, although the X atoms occupy octahedral positions. Unlike graphene and transition metal dichalcogenides (TMDs), which have Van der Waals bonds between the layers, the M–X bonds are ferro/covalent/ionic. A research team from Drexel University in Philadelphia, led by researchers Gogotsi and Barsoum, discovered the first MXene fabric, $Ti_3C_2T_x$ [9]. These transition steel carbides and nitrides are generally defined by the formula $Mn_1X_nT_n$, where M represents a transition metal, X represents carbon and/or nitrogen, and T represents a surface end group such as –OH, –F, or –O. The conductive properties and charge-switching capabilities of these materials, which can be attributed to the various oxidation states of the transition metal M and the layered structure of MXenes, make them ideal for electrochemical packages, especially energy storage devices. Nevertheless, MXenes are challenged with numerous obstacles that limit developing them into more feasible supercapacitor electrodes than other carbons and metal materials. MXene's chemistry and complexity are higher, and therefore its energy storage mechanism in different electrolytes has not yet been entirely disclosed. Though various synthesis methodologies for MXenes are established, controlled surface chemistry and structure are still challenging problems; lastly, the agglomeration issue frequently encountered by 2D materials in real applications still needs further studies. Furthermore, MXenes exhibit target layer end groups, ferroelectric conductivity, fast redox reactions, and higher charge capacities compared to other materials due to their tunable electrical and magnetic properties [4, 10]. The distinctive layered architecture of MXenes facilitates both the transport of electrolyte ions and the availability of active redox sites from transition metals on their surfaces. These characteristics render MXenes especially advantageous for use as high-performance electrodes in electrochemical capacitors, commonly known as supercapacitors.

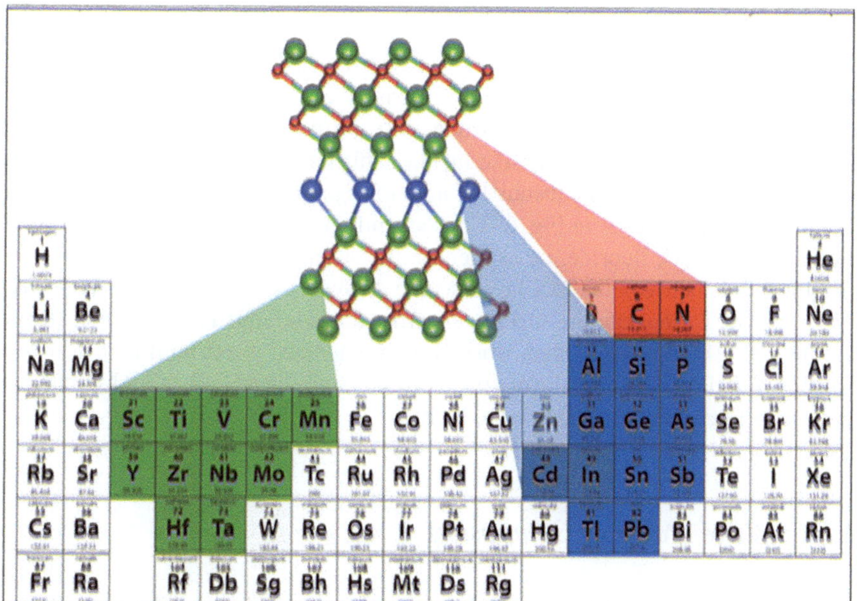

Fig. 1 Periodic table showing the MAX phases [11]. Reproduced with permission @ Copyright (2023) Elsevier

This chapter presents the properties of MXenes and the key elements that significantly affect the energy storage capabilities of supercapacitors. It also offers an intensity recent advances in the packages of MXenes and their hybrids to substances including carbonaceous materials, engaging in polymers, (TMDs), and transition steel oxides (TOs). Moreover, the chapter discusses present research gaps in MXene-based electrodes for device packages, their relevance in the international electricity marketplace, and diverse strategies for improving electrochemical properties and lowering charges, all of which can be discussed in phrases of future views.

2 Synthesis Method

During the past decades, MXene composite development for energy storage devices has become an attractive technique for the preparation of strong, versatile material. MXene, with its two-dimensional morphology, has stacking and remarkable flexibility, making it unique properties to fabricate the multifunctional composites. A various types of new composites have been fabricated by simply mixing MXene with polymers, metal oxides, and carbon nanotubes [12, 13]. The 2D shapes with a multilayered structure demonstrating the good electrical conductivity with high mechanical strength. The M–A bond typically exhibits a metallic character in the MAX phase, whereas the M–X bond exhibits a combination of ionic, metallic, and covalent characteristics. In contrast to graphite and TMDs, where a weak van der Waals force holds the layers together, mechanical exfoliation or sharing of these bonds has not been feasible due to the strong interlayer bonding in the MAX phases. MAX phases have a unique combination of characteristics that influence metals and ceramics. Like ceramics, they are strong, stiff, brittle, and thermally stable. They also produce heat and electricity like metals. This method involves stirring layered MAX phase powders with aqueous HF acid for a predetermined amount of time at room temperature shown in Fig. 2. Consequently, weak bonds of surface terminations like hydroxyl, fluoride, or oxygen on the surface of ML-MXene replace the metallic bonds between the MX layers as the A layers of the MAX phase are selectively etched. To separate the supernatant from the solid, this is further centrifuged and filtered. The mixture is then washed with deionized water until its pH is kept between 4 and 6. Thus, few-layer (FL) MXene is produced. When an MXene has fewer than five layers, it is deemed to be FL.

2.1 MXene–Polymer Composites

These include the use of MXenes with enhanced properties, hydrophilic components, and metal components to enhance electrical and thermal properties in polymers when formed into composites. Single-layered MXene can easily offer good hydrophilicity and polymer compatibility as compared to multilayered MXene. As Ling et al. [7]

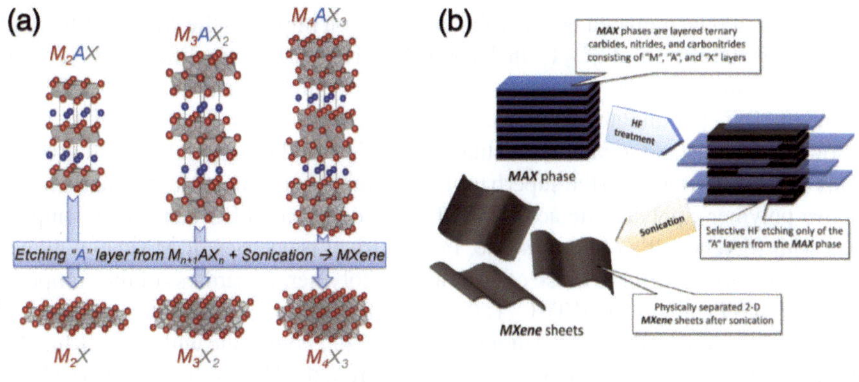

Fig. 2 Schematic of MXene synthesis [14]. Reproduced with permission @ Copyright (2019) Elsevier

explained his recent key findings, the resulting composite of $Ti_3C_2T_x$–polyvinyl alcohol was obtained by mixing the colloidal solutions of the films with the aqueous solution of PVA; PVA is distributed over the sheets of $Ti_3C_2T_x$, creating the coating process for composite material $Ti_3C_2T_x$–PVA. The formation of composite material was without any agglomeration. Naguib and colleagues fabricated $Ti_3C_2T_x$–PAM composites using a two-step process. To improve the $Ti_3C_2T_x$ layer separation and achieve total delamination of the $Ti_3C_2T_x$ layers, Dimethyl sulfoxide (DMSO) is an organosulfur was first introduced into the $Ti_3C_2T_x$ interlayers [15–17]. They enhance the polymer and thermal properties in transitioning from the polymers to composite. Single-layer MXene is more hydrophilic when compared to multi-layer MXene; it has a richer composition of polymer. Normally, MXene is divided into many layers before adding into the polymers. Ling et al. Single-layered composites followed from the mixture of colloidal films of $Ti_3C_2T_x$ with the aqueous solutions of PVA. In the resulting composite $Ti_3C_2T_x$/PVA, there is a layer of PVA dispersed over a layer of $Ti_3C_2T_x$. The very strong hydrogen bonding between PVA and hydrophilic $Ti_3C_2T_x$ imparts good thermal stability to the mix. Composites of $Ti_3C_2T_x$/PVA possess better tensile and compressive characteristics and turn out to be more flexible than the individual $Ti_3C_2T_x$ and PVA. That is impressive, as the tensile strength recorded for the PVA containing 40 wt% $Ti_3C_2T_x$ was 91 MPa, around four times that of the neat $Ti_3C_2T_x$ [13, 18]. Besides, the addition of Ti could significantly enhance the electrical conductivity of composites. DMSO for expanding the layer and making every layer fully delaminate. Mix well the solution of $Ti_3C_2T_x$ and PAM, then dry in a room temperature environment over four to five days. When the content of $Ti_3C_2T_x$ reaches 6 wt%, the composite exhibits ideal conductivity and the highest electrical conductivity of 3.3×10^{-2} S/m. According to SEM, the fracture of the $Ti_3C_2T_x$/PAM sample with the content of 6 wt% is very severe, which is the quality of the bone fracture pattern. It was also documented in literature studies that UHMWPE (Ultra-high-molecular-weight polyethylene) can also form composites with $Ti_3C_2T_x$ [19]. Before producing $Ti_3C_2T_x$/UHMWPE composites, surface-modified $Ti_3C_2T_x$

powder should be prepared because the modification improves the compatibility and dispersion stability of $Ti_3C_2T_x$ in UHMWPE. The surface-modified $Ti_3C_2T_x$ and UHMWPE mixture is then shaped using a high-pressure furnace, heated to 220 °C at a rate of 10 °C per minute and held at 10 MPa for 30 min. The flexibility of $Ti_3C_2T_x$ allows it to withstand greater stress than UHMWPE, reducing cracking and growth. This device contributes to the superb tools discovered in this mixture. Similarly to the above polymers, polyacrylic acid, PVP, PEO and alginate/PEO can shape composites with MXenes. It's obvious sense, the electrical properties of MXene–polymer composites are superior to those of MXene and polymer. Regardless of the composition of $Ti_3C_2T_x$, all prepared $Ti_3C_2T_x$/UHMWPE composites have better mechanical strength than pure UHMWPE. Among them, $Ti_3C_2T_x$ composite with 10 wt% has the highest elongation, yield strength, and tensile strength. The strain used in $Ti_3C_2T_x$/UHMWPE composites improves the mechanical properties of UHMWPE. Because of the composite structure, the stress in UHMWPE can be efficiently transferred to $Ti_3C_2T_x$ through the polymer chains [20].

2.2 Composites of Metal/Oxide and MXene

The composites of MXene with metals/metal oxides illustrate high electrical conductivity and good electrochemical stability. Rashid et al. [21] presented the Au/$Ti_3C_2T_x$ nanocomposite was synthesized through chemical reduction. Along with the addition of a reducing agent to assist in lowering the solution containing the metal, stirring was done. Addition of $Ti_3C_2T_x$ MXene was then made to the previous mixture. After 30 min, ultrasound treatment was given to the chemical. After the reaction was completed three times, the solution was drained with deionized water using cellulose filter paper with a 0.1 mm pore size. The filtered material was then dried under vacuum in an oven at 80 C for two hours. Wang et al. developed $Li_4Ti_5O_{12}$–$Ti_3C_2T_x$ composite [15] where the two-step synthesis procedure was showing great potential for successfully compositions. $Ti_3C_2T_x$, different hydrogen peroxides, and $LiOH{\cdot}H_2O$ can be mixed with deionized (DI) water to produce lithium ions that can first permeate the $Ti_3C_2T_x$ layer and separate them from one another [15, 22]. To obtained products repeatedly cleaned in DI water, dried for 24 h at 40 C, and finally calcined for 4 h at 550 C in a N_2 environment. TiO_2–$Ti_3C_2T_x$ nanocomposites were synthesized in a hydrothermal setting by Zhu et al. The sample is heated at 500 C for 4 h to make the TiO_2–$Ti_3C_2T_x$ nanocomposite. This occurs after tetra butyl titanate (TBOT) interacts with $Ti_3C_2T_x$ to produce precipitates of TiO_2 particles have potential for the composition after the surface modifications [23–25].

2.3 MXene-Based Composite Materials for Supercapacitor

MXene provides an ideal surface for energy storage due to its electrochemically active surface, specific surface area, and interlayer structure. However, problems such as ridge repair and poor quality of materials hinder their future development. A feasible and reliable strategy to improve the performance of supercapacitors is to create composite structures or heterostructures that solve the problems associated with MXene as an electrode material. The classification of supercapacitor given below with work mechanism.

Electrostatic double layer capacitors (EDLCs)—These are based on the physical double layering of charge at the electrode–electrolyte interface and utilize an electric double layer. They employ carbon-based materials as electrodes and are characterized by high power density and long cycle life.

Pseudo capacitors—These store charge by rapid and reversible redox reaction on the electrode surface. They tend to employ metal oxides or conductive polymers, which deliver greater energy density than EDLCs.

Hybrid Supercapacitors—These feature the characteristics of both EDLCs and pseudo capacitors, combining various electrode materials to minimize energy and maximize the power density. Supercapacitors build energy using electrostatic and electrochemical procedures. Contrary to batteries that utilize slow chemical reactions, supercapacitors charge, and discharge quickly. The energy is stored at the electrode–electrolyte interface where ions build and form capacitance. This allows them to supply bursts of energy effectively and are best for uses which demand instant energy discharge, for instance, regenerative breaking in electric motors. The incorporation of MXenes into supercapacitor devices has the potential to maximize its performance by increasing conductivity, stability, and charge storage. Nevertheless, its applicability is faced with challenges such as surface chemistry management and agglomeration.

The MXenes provide high capacitance, outstanding rate capability, and strong long-term stability by striking a convincing balance between the pseudocapacitive behavior of metal oxides and the high conductivity of carbon-based materials shown in Table 1. Their special qualities make them excellent candidates for next-generation high-performance supercapacitors, particularly in flexible and wearable electronics, even though they are not yet the most affordable or scalable option.

MXene is a suitable filler for composites due to its reliable ratio, tunability of surface end groups, and high elastic modulus. With the development of composite materials, researchers have decided to establish elastic, wearable, and portable electronic devices for supercapacitor [26]. The work Qin et al. [27] describes a simple-state supercapacitor which consists of MXene and poly(3,4-ethylenedioxythiophene) poly(styrenesulfonic acid) layers. It uses concentrated H_2SO_4. Such a revolutionary supercapacitor offers the maximum capacitance of 568 F/cm^3, energy density of 33.2 mWh/cm^3, and energy density of 19,470 mW/cm^3. When the gap MoC MXene layer interacts with the PEDOT material, it forms nearly continuous triangular current charge discharge (GCD) curve. That means the redox reaction is reversible

Table 1 Parameters of the supercapacitor

Material	Specific capacitance (F/g)	Power density (W/kg)	Cycling stability
MXenes (e.g., $Ti_3C_2T_x$)	300–1500	Up to 10,000	>90% after 10,000+
Activated carbon	100–300	1,000–5,000	~95% after 10,000+
Graphene	200–500	5,000–10,000	90–95% after 5,000–10,000
Carbon nanotubes (CNTs)	100–300	10,000+	Excellent (>95%)
Transition metal oxides (e.g., MnO_2, RuO_2)	300–1000	500–2,000	Moderate (70–90%)
Conducting polymers (e.g., PANI, PEDOT)	200–700	500–2,000	Low to moderate (50–80%)

and the hybrid material has high Coulombic efficiency. Li et al., in the second work, used the energy storing abilities of CNTs, MXene, and PANI to fabricate an asymmetric supercapacitor by using $Ti_3C_2T_x$ MXene as the negative electrode and rGO/CNT/PANI as the well electrode. Fabrication of solid-state supercapacitor by concentrated H_2SO_4 determination in the coating process of MXene and poly(3,4-ethylenedioxythiophene) poly(styrene sulfonic acid). The indications of the new supercapacitor are that the maximum capacitance was 568 F/cm^3, energy density at 33.2 mWh/cm^3, and energy density at 19,470 mW/cm^3. In fact, the almost triangular-shaped GCD curve that forms when the space-cut MoC MXene layer interacts with the PEDOT material is an indication of good performance compared to Coulombic and reverse redox reactions. Li et al. developed a second asymmetric supercapacitor based on energy-storing capabilities of CNT, MXene, and PANI. The negative electrode was $Ti_3C_2T_x$ MXene, while the positive electrode was rGO/CNT/PANI. Jian et al. [28] developed a carambola-inspired MXene–Polypyrrole (PPy) composite utilizing a one-step co-electrodeposition technique. The presence of functional groups (–F, –OH, or –O) on the surface of MXene makes it a core polymer, but the introduction of monomer radical cations from pyrrole monomer enhances surface polymerization. This strategy upgrades particle distribution and electron transmission by changing the two-dimensional structure into a three-dimensional shape. At a current thickness of 0.5 A/g, the composite incorporates a gravimetric capacitance of 416 F/g. It confirmed the pseudo-rectangular and symmetric triangular cyclic voltammetry and galvanostatic rate-discharge bends one by one, which provided a reversible charging and discharging take care of with fantastic capacitive execution. The symmetric supercapacitor joined on ITO glass the use of MXene/PPy composite displays very incredible cyclic solidness: a selected capacitance of 184 F/g at a test price of 10 mV/s, that's able to keep 86.4% capacitance inside 5000 cycles at 5 A/g. Currently, the electronics enterprise is dealing with an boom in the utilization of those planar supercapacitor devices as a result of their gift shape for flexibility, portability, and safety. Zhang et al. [29] have been able to efficaciously fabricate a planar supercapacitor device using Ti_3C_2/PPy added by using electrochemical polymerization and electrophoretic deposition. A price of 109 four mF/cm^2 changed into recorded the usage of a 3-electrode configuration in a 2 M electrolyte solution shown in Fig. 3. An 86.7 mF/cm^2 and 1.5 mA/cm^2 current thickness solid-state electrolyte in PVA/H_2SO_4 was used. After the belief of a planar $Ti_3C_2T_x$/PPy structure with the aid of laser ablation, an interdigital design is engraved on ITO glass. The $Ti_3C_2T_x$ portions are electrophoretically retained for 2 min on the potential of 30 V at the ITO glass floor. On this identify, chronoamperometry technique has been used for the deposition of PPy at the $Ti_3C_2T_x$/ITO by way of the assistance of electrochemical polymerization at 0.8 V. Other materials for redox crests, which recommend their tall houses. PVA/H_2SO_4 use solid-nation electrolyte along with 86.7 mF/cm^2 and current thickness 1.5 mA/cm^2. On this approach, following the preparation of a planar $Ti_3C_2T_x$/PPy, the shape is reduced into an interdigital sample on ITO glass through laser ablation. Later, the $Ti_3C_2T_x$ chips are electrophoretically deposited for 2 min at an ability of 30 V onto the ITO glass floor. A planar supercapacitor with an aerial control thickness of 0.0845 mW/cm^2 provides a floating energy density in the air of three 398 μWh/cm^2. The tool's integration with a solar-powered cell showed its utility. The conjugated microporous polymers (CMPs) are exciting natural porous substances made up of rigid covalently bonded devices prepared in a 3D shape. While polymers, CMPs are totally exclusive and may be prepared inexpensively by more than one methodology, rendering them beneficial in fuel adsorption, catalysis, and strength applications. Conjugated microporous polymer based totally on MXene, *p*-iodophenyl functionalized MXene become synthesized with the aid of Yang et al. Restacking and oxidation of MXene have been effectively expected with the aid of conjugated microporous polymers generated from MXene, which had higher surface areas and -dimensional architectures. In comparison to its substance, the final cathode material has a bigger capacitance. The synergistic substance that integrates MXene with one-of-a-kind sections of the microporous polymer network is what is answerable for this enhancement [28, 30, 31] (Table 2).

In recent year, MXene-based interconnected microelectrodes have gained large-scale applications in flexible and robust substrates for MSCs, allowing high precision control of the distance between microelectrodes from tens to several micrometers. For example, Peng et al. fabricated all-solid-state MXene-MSCs using coordinated laser ablation. Smaller $Ti_3C_2T_x$ nanosheets (s-$Ti_3C_2T_x$) were subsequently deposited on the substrate, where larger $Ti_3C_2T_x$ nanosheets (L-$Ti_3C_2T_x$) served as the final collector layer. Interdigitated microelectrodes with a thickness of about 3 μm were fabricated using 10.6 μm CO_2 laser ablation. The L-s-MSCs exhibited excellent electrochemical performances, with an air capacitance of 27.3 mF cm^{-2} and a volumetric capacitance of 356.8 F cm^{-3}. The L-s-$Ti_3C_2T_x$ MSCs exhibited an air capacitance of 27.3 mF cm^{-2} and a volumetric capacitance of 356.8 F cm^{-3}. At an experimental rate of 50 mV s^{-1}, the capacitance was found to be very well maintained even after 10,000 cycles. Laser etching also has the advantage of generating MSCs on both sides of the substrate, which can ultimately double the capacity. A solid polyethylene terephthalate (PUP) substrate was directly cleaved using a femtosecond laser into an ultra-precise waist section of double-sided MSCs (Fig. 4a–d). The cross-sectional

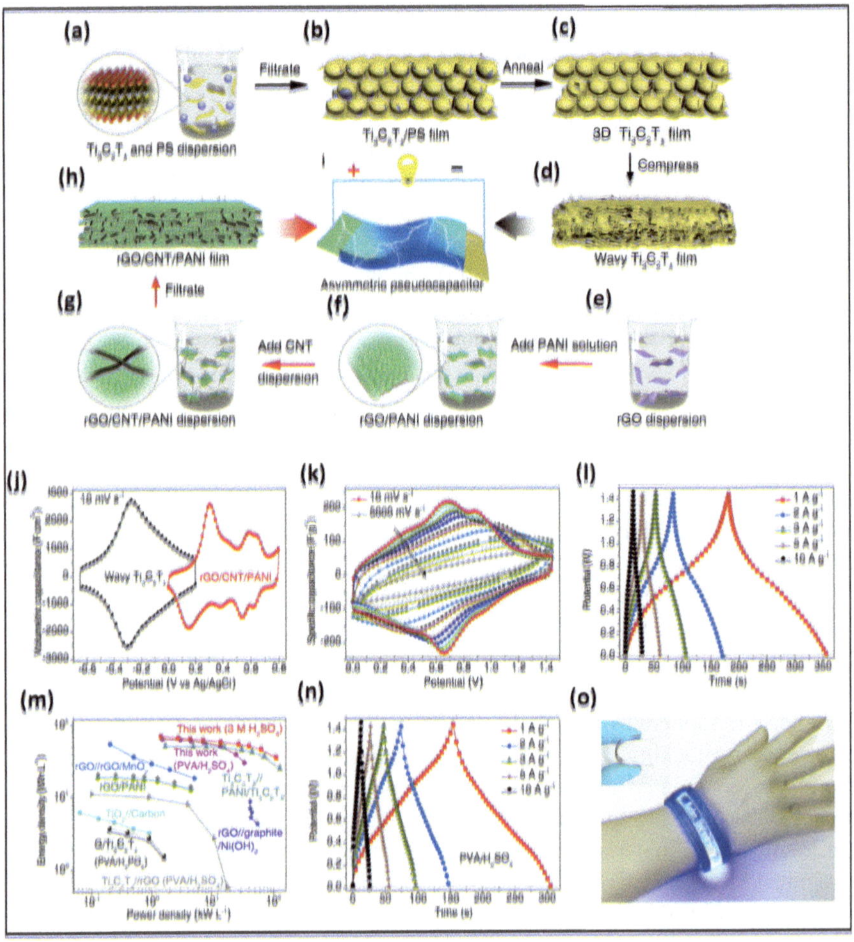

Fig. 3 Showing the preparation for fabricating MXene//rGO/CNT/PANI devices with wavy shape includes the following steps: **a** mixing $Ti_3C_2T_x$ and dispersing polystyrene in water; **b** forming unsupported MXene/polystyrene films by filtration; **c** placing the MXene films formed by curing to break the polystyrene circles; **d** compressing the formed permeable films within the MXene corrugated films; **e** planning the dispersion of GO in *N*-methylpyrrolidine; **f** improving the scattering of rGO/PANI by incorporating graphene and PANI; **g** measuring the scattering of rGO/CNT/PANI by incorporating CNT to form rGO/CNT/PANI films; **h** supporting scheme of MXene//rGO/CNT/PANI hilter-kilter wavy devices; **i** cyclic voltammetry (CV) bending for cathode and anode electrodes at a test rate of 10 mV/s; **j** CV bending of the reversed devices at various filtration rates of 10, 20, 50, 100, 200, 500, 1000, 2000, 3000, and 5000 mV/s; **k** galvanostatic charge–discharge (GCD) bending of the Hilter Kilter devices in gelled electrolyte H_2SO_4; **l** Ragone plot comparing the performance of the rejected devices with other cathode materials. **m** Bending of the GCD gadget in PVA/H_2SO_4 gel electrolyte; **n** expanded image of the blue Driven in layout with an insert showing the adaptive rejection gadget [45]. Reproduce with permission @ Copyright (2020) Elsevier

Table 2 A comprehensive overview of supercapacitor electrodes that incorporate MXene and its composite for supercapacitor

Sl. No.	Materials	MXene synthesis	Electrolyte	Specific capacitance	Refs.
1	*n*-Butyllithium treated $Ti_3C_2T_x$	Etching	1 M H_2SO_4	523 F/g	[32]
2	$MoS_2/Ti_3C_2T_x$	Etching	1 M H_2SO_4	342 F/g	[33]
3	CNTs/$Ti_3C_2T_x$	Etching	1 M H_2SO_4	300 F/g	[34]
4	rGO/$Ti_3C_2T_x$	Etching	1 M H_2SO_4	195 F/g	[35]
5	Folded structure $Ti_3C_2T_x$	Etching	3 M H_2SO_4	345 F/g	[36]
6	LDH/$Ti_3C_2T_x$	Etching	6 M KOH	655 F/g	[37]
7	$BaTiO_3/Ti_3C_2T_x$	Etching	0.1 M H_2SO_4	254 F/g	[38]
8	$Ti_3C_2T_x/Ni_2CO_3(OH)_2$	Etching	6 M KOH	173 F/g	[39]
9	N_2H_4 modified $Ti_3C_2T_x$	Etching	1 M H_2SO_4	250 F/g	[40]
10	$Ti_3C_2T_x$/PPy/PVA	Etching	PVA/H_2SO_4	614 F/g	[41]
11	CTAB-Ti_3C_2@V_2O_5	Etching	Na_2SO_4	477 mF/cm^3	[42]
12	Nb_2CT_x/CNT	Etching	H_2SO_4	462 mF/cm^3	[43]
13	$NiMoO_4/V_2CT_x$@rGO	Electrostatic	PVA–KOH	346 F/g	[44]
14	$Ti_3C_2T_x$/N-doped CNT	Etching	H_2SO_4	299.52 F/g	[45]
15	$Ti_3C_2T_x$/cellulose nanofiber/porous carbon film	Vacuum filtration	PVA/KOH	143 mF/cm^2	[46]

MSCs demonstrated a volumetric capacitance of 308 F cm^{-3} at a thickness of 10 μm. In addition, as shown in Fig. 4e, the output voltage increased dynamically from 0.6 to 4.8 V, while the charging and discharging times of the blocks coupled to the assembly remained almost unchanged. Flexible MXene-MSCs comprising MSC winders in series can provide an operating voltage up to 7.2 V. Alsharif et al. fabricated flexible and twisted MXene-on-paper coplanar MSCs with thicknesses of 100–150 μm using Meyer rods. Although the devices cannot effectively change the thickness and precision, they provide tremendous integration results. The figure shows the state-of-the-art laser-cut MXene-MSC all-in-one Si electronic device configuration. Figure 4g–j demonstrates an extremely high volumetric energy density of 75 mWh cm^{-2} and an extremely high operating voltage of 1.2 V. This high level of integration will undoubtedly further accelerate the development of high-performance Si laser devices using MXene. In any case, the edges of MXene interdigitated microelectrodes are prone to oxidation during high-temperature laser processing. To overcome this problem, Huang et al. developed a novel UV laser stamping technique to pattern MXene thin films using a low-temperature UV laser. The low pulse width of the low-temperature UV laser allows the material to be removed without significant surface heating, which reduces oxidation during the laser etching process and prevents edge defects caused by localized high temperatures. Secondly, the laser can create microchannels that enhance particle transport, which improves the electrochemical performance of MSC devices. In this regard, Jiao et al. presented self-supporting MXene-MSCs using PVA/H_2SO_4 gel as electrolyte, which exhibited remarkable mechanical adaptability with a 60° bending curvature and remarkable electrochemical performances, such as achieving an air thickness of 43.5 mWh cm^2 and a fixed areal capacitance of 340 mF cm^2. With respect to the MXene nanosheets in the composite paper, the BCF plays two roles: as a stabilizer to enhance the mechanical properties and as a separator to reduce re-agglomeration, establish interlayer dispersion, promote particle transport, and increase the accessibility of the active sites. The 100% plastically strained twisted and twisted MXene-BCF MSCs exhibited an ether capacitance of 111.5 mF cm^2 without any obvious capacitance degradation, as shown in Fig. 3k. In addition, the MXene-BCF MSC composite block exhibited 132% higher air energy thickness than that of pure MXene. The enhanced particle transport channels between the multiple layers of MXene increased the exchange capacity of divalent Zn^2 and improved the overall electrochemical performance of Zn ion MSCs. More importantly, in terms of specifications with circular financial systems and green technology knowledge, these composite stretchable MXene-BCF MSCs have great potential to enhance the utilization of flexible and stretchable energy sources, making them a possible alternative energy source for wearable devices. In addition, laser engraving has been used not only for the fabrication of symmetrical full MXene-based MSCs but also for the fabrication of cross-MSCs, especially HMSCs and inverted MSCs, especially AMSCs, which can operate at much higher voltages than symmetrical MXene-based ones. MSC. Zinc-ion HMSC using polyacrylamide/$ZnSO_4$ hydrogel electrolyte showed high operating voltage of 1.65 V, capacitance of 129 mF cm^2, and energy density of 48.9 μWh cm^2. This is achieved by laser etching the interdigitated structure using MXene as capacitor anode and V_2O_5 as battery anode. Wu et al. aligned the horrible MXene anode with the high-quality CNT@PPy cathode using vacuum filtration and laser reduction. The result is an AMSC with an air capacity of 150.2 mF cm^2 and an operating voltage of 1.4 V. The AMSC islands were connected and folded to fabricate a spiral-shaped version of the stretchable AMSC. Within the scan range, it has linear/areal capacitances of 1.5–29.4 mF cm^2 and 1.4–26.8 mF cm^2, and linear/areal current densities of 7.9 mF cm^2 and 7.2 μWh cm^2, respectively. The device maintained significant capacitance at specific measurements of 100% strain, 180° bending, and 360° twist. Finally, an elastic string-like AMSC (WSS-AMSC) model was fabricated by bonding and folding the islands of AMSC, exhibiting various line/area shapes using a flexible silicone bonding substrate with stress-absorbing and water-resistant functions. The WSS-AMSCs showed uncommon washability and longevity, retaining 87% and 84% capacitance for over 1000 cycles at 100% prolongation interims (50 cycles), and after thorough cleaning (10 instances), in my opinion. This creative rib topology design propels the key detail of creating development in seclusion and encourages the enhancement of complex straight adaptable power ability devices decided from littler subsystems, coming about in a direct, stretchable, wearable, and regular micropower supply with bendy yield voltage/modern, customizable vitality thickness, and incredible deformability and washability [47–51].

In addition, except for the aqueous method assisted by the chemical reduction of graphene oxide with the use of a reducing agent, the instinct of stacking two-dimensional graphene nanosheets drives 3D self-assembly of graphene hydrogel, which could be obviously observed. However, residual oxygen-containing groups do induce re-aggregation of graphene, leading to the decline of electric conductivity and further influencing the electrochemical performance. The restacking problem is improved upon the incorporation of MXene into graphene nanosheets. Likewise, the flexible, free-standing MXene/holey graphene varied film was prepared by Fan et al. via filtering alkalized MXene and holey graphene oxide dispersions, then mildly reducing them. The water-soluble base cloth destroys the charge balance of MXene, thus accelerating the conversion of –F terminals to –OH terminations. The subsequent termination of –OH terminals from reduction will raise the proportional content of titanium atoms, thus achieving higher pseudocapacitive activity. Figure 5 represents the fabrication method of the MXene-holey graphene film and symmetric supercapacitor effects. Figure 5a–e contains a CV curve that describes a huge amount of capacitance formed about pseudo capacitance harnessed into the atoms of titanium which promotes long-term and reversible redox states. The prepared fabric exhibits a volumetric capacitance of 1455 F/cm^3, while maintaining a scan rate of 2 mV/s and capacitance retention of sixty nine percent at 500 mV/s. Additionally, the fabricated symmetric supercapacitor exhibits a volumetric strength density of 38.6 Wh/L with an energy density of 206 W/L. Assuming the findings to be consistent, by increasing volumetric electricity density, the MXene-holey graphene movie symmetric supercapacitor device maintains a regular volumetric strength density compared with the pure MXene supercapacitor tool. Therefore, it would appear to intimate that the virgin terminal material has extensive capability for software in the fate of power storage, particularly in wearable devices. Zhou et al. [43, 53–55] successfully synthesized MXene-functionalized GO platelets through covalent bonding. C 1s CPs of conjugated 1-amino pyrene-disuccinimidyl substrates enable π–π bridging contacts between rGO platelets for emergent interfacial synergism and exceptional toughness. Operative current density of 1 A/cm^2 yields a volumetric capacitance of 645 F/cm^3 superior to the sum of the parts.

Since Mxene showed the good electrical and thermal properties which makes its suitable for the supercapacitors the rapid ion transport and effective charge storage are possible because of their hydrophilic nature, high electrical conductivity, and large surface area. Furthermore, functionalization happened because of by their adjustable surface chemistry, which enhances their electrochemical performance which is beneficial for the functional materials based on MXene. Although MXenes on their own have great qualities, MXene-based functional materials go one step further. Researchers can improve the stability, capacitance, and cycling performance of MXenes by mixing them with other materials. Among the benefits are, the better structural integrity, hybrid materials reduce the agglomeration problem that 2D materials frequently have, the improved electrochemical characteristics Functionalization with metal oxides or polymers can increase the amount of charge that can be stored.

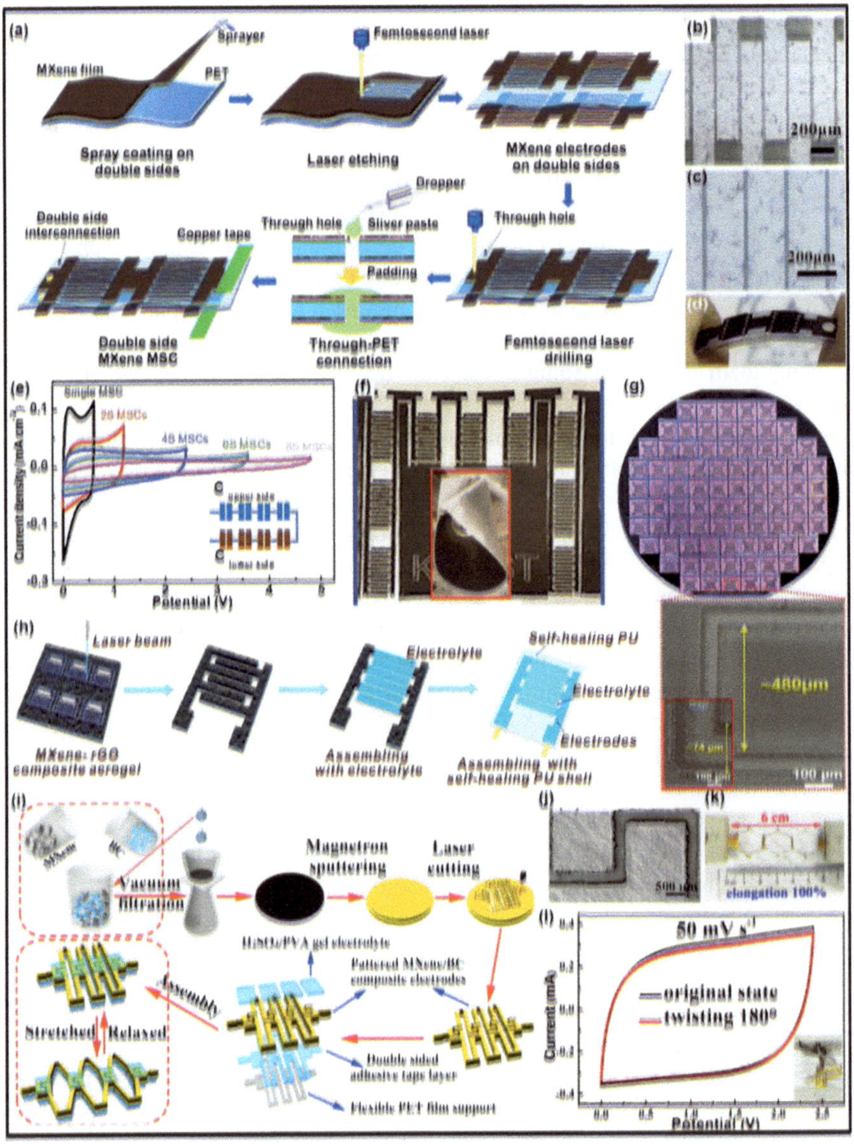

Fig. 4 **a** Microfabrication method of double-sided MXene-MSCs on a puppy substrate. **b, c** Interdigital MXene microelectrodes placed on both sides of the substrate. **d** A virtual photograph showing series connection of MXene-MSCs. **e** CV curves recorded at the price of one hundred mV s^{-1} in collection connection below a modern density of 0.2 mA cm^{-2}. **f** Meyer rod coating and laser patterning process of MXene on printing paper; **g** a virtual image presents the scalable on-chip MXene-MSCs in a silicon chip and SEM pics of laser-cutting finger electrodes, **h** the fabrication manner of self-restoration MXenerGO MSCs; **i** a step-through-step flowchart of all-strong-nation stretchable MXene-BCF MSCs production. **j** SEM photo supplying the symmetric interdigital microelectrodes of MXene-BCF MSCs. **k** Optical photographs of collection-related MXene-BCF MSCs for the duration of one hundred% stretching, and **l** the corresponding CV curves [52]. Reproduced with permission @ Copyright (2022) Elsevier

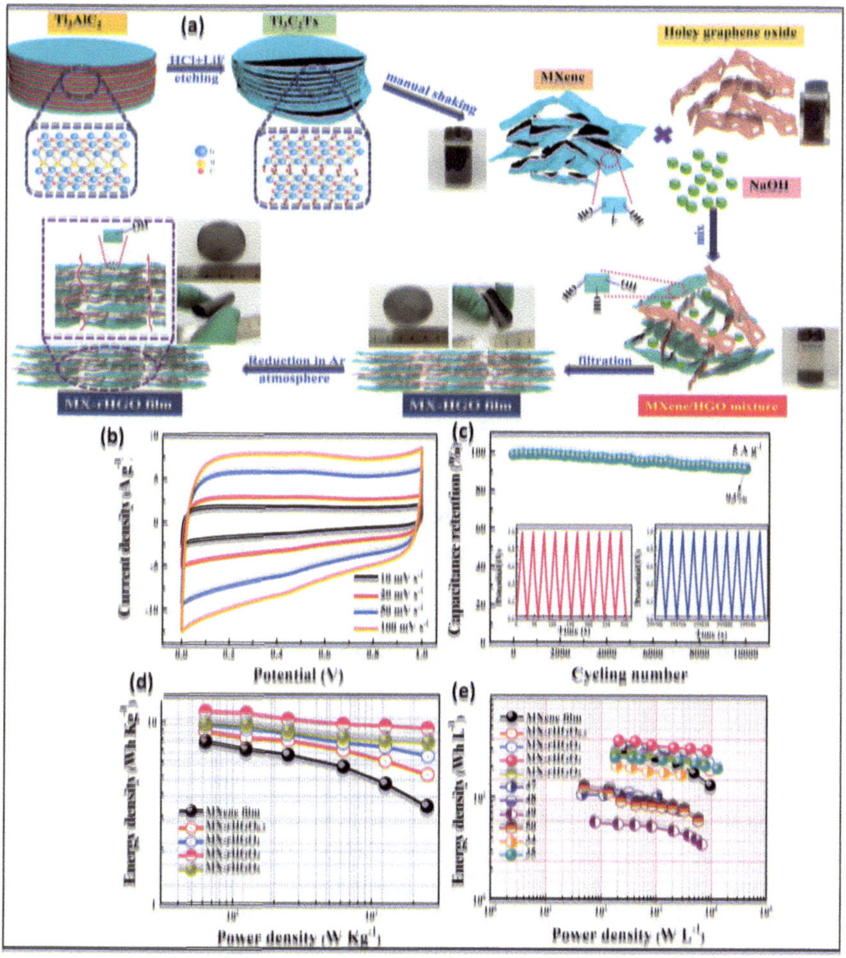

Fig. 5 Illustrates the following: **a** a visual depiction of the manner for growing MXene-holey graphene movie; **b** the cyclic voltammetry (CV) curve of the MXene-holey graphene film applied in a symmetric supercapacitor at unique scan quotes; **c** the capacitance retention of the symmetric supercapacitor working at a modern-day density of 5 A/g over 10,000 cycles, with the steady modern-day charging and discharging curves for the primary ten and final ten cycles provided as an inset; **d** the gravimetric power density; and **e** the volumetric power and power density of the MXene-holey graphene based totally symmetric supercapacitor, as compared to preceding studies [56]. Reproduce with permission @ Copyright (2022) Elsevier

3 Challenges

MXene-based composites have great potential for supercapacitor applications owing to high electrical conductivity and pseudocapacitive characteristics. A series of obstructions, however, influence their optimal performance and scale-up. One major disadvantage is restacking of MXene nanosheets, which limits ion diffusion and reduces active surface area and thus lower capacitance. Additionally, their small voltage window, especially in aqueous electrolytes, constrains accessible energy density. Surface terminations on MXenes (–OH, –F, –O) are generally impulsive, though these cause unpredictable electrochemical response. Structural instability in cyclic durability tests results from mechanical pressure by ion intercalation, and thus there is a decay of capacitance. Additionally, poor functionality at high rates of charge discharge is a characteristic manifested by MXene-based electrodes because of limited transport pathways by ions. On a production scale, commercial production of high-quality MXenes is a challenge, especially using a deleterious and expensive HF-based etching technique. Lastly, compatibility with electrolyte is limited as specific MXene materials disintegrate with some of their vicinal environments. Nevertheless, several key challenges have been identified as follows:

- Limited availability of precursors
- High synthesis costs
- Toxicity and environmental safety concerns
- Decreased chemical and thermal stability
- Insufficient availability of accurate theoretical simulations.

4 Conclusion and Future Perspectives

The challenges related with MXenes are various such as restricting, uniform structure, and can be enhance the electrical conductivity. Addressing these challenges and establishing future objectives will facilitate the enhancement of MXene-based supercapacitors, thereby improving the ongoing energy crisis and the limitations of existing energy storage solutions. This chapter gives detailed information of the fundamental characteristics of MXenes, fabrication of MXene with composite structure. This chapter also showed a MXene composite system for understanding the charge capacity components, the variables affecting their supercapacitor performance, and the different categories of supercapacitor devices utilizing MXenes. To check the restacking caused by an abundance of utilitarian bunches and to upgrade the execution of MXene-based supercapacitors, this work investigates the capacitive execution of MXene types with materials such as carbonaceous substances, move metal oxides (TMOs), and move metal dichalcogenides (TMDs). Our investigation indicates that the improvement of crossover materials could be a noteworthy methodology for making strides in the supercapacitor execution of MXene cathodes, although a comprehensive understanding of these materials is basic. Furthermore, elective procedures such as stage designing, deformity creation, and the application of attractive areas may demonstrate advantages in improving the charge storage capabilities of MXenes. Eventually, optimizing generation costs is significant for making MXene-based cathodes broadly accessible within the worldwide supercapacitor.

References

1. Kim, H.-J., Krishna, T., Zeb, K., Rajangam, V., Gopi, C.V.M., Sambasivam, S., Raghavendra, K.V.G., Obaidat, I.M.: A comprehensive review of Li-ion battery materials and their recycling techniques. Electronics **9**(7), 1161 (2020)
2. Béguin, F., Presser, V., Balducci, A., Frackowiak, E.: Carbons and electrolytes for advanced supercapacitors. Adv. Mater. **26**(14), 2219–2251 (2014)
3. Soonmin, H., Nandi, P., Paulraj, I., Ali, D., Sonker, R.K.: Chalcogenides and their nanocomposites: fundamental, properties and applications. In: Metal-Chalcogenide Nanocomposites, pp. 1–27. Woodhead Publishing (2024)
4. Zhu, Y., Ma, J., Das, P., Wang, S., Wu, Z.S.: High-voltage MXene-based supercapacitors: present status and future perspectives. Small Methods **7**(8), 2201609 (2023)
5. Zhu, Y., Ma, J., Das, P., Wang, S., Wu, Z.S.: High-voltage MXene-based supercapacitors: present status and future perspectives. Small Methods **7**(8), 2370046 (2023)
6. Zhu, Y., Zhang, Q., Ma, J., Das, P., Zhang, L., Liu, H., Wang, S., Li, H., Wu, Z.S.: Three-dimensional (3D)-printed MXene high-voltage aqueous micro-supercapacitors with ultrahigh areal energy density and low-temperature tolerance. Carbon Energy e481 (2024)
7. Assad, H., Fatma, I., Sharma, P.K., Kumar, A.: Structural, morphological and tribological properties of polymer/MXene composites. In: MXene Reinforced Polymer Composites: Fabrication, Characterization and Applications, pp. 221–255 (2024)
8. Hu, M., Chen, L., Jing, Y., Zhu, Y., Dai, J., Meng, A., Sun, C., Jia, J., Li, Z.: Intensifying electrochemical activity of $Ti_3C_2T_x$ MXene via customized interlayer structure and surface chemistry. Molecules **28**(15), 5776 (2023)
9. Xie, Y., Naguib, M., Mochalin, V.N., Barsoum, M.W., Gogotsi, Y., Yu, X., Nam, K.-W., Yang, X.-Q., Kolesnikov, A.I., Kent, P.R.: Role of surface structure on Li-ion energy storage capacity of two-dimensional transition-metal carbides. J. Am. Chem. Soc. **136**(17), 6385–6394 (2014)
10. Huang, H., Yang, W.: MXene-based micro-supercapacitors: ink rheology, microelectrode design and integrated system. ACS Nano **18**(6), 4651–4682 (2024)
11. Akhter, R., Maktedar, S.S.: MXenes: a comprehensive review of synthesis, properties, and progress in supercapacitor applications. J. Mater. (2023)
12. Ling, Z., Ren, C.E., Zhao, M.-Q., Yang, J., Giammarco, J.M., Qiu, J., Barsoum, M.W., Gogotsi, Y.: Flexible and conductive MXene films and nanocomposites with high capacitance. Proc. Natl. Acad. Sci. **111**(47), 16676–16681 (2014)
13. Naguib, M., Saito, T., Lai, S., Rager, M.S., Aytug, T., Paranthaman, M.P., Zhao, M.-Q., Gogotsi, Y.: $Ti_3C_2T_x$ (MXene)–polyacrylamide nanocomposite films. RSC Adv. **6**(76), 72069–72073 (2016)
14. Salim, O., Mahmoud, K.A., Pant, K.K., Joshi, R.K.: Introduction to MXenes: synthesis and characteristics. Mater. Today Chem. **14**, 100191 (2019)
15. Wang, J., Dong, S., Li, H., Chen, Z., Jiang, S., Wu, L., Zhang, X.: Facile synthesis of layered $Li_4Ti_5O_{12}$-$Ti_3C_2T_x$ (MXene) composite for high-performance lithium ion battery. J. Electroanal. Chem. **810**, 27–33 (2018)
16. Zhang, C., Kim, S.J., Ghidiu, M., Zhao, M.Q., Barsoum, M.W., Nicolosi, V., Gogotsi, Y.: Layered orthorhombic Nb_2O_5@$Nb_4C_3T_x$ and TiO_2@$Ti_3C_2T_x$ hierarchical composites for high performance Li-ion batteries. Adv. Func. Mater. **26**(23), 4143–4151 (2016)
17. Mohajer, F., Ziarani, G.M., Badiei, A., Iravani, S., Varma, R.S.: MXene-carbon nanotube composites: properties and applications. Nanomaterials **13**(2), 345 (2023)
18. Zhi, W., Xiang, S., Bian, R., Lin, R., Wu, K., Wang, T., Cai, D.: Study of MXene-filled polyurethane nanocomposites prepared via an emulsion method. Compos. Sci. Technol. **168**, 404–411 (2018)
19. Zhang, H., Wang, L., Chen, Q., Li, P., Zhou, A., Cao, X., Hu, Q.: Preparation, mechanical and anti-friction performance of MXene/polymer composites. Mater. Des. **92**, 682–689 (2016)
20. Levitt, A.S., Alhabeb, M., Hatter, C.B., Sarycheva, A., Dion, G., Gogotsi, Y.: Electrospun MXene/carbon nanofibers as supercapacitor electrodes. J. Mater. Chem. A **7**(1), 269–277 (2019)

21. Protyai, M.I.H., Rashid, A.B.: A comprehensive overview of recent progress in MXene-based polymer composites: their fabrication processes, advanced applications, and prospects. Heliyon (2024)
22. Mohajer, F., Ziarani, G., Badiei, A., Iravani, S., Varma, R.: MXene-carbon nanotube composites: properties and applications. Nanomaterials **13**(345), 2023 (2023)
23. Wang, H.-W., Naguib, M., Page, K., Wesolowski, D.J., Gogotsi, Y.: Resolving the structure of $Ti_3C_2T_x$ MXenes through multilevel structural modeling of the atomic pair distribution function. Chem. Mater. **28**(1), 349–359 (2016)
24. Brady, A., Liang, K., Vuong, V.Q., Sacci, R., Prenger, K., Thompson, M., Matsumoto, R., Cummings, P., Irle, S., Wang, H.-W.: Pre-sodiated $Ti_3C_2T_x$ MXene structure and behavior as electrode for sodium-ion capacitors. ACS Nano **15**(2), 2994–3003 (2021)
25. Prenger, K., Sun, Y., Ganeshan, K., Al-Temimy, A., Liang, K., Dun, C., Urban, J.J., Xiao, J., Petit, T., Van Duin, A.C.: Metal cation pre-intercalated $Ti_3C_2T_x$ MXene as ultra-high areal capacitance electrodes for aqueous supercapacitors. ACS Appl. Energy Mater. **5**(8), 9373–9382 (2022)
26. Pomerantseva, E., Bonaccorso, F., Feng, X., Cui, Y., Gogotsi, Y.: Energy storage: the future enabled by nanomaterials. Science **366**(6468), eaan8285 (2019)
27. Qin, L., Jiang, J., Tao, Q., Wang, C., Persson, I., Fahlman, M., Persson, P.O., Hou, L., Rosen, J., Zhang, F.: A flexible semitransparent photovoltaic supercapacitor based on water-processed MXene electrodes. J. Mater. Chem. A **8**(11), 5467–5475 (2020)
28. Zhang, C., Xu, S., Cai, D., Cao, J., Wang, L., Han, W.: Planar supercapacitor with high areal capacitance based on Ti_3C_2/polypyrrole composite film. Electrochim. Acta **330**, 135277 (2020)
29. Gürbüz, B., Ciftci, F.: Bio-electric-electronics and tissue engineering applications of MXenes wearable materials: a review. Chem. Eng. J. 151230 (2024)
30. Yang, W., Huang, B., Li, L., Zhang, K., Li, Y., Huang, J., Tang, X., Hu, T., Yuan, K., Chen, Y.: Covalently sandwiching MXene by conjugated microporous polymers with excellent stability for supercapacitors. Small Methods **4**(10), 2000434 (2020)
31. Song, Q., Zhan, Z., Chen, B., Zhou, Z., Lu, C.: Biotemplate synthesis of polypyrrole@bacterial cellulose/MXene nanocomposites with synergistically enhanced electrochemical performance. Cellulose **27**, 7475–7488 (2020)
32. Chen, X., Zhu, Y., Zhang, M., Sui, J., Peng, W., Li, Y., Zhang, G., Zhang, F., Fan, X.: N-Butyllithium-treated $Ti_3C_2T_x$ MXene with excellent pseudocapacitor performance. ACS Nano **13**(8), 9449–9456 (2019)
33. Chandran, M., Thomas, A., Raveendran, A., Vinoba, M., Bhagiyalakshmi, M.: MoS_2 confined MXene heterostructures as electrode material for energy storage application. J. Energy Storage **30**, 101446 (2020)
34. Chen, H., Yu, L., Lin, Z., Zhu, Q., Zhang, P., Qiao, N., Xu, B.: Carbon nanotubes enhance flexible MXene films for high-rate supercapacitors. J. Mater. Sci. **55**, 1148–1156 (2020)
35. Wang, Z., Chen, Y., Yao, M., Dong, J., Zhang, Q., Zhang, L., Zhao, X.: Facile fabrication of flexible rGO/MXene hybrid fiber-like electrode with high volumetric capacitance. J. Power Sources **448**, 227398 (2020)
36. Fan, Z., Wang, J., Kang, H., Wang, Y., Xie, Z., Cheng, Z., Liu, Y.: A compact MXene film with folded structure for advanced supercapacitor electrode material. ACS Appl. Energy Mater. **3**(2), 1811–1820 (2020)
37. Wang, Y., Dou, H., Wang, J., Ding, B., Xu, Y., Chang, Z., Hao, X.: Three-dimensional porous MXene/layered double hydroxide composite for high performance supercapacitors. J. Power Sources **327**, 221–228 (2016)
38. Pani, J., Borkar, H.: Enhanced electrochemical energy storage performance by mediating $BaTiO_3$ nanoparticles into the multilayers of $Ti_3C_2T_x$ MXene. J. Electroanal. Chem. **956**, 118092 (2024)
39. Yao, P., Yu, J., Zhou, J., Zhang, S., Zhang, M., Liu, H., Yang, B., Zhang, T., Zhu, C., Xu, J.: Engineering 3D electron and ion transport channels by constructing sandwiched holey quaternary metal oxide nanosheets for high-performance flexible energy storage. Sci. China Mater. **63**(9), 1719–1730 (2020)

40. Mashtalir, O., Lukatskaya, M.R., Kolesnikov, A.I., Raymundo-Pinero, E., Naguib, M., Barsoum, M., Gogotsi, Y.: The effect of hydrazine intercalation on the structure and capacitance of 2D titanium carbide (MXene). Nanoscale **8**(17), 9128–9133 (2016)
41. Zhang, W., Ma, J., Zhang, W., Zhang, P., He, W., Chen, J., Sun, Z.: A multidimensional nanostructural design towards electrochemically stable and mechanically strong hydrogel electrodes. Nanoscale **12**(12), 6637–6643 (2020)
42. Zhang, Z., Guo, M., Tang, Y., Liu, C., Zhou, J., Yuan, J., Gu, J.: High areal capacitance of vanadium oxides intercalated Ti_3C_2 MXene for flexible supercapacitors with high mass loading. Nanotechnology **31**(16), 165403 (2020)
43. Zhou, T., Wu, C., Wang, Y., Tomsia, A.P., Li, M., Saiz, E., Fang, S., Baughman, R.H., Jiang, L., Cheng, Q.: Super-tough MXene-functionalized graphene sheets. Nat. Commun. **11**(1), 2077 (2020)
44. Sun, Y., Yi, R., Zhao, Y., Liu, C., Yuan, Y., Geng, X., Li, W., Feng, Z., Mitrovic, I., Yang, L.: Improved pseudocapacitances of supercapacitors based on electrodes of nitrogen-doped $Ti_3C_2T_x$ nanosheets with in-situ growth of carbon nanotubes. J. Alloy. Compd. **859**, 158347 (2021)
45. Li, K., Wang, X., Wang, X., Liang, M., Nicolosi, V., Xu, Y., Gogotsi, Y.: All-pseudocapacitive asymmetric MXene-carbon-conducting polymer supercapacitors. Nano Energy **75**, 104971 (2020)
46. Chen, W., Zhang, D., Yang, K., Luo, M., Yang, P., Zhou, X.: Mxene ($Ti_3C_2T_x$)/cellulose nanofiber/porous carbon film as free-standing electrode for ultrathin and flexible supercapacitors. Chem. Eng. J. **413**, 127524 (2021)
47. Xu, T., Du, H., Liu, H., Liu, W., Zhang, X., Si, C., Liu, P., Zhang, K.: Advanced nanocellulose-based composites for flexible functional energy storage devices. Adv. Mater. **33**(48), 2101368 (2021)
48. Li, T., Chen, C., Brozena, A.H., Zhu, J., Xu, L., Driemeier, C., Dai, J., Rojas, O.J., Isogai, A., Wågberg, L.: Developing fibrillated cellulose as a sustainable technological material. Nature **590**(7844), 47–56 (2021)
49. Zhao, D., Zhu, Y., Cheng, W., Chen, W., Wu, Y., Yu, H.: Cellulose-based flexible functional materials for emerging intelligent electronics. Adv. Mater. **33**(28), 2000619 (2021)
50. Tu, H., Zhu, M., Duan, B., Zhang, L.: Recent progress in high-strength and robust regenerated cellulose materials. Adv. Mater. **33**(28), 2000682 (2021)
51. Kurra, N., Ahmed, B., Gogotsi, Y., Alshareef, H.N.: MXene-on-paper coplanar microsupercapacitors. Adv. Energy Mater. **6**(24), 1601372 (2016)
52. Zhu, Y., Wang, S., Ma, J., Das, P., Zheng, S., Wu, Z.-S.: Recent status and future perspectives of 2D MXene for micro-supercapacitors and micro-batteries. Energy Storage Mater. **51**, 500–526 (2022)
53. Zhou, Z., Panatdasirisuk, W., Mathis, T.S., Anasori, B., Lu, C., Zhang, X., Liao, Z., Gogotsi, Y., Yang, S.: Layer-by-layer assembly of MXene and carbon nanotubes on electrospun polymer films for flexible energy storage. Nanoscale **10**(13), 6005–6013 (2018)
54. Fan, Z., Wang, Y., Xie, Z., Wang, D., Yuan, Y., Kang, H., Su, B., Cheng, Z., Liu, Y.: Modified MXene/holey graphene films for advanced supercapacitor electrodes with superior energy storage. Adv. Sci. **5**(10), 1800750 (2018)
55. Song, L., Jin, X., Dai, C., Han, Y., Zhang, J., Zhang, Z., Qu, L.: Multifunctional devices based on planar microsupercapacitors: progress and challenges. Sci. China Mater. **65**(12), 3202–3228 (2022)
56. Thomas, S.A., Patra, A., Al-Shehri, B.M., Selvaraj, M., Aravind, A., Rout, C.S.: MXene based hybrid materials for supercapacitors: recent developments and future perspectives. J. Energy Storage **55**, 105765 (2022)

MIX
Papier aus verantwortungsvollen Quellen
Paper from responsible sources
FSC® C105338

If you have any concerns about our products,
you can contact us on
ProductSafety@springernature.com

In case Publisher is established outside the EU,
the EU authorized representative is:
Springer Nature Customer Service Center GmbH
Europaplatz 3, 69115 Heidelberg, Germany

Printed by Libri Plureos GmbH
in Hamburg, Germany